Parametric Modeling

with Creo Parametric 10.0

Randy H. Shih

Oregon Institute of Technology

SDC
PUBLICATIONS

SDC Publications
P.O. Box 1334
Mission, KS 66222
913-262-2664
www.SDCpublications.com
Publisher: Stephen Schroff

ISBN-13: 978-1-63057-620-2
ISBN-10: 1-63057-620-4

Printed and bound in the United States of America.

Preface

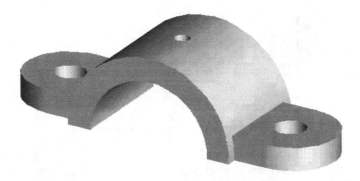

The primary goal of *Parametric Modeling with Creo Parametric 10.0* is to introduce the aspects of **Solid Modeling** and **Parametric Modeling**. This text is intended to be used as a training guide for students and professionals. This text covers Creo Parametric and the lessons proceed in a pedagogical fashion to guide you from constructing basic shapes to building intelligent solid models and creating multi-view drawings. This text takes a hands-on, exercise-intensive approach to all the important *Parametric Modeling* techniques and concepts. This textbook contains a series of ten tutorial style lessons designed to introduce beginning CAD users to Creo Parametric. This text is also helpful to Creo Parametric users upgrading from a previous release of the software. The solid modeling techniques and concepts discussed in this text are also applicable to other parametric feature-based CAD packages. The basic premise of this book is that the more designs you create using Creo Parametric, the better you learn the software. With this in mind, each lesson introduces a new set of commands and concepts, building on previous lessons. This book does not attempt to cover all of Creo Parametric's features, only to provide an introduction to the software. It is intended to help you establish a good basis for exploring and growing in the exciting field of **Computer Aided Engineering**.

Acknowledgments

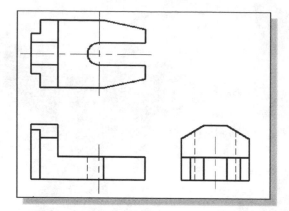

This book would not have been possible without a great deal of support. First, special thanks to two great teachers, Prof. George R. Schade of University of Nebraska-Lincoln and Mr. Denwu Lee from Taiwan, who taught me the fundamentals, the intrigue, and the sheer fun of Computer Aided Engineering.

The effort and support of the editorial and production staff of SDC Publications is gratefully acknowledged. I would especially like to thank Stephen Schroff for the support during this project.

I am grateful that the Mechanical Engineering Department of Oregon Institute of Technology has provided me with an excellent environment in which to pursue my interests in teaching and research activities.

Finally, truly unbounded thanks are due to my wife Hsiu-Ling and our daughter Casandra for their understanding and encouragement throughout this project.

Randy H. Shih
Klamath Falls, Oregon
Spring, 2023

Table of Contents

Chapter 2
Constructive Solid Geometry Concepts

Chapter 3
Model History Tree

Chapter 4
Constraints and Parametric Relations

Chapter 5
Parent/Child Relationships

Chapter 6
Datum Features, 3D Annotation and Part Drawings

Chapter 7
Introduction to 3D Printing

Chapter 8
Symmetrical Features in Designs

Chapter 9
Three Dimensional Construction Tools

Chapter 10
Basic Sheet Metal Designs

Chapter 11
Advanced Modeling Techniques

Chapter 12
Assembly Modeling – Putting It All Together

Chapter 13
Advanced Assembly Modeling and Animation

APPENDIX

INDEX

Introduction

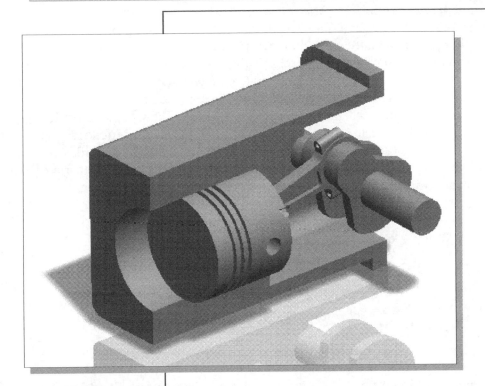

Learning Objectives

♦ **Development of Computer Geometric Modeling**

♦ **Feature-Based Parametric Modeling**

♦ **Getting Started with Creo Parametric**

♦ **Startup Options and Units Setup**

♦ **Creo Parametric Screen Layout**

♦ **Mouse Buttons**

♦ **Creo Parametric On-Line Help**

Introduction

The rapid changes in the field of **Computer Aided Engineering** (CAE) have brought exciting advances in the engineering community. Recent advances have made the long-sought goal of **concurrent engineering** closer to a reality. CAE has become the core of concurrent engineering and is aimed at reducing design time, producing prototypes faster, and achieving higher product quality. **Creo Parametric**, previously known as **Pro/Engineer**, is an integrated package of Mechanical Computer Aided Engineering software tools developed by *Parametric Technology Corporation* (**PTC**). Creo Parametric is a tool that facilitates a concurrent engineering approach to the design, analysis, and manufacturing of mechanical engineering products. The Creo Parametric software allows us to quickly create three-dimensional solid models. Real-life loads can be simulated on the computer to predict the behaviors of the designs under specific operating conditions. The computer models can also be used directly by manufacturing equipment such as machining centers, lathes, mills, or rapid prototyping machines to manufacture the product.

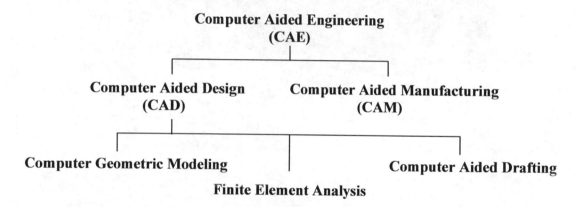

Development of Computer Geometric Modeling

Computer geometric modeling is a relatively new technology and its rapid expansion in the last fifty years is truly amazing. Computer-modeling technology advanced along with the development of computer hardware. The first-generation CAD programs, developed in the 1950s, were mostly non-interactive; CAD users were required to create program-codes to generate the desired two-dimensional (2D) geometric shapes. Initially, the development of CAD technology occurred mostly in academic research facilities. The Massachusetts Institute of Technology, Carnegie-Mellon University, and Cambridge University were the lead pioneers at that time. The interest in CAD technology spread quickly and several major industry companies, such as General Motors, Lockheed, McDonnell, IBM, and Ford Motor Co., participated in the development of interactive CAD programs in the 1960s. Usage of CAD systems was primarily in the automotive industry, aerospace industry, and government agencies that developed their own programs for their specific needs. The 1960s also marked the beginning of the development of finite element analysis methods for computer stress analysis and computer aided manufacturing for generating machine toolpaths.

The 1970s are generally viewed as the years of the most significant progress in the development of computer hardware, namely the invention and development of **microprocessors**. With the improvement in computing power, new types of 3D CAD programs that were user-friendly and interactive became reality. CAD technology quickly expanded from very simple **computer aided drafting** to very complex **computer aided design**. The use of 2D and 3D wireframe modelers was accepted as the leading-edge technology that could increase productivity in industry. The development of surface modeling and solid modeling technology were taking shape by the late 1970s, but the high cost of computer hardware and programming slowed the development of such technology. During this time period, the available CAD systems all required expensive room-sized mainframe computers that were extremely high in cost.

In the 1980s, improvements in computer hardware brought the power of mainframes to the desktop at less cost and with more accessibility to the general public. By the mid-1980s, CAD technology had become the main focus of a variety of manufacturing industries and was very competitive with traditional design/drafting methods. It was during this period of time that 3D solid modeling technology had major advancements, which boosted the usage of CAE technology in industry.

The introduction of the *feature-based parametric solid modeling* approach, at the end of the 1980s, elevated CAD/CAM/CAE technology to a new level. In the 1990s, CAD programs evolved into powerful design/manufacturing/management tools. CAD technology has come a long way, and during these years of development, modeling schemes progressed from two-dimensional (2D) wireframe to three-dimensional (3D) wireframe, to surface modeling, to solid modeling and, finally, to feature-based parametric solid modeling.

The first-generation CAD packages were simply 2D **Computer Aided Drafting** programs, basically the electronic equivalents of the drafting board. For typical models, the use of this type of program would require that several views of the objects be created individually as they would be on the drafting board. The 3D designs remained in the designer's mind, not in the computer database. The mental translation of 3D objects to 2D views is required throughout the use of the packages. Although such systems have some advantages over traditional board drafting, they are still tedious and labor intensive. The need for the development of 3D modelers came quite naturally, given the limitations of the 2D drafting packages.

The development of three-dimensional modeling schemes started with three-dimensional (3D) wireframes. Wireframe models are models consisting of points and edges, which are straight lines connecting between appropriate points. The edges of wireframe models are used, similar to lines in 2D drawings, to represent transitions of surfaces and features. The use of lines and points is also a very economical way to represent 3D designs.

The development of the 3D wireframe modeler was a major leap in the area of computer geometric modeling. The computer database in the 3D wireframe modeler contains the locations of all the points in space coordinates and it is typically sufficient to create just one model rather than multiple views of the same model. This single 3D model can then be viewed from any direction as needed. Most 3D wireframe modelers allow the user to create projected lines/edges of 3D wireframe models. In comparison to other types of 3D modelers, 3D wireframe modelers require very little computing power and generally can be used to achieve reasonably good representations of 3D models. However, because surface definition is not part of a wireframe model, all wireframe images have the inherent problem of ambiguity. Two examples of such ambiguity are illustrated.

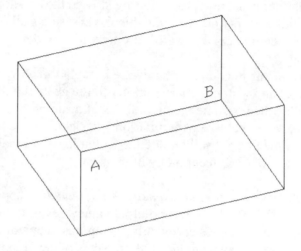

Wireframe Ambiguity: Which corner is in front, A or B?

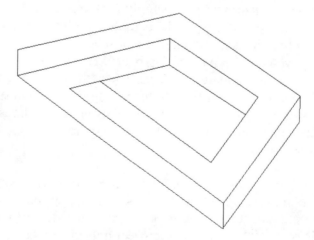

A non-realizable object: Wireframe models contain no surface definitions.

Surface modeling is the logical development in computer geometry modeling to follow the 3D wireframe modeling scheme by organizing and grouping edges that define polygonal surfaces. Surface modeling describes the part's surfaces but not its interiors. Designers are still required to interactively examine surface models to ensure that the various surfaces on a model are contiguous throughout. Many of the concepts used in 3D wireframe and surface modelers are incorporated in the solid modeling scheme, but it is solid modeling that offers the most advantages as a design tool.

In the solid modeling presentation scheme, the solid definitions include nodes, edges, and surfaces, and it is a complete and unambiguous mathematical representation of a precisely enclosed and filled volume. Unlike the surface modeling method, solid modelers start with a solid or use topology rules to guarantee that all of the surfaces are stitched together properly. Two predominant methods for representing solid models are **constructive solid geometry** (CSG) representation and **boundary representation** (B-rep).

The CSG representation method can be defined as the combination of 3D solid primitives. What constitutes a "primitive" varies somewhat with the software but typically includes a rectangular prism, a cylinder, a cone, a wedge, and a sphere. Most solid modelers also allow the user to define additional primitives, which are shapes typically formed by the basic shapes. The underlying concept of the CSG representation method is very straightforward; we simply **add** or **subtract** one primitive from another. The CSG approach is also known as the machinist's approach as it can be used to simulate the manufacturing procedures for creating the 3D object.

In the B-rep representation method, objects are represented in terms of their spatial boundaries. This method defines the points, edges, and surfaces of a volume, and/or issues commands that sweep or rotate a defined face into a third dimension to form a solid. The object is then made up of the unions of these surfaces that completely and precisely enclose a volume.

By the 1980s, a new paradigm called *concurrent engineering* had emerged. With concurrent engineering, designers, design engineers, analysts, manufacturing engineers, and management engineers all work closely together right from the initial stages of the design. In this way, all aspects of the design can be evaluated, and any potential problems can be identified right from the start and throughout the design process. Using the principles of concurrent engineering, a new type of computer modeling technique appeared. The technique is known as the *feature-based parametric modeling technique.* The key advantage of the feature-based parametric modeling technique is its capability to produce very flexible designs. Changes can be made easily, and design alternatives can be evaluated with minimum effort. Various software packages offer different approaches to feature-based parametric modeling, yet the end result is a flexible design defined by its design variables and parametric features.

Feature-Based Parametric Modeling

One of the key-elements in the *Creo Parametric* solid modeling software is its use of the **feature-based parametric modeling technique**. The feature-based parametric modeling approach has elevated solid modeling technology to the level of a very powerful design tool. Parametric modeling automates design and revision procedures by the use of parametric features. Parametric features control the model geometry by the use of design variables. The word ***parametric*** means that the geometric definitions of the design, such as dimensions, can be varied at any time in the design process. Features are predefined parts or construction tools in which users define the key parameters. A part is described as a sequence of engineering features, which can be modified/changed at any time. The concept of parametric features makes the modeling more closely match the actual design-manufacturing process than the mathematics of a solid modeling program. In parametric modeling, models and drawings are updated automatically when the design is refined.

Parametric modeling offers many benefits:

- **We begin with simple, conceptual models with minimal detail; this approach conforms to the design philosophy of "shape before size."**

- **Geometric constraints, dimensional constraints, and relational parametric equations can be used to capture design intent.**

- **The ability to update an entire system, including parts, assemblies and drawings after changing one parameter of complex designs.**

- **We can quickly explore and evaluate different design variations and alternatives to determine the best design.**

- **Existing design data can be reused to create new designs.**

- **Quick design turn-around.**

The feature-based parametric modeling technique enables the designer to incorporate the original **design intent** into the construction of the model, and the individual features control the geometry in the event of a design change. As features are modified, the system updates the entire part by re-linking the individual features of the model.

Creo Parametric is the twenty-fifth release, with many added features and enhancements, of the original *Creo Parametric* software produced by Parametric Technology Corporation (PTC). *Creo Parametric* is considered the industry leader in setting the standard for the feature-based modeling paradigm. *Creo Parametric's* **Behavioral Modeling** and **Intent Referencing** allow users to concentrate on the design without depending on the associated parameters or constraints. Users can specify how features interact with each other and *Creo Parametric* will automatically adjust sizes and positions as changes are made.

Getting Started with *Creo Parametric*

- *Creo Parametric* is composed of several application software modules (these modules are called *applications*), all sharing a common database. In this text, the main concentration is placed on the solid modeling modules used for part design. The general procedures required in creating solid models, engineering drawings, and assemblies are illustrated.

- The program takes a while to load, so be patient. The tutorials in this text are based on the assumption that you are using *Creo Parametric's* default settings. If your system has been customized for other uses, some of the settings may appear differently and not work with the step-by-step instructions in the tutorials. Contact your instructor and/or technical support personnel to restore the default software configuration.

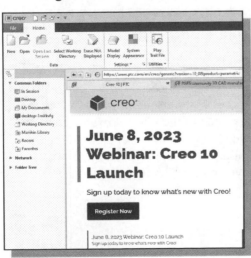

The **Explore Creo** window provides general information regarding changes in the new version of **PTC Creo** software; you are encouraged to select and view the different options available within this window.

Creo Parametric Screen Layout

The default *Creo Parametric drawing screen* contains the *Quick Access* toolbar, the *Ribbon* toolbar, the *Navigator* area, the *web browser*, the *message area*, the *Status Bar,* and the *Folder Tree.* A line of quick text appears next to the icon as you move the *mouse cursor* over different icons. You may resize the *Creo Parametric* drawing window by clicking and dragging at the edges of the window, or relocate the window by clicking and dragging at the window title area.

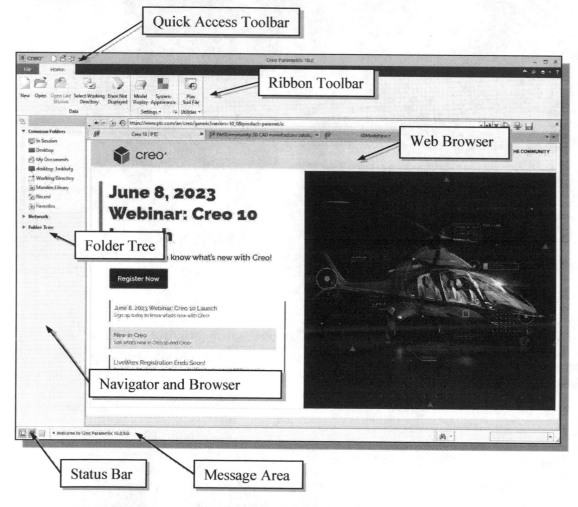

❖ *Creo Parametric* uses the **context sensitive menus** approach, which means additional selection menus and option windows will only be available when they are applicable to the current task. The appearance of the *Creo Parametric* main window and toolbars, as shown in the figure above, remains the same on the screen throughout the different phases of modeling. Note that the menu items and icons of the non-applicable options are grayed out, which means they are temporarily disabled.

- **Ribbon toolbar**

  The *Ribbon* toolbar at the top of the main window contains operations that you can use for all modes of the system. The *Ribbon* toolbar contains command buttons organized in a set of tabs. On each tab, related buttons are grouped. You can minimize the *Ribbon* to make more space available on your screen. You can customize the *Ribbon* by adding, removing, or moving buttons.

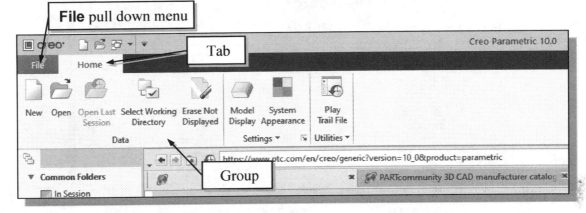

- **Quick Access toolbar**

  The *Quick Access* toolbar located at the top of the main window allows us quick access to frequently used commands, for example, the *basic file related* commands, such as **New file, Open, Save** and common tools such as **Undo, Redo**. Note the toolbar can be customized by adding and removing sets of options or individual commands.

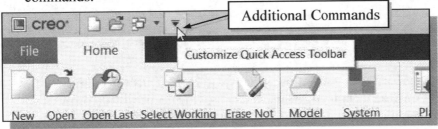

- **Message Area**

  The message provides both an interface to the commands and also provides a line of quick help on the activated command.

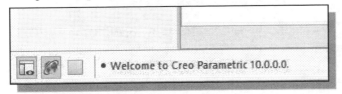

- **Graphics display area**

  The graphics display area is the area where parts, assemblies, and drawings are displayed. The mode's display is controlled by the *environment settings*. The *Display Control* toolbar is also located near the top of the graphics area.

- ## Navigator and Browser
  The tabs across the top of the *Navigator* provide direct access to four groups of options. The initial content of the *Navigator* is the *Folder* information where *Creo Parametric* is launched. The *Navigator* includes the *Model Tree, Layer Tree, Detail Tree, Folder browser*, and *Favorites*.

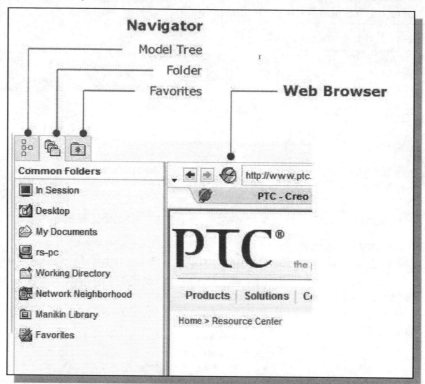

- ## Web Browser
  The *Creo Parametric web browser* allows the user to quickly access **model information** and **on-line documentation**. It also provides general web browsing capabilities. The initial content of the *Creo Parametric* web browser is the *Creo Parametric Web Tools*. The web browser will close automatically when you open a model; it can also be opened or closed by toggling the *Browser Sash* with the left-mouse-button.

- ## Navigator Display Controls
  The *Navigator* and the *web browser* can be opened or closed by toggling the *Display Controls* with the left-mouse-button.

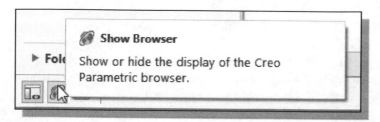

Basic Functions of Mouse Buttons

Creo Parametric utilizes the mouse buttons extensively. In learning *Creo Parametric's* interactive environment, it is important to understand the basic functions of the mouse buttons. A three-button mouse is highly recommended with *Creo Parametric* since the package uses all three buttons for various functions. If a two-button mouse is used instead, some of the operations, such as the *Dynamic Viewing* functions, will not be available.

- **Left mouse button**

  The left mouse button is used for most operations, such as selecting menus and icons, or picking graphic entities. One click of the button is used to select icons, menus and form entries, and to pick graphic items.

- **Middle mouse button**

  The middle button is often used to accept the default setting of a prompt or to end a process. It is also used as the shortcut to perform the *Dynamic Viewing* functions: Rotate, Zoom, and Pan.

- **Right mouse button**

  The right mouse button is used to query a selection and is also used in the *Creo Parametric Sketcher* to create tangent arcs.

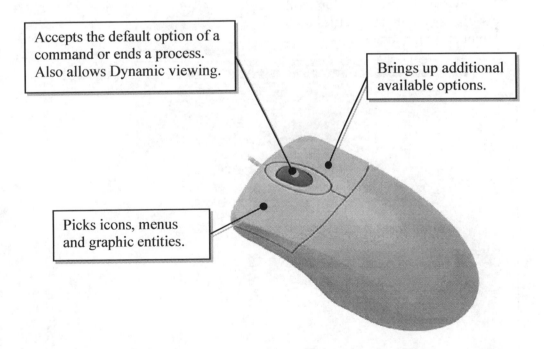

Accepts the default option of a command or ends a process. Also allows Dynamic viewing.

Brings up additional available options.

Picks icons, menus and graphic entities.

Model Tree window and Feature toolbars

❖ The *Creo Parametric* **Model Tree** window and **Feature** toolbars are two of the most important *Creo Parametric* interfaces; they will appear on the screen when we begin a modeling session.

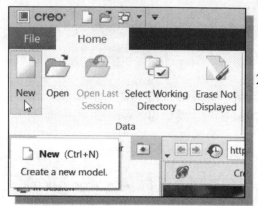

1. Click on the **New** icon, the first icon in the *Ribbon* toolbar, as shown.

2. In the *New* dialog box, confirm the model's Type is set to **Part** (and the Sub-type is set to Solid).

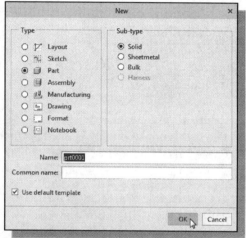

3. Click on the **OK** button to accept the default settings and enter the *Creo Parametric Part Modeling* mode.

❖ The **Navigator** and the **Ribbon** toolbar now display commands related to the current part modeling task.

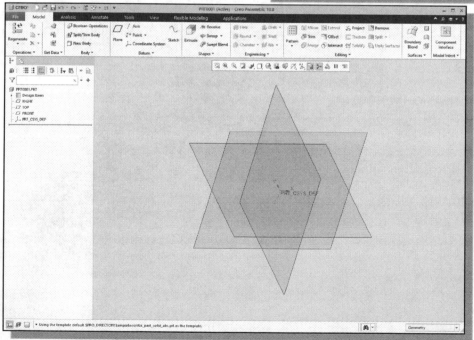

Online Help

❖ Several types of online *Help* are available at any time during a *Creo Parametric* session. *Creo Parametric* provides many help functions, such as:

- **Creo Parametric Help system**: The multiple Creo help system can be accessed through the File tab in the Ribbon of the *Creo Parametric* window. Note that many of the Help options use the *Creo Parametric Web Tools*, which allow us to access the *Creo Parametric* resource center available at the PTC website through the internet.

- **Creo Parametric Help Center**: Click on the **Help** or the **Command Search** option in the *Ribbon* toolbar to access the *Creo Parametric Help Center*.

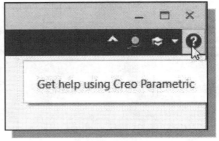

- **Context Sensitive Help**: The context sensitive help can be used to quickly access the Help menu on specific tasks or icons displayed on the screen.

Leaving *Creo Parametric*

1. To leave *Creo Parametric*, use the left-mouse-button and click on the **File** tab in the *Creo Parametric Ribbon* toolbar, then choose **Exit** from the pull-down menu.

2. In the confirmation dialog box, click on the **Yes** button to exit *Creo Parametric*.

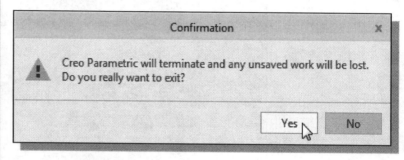

Creating a CAD files folder

It is a good practice to create a separate folder to store your CAD files. You should not save your CAD files in the same folder where the *Creo Parametric* application is located. It is much easier to organize and back up your project files if they are in a separate folder. Making folders within this folder for different types of projects will help you organize your CAD files even further. When creating CAD files in *Creo Parametric*, it is strongly recommended that you *save* your CAD files on the hard drive.

- To create a new folder with most *Windows* systems:

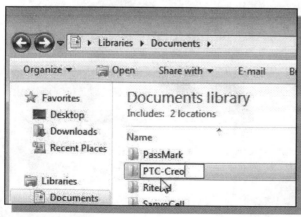

1. In *My Computer*, or start the **Windows Explorer** under the *Start* menu, open the **Documents** folder.

2. **Right-mouse-click** once and select **New-Folder**.

3. Enter the new name of the folder, and press **ENTER**.

Chapter 1
Parametric Modeling Fundamentals

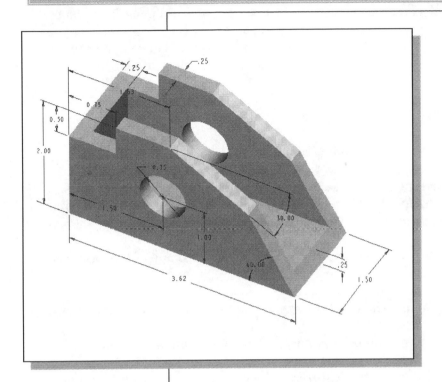

Learning Objectives

- ◆ **Create Simple Extruded Solid Models**
- ◆ **Understand the Basic Parametric Modeling Process**
- ◆ **Create 2D Sketches**
- ◆ **Understand the "Shape before Size" approach**
- ◆ **Use the Dynamic Viewing commands**
- ◆ **Create and Modify Parametric Dimensions**

Introduction

The **feature-based parametric modeling** technique enables the dimensioner to incorporate the original **design intent** into the construction of the model. The word *parametric* means the geometric definitions of the design, such as dimensions, can be varied at any time in the design process. Parametric modeling is accomplished by identifying and creating the key features of the design with the aid of computer software. The design variables, described in the sketches and features, can be used to quickly modify/update the design.

In *Creo Parametric*, the parametric part modeling process involves the following steps:

1. **Set up Units and Basic Datum Geometry.**

2. **Determine the type of the base feature, the first solid feature, of the design. Note that *Extrude*, *Revolve*, or *Sweep* operations are the most common types of base features.**

3. **Create a rough two-dimensional sketch of the basic shape of the base feature of the design.**

4. **Apply/modify constraints and dimensions to the two-dimensional sketch.**

5. **Transform the two-dimensional parametric sketch into a 3D feature.**

6. **Add additional parametric features by identifying feature relations and complete the design.**

7. **Perform analyses/simulations, such as finite element analysis (FEA) or cutter path generation (CNC), on the computer model and refine the design as needed.**

8. **Document the design by creating the desired 2D/3D drawings.**

The approach of creating three-dimensional features using two-dimensional sketches is an effective way to construct solid models. Many designs are in fact the same shape in one direction. Computer input and output devices we use today are largely two-dimensional in nature, which makes this modeling technique quite practical. This method also conforms to the design process that helps the designer with conceptual design along with the capability to capture the *design intent*. Most engineers and designers can relate to the experience of making rough sketches on restaurant napkins to convey conceptual design ideas. Note that *Creo Parametric* provides many powerful modeling and design tools, and there are many different approaches to accomplish modeling tasks. The basic principle of **feature-based modeling** is to build models by adding simple features one at a time. In this chapter, a very simple solid model with extruded features is used to introduce the general **feature-based parametric** modeling procedure.

The *Adjuster* design

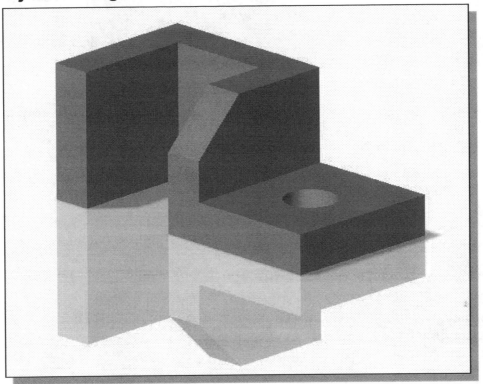

Starting *Creo Parametric*

How to start *Creo Parametric* depends on the type of workstation and the particular software configuration you are using. With most *Windows* and *UNIX* systems, you may select **Creo Parametric** on the *Start* menu or select the **Creo Parametric** icon on the desktop. Consult your instructor or technical support personnel if you have difficulty starting the software.

1. Select the **Creo Parametric** option on the *Start* menu or select the **Creo Parametric** icon on the desktop to start *Creo Parametric*. The *Creo Parametric* main window will appear on the screen.

2. Click on the **New** icon, located in the *Ribbon* toolbar as shown.

3. In the *New* dialog box, confirm the model's Type is set to **Part** (**Solid** Sub-type).

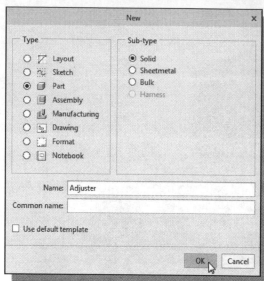

4. Enter **Adjuster** as the part Name as shown in the figure.

5. Turn *off* the **Use default template** option.

6. Click on the **OK** button to accept the settings.

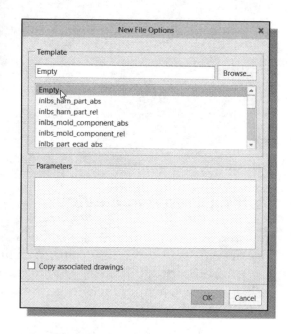

7. In the *New File Options* dialog box, select **EMPTY** in the option list to not use any template file.

8. Click on the **OK** button to accept the settings and enter the *Creo Parametric Part Modeling* mode.

❖ Note that the part name, *Adjuster.prt*, appears in the *Navigator Model Tree* window and the title bar area of the main window.

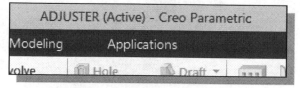

Step 1: Units and Basic Datum Geometry Setups

◆ Units Setup

When starting a new model, the first thing we should do is to choose the set of units we want to use.

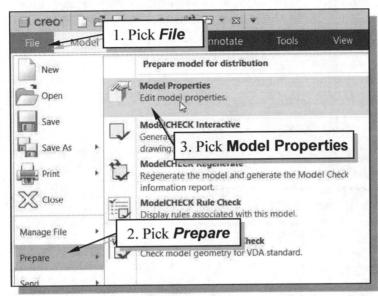

1. Use the left-mouse-button and select **File** in the pull-down menu area.

2. Use the left-mouse-button and select **Prepare** in the **pull-down list** as shown.

3. Select **Model Properties** in the *expanded list* as shown.

➢ Note that the *Creo Parametric* menu system is context-sensitive, which means that the menu items and icons of the non-applicable options are grayed out (temporarily disabled).

4. Select the **Change** option that is to the right of the **Units** option in the *Model Properties* window.

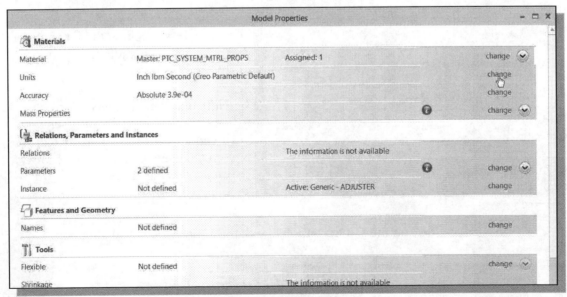

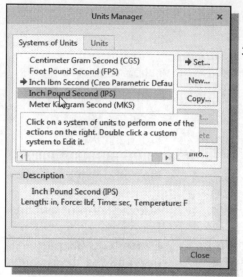

5. In the **Units Manager – Systems of Units** form, the *Creo Parametric* default setting **Inch Ibm Second** is displayed. The set of units is stored with the model file when you save. Pick **Inch Pound Second (IPS)** by clicking in the list window as shown.

6. Click on the **Set** button to accept the selection. Notice the arrow in the Units list now points toward the **Inch Pound Second (IPS)** units set.

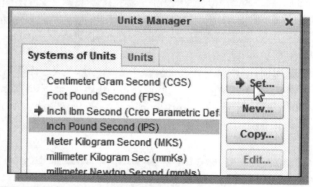

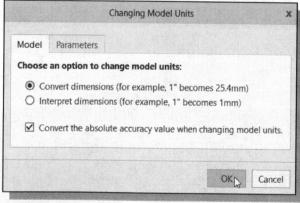

7. In the *Changing Model Units* dialog box, click on the **OK** button to accept the default option to change the units.

➢ Note that *Creo Parametric* allows us to change model units even after the model has been constructed; we can change the units by (1) *Convert dimensions* or (2) *Interpret dimensions*.

8. Click on the **Close** button to exit the *Units Manager* dialog box.

9. Pick **Close** to exit the *Model Properties* window.

➤ Adding the First Part Features – Datum Planes

❖ *Creo Parametric* provides many powerful tools for model creation. In doing feature-based parametric modeling, it is a good practice to establish three reference planes to locate the part in space. The reference planes can be used as location references in feature constructions.

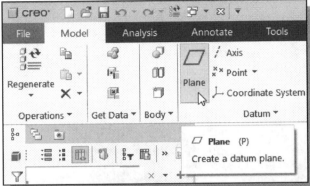

1. Move the cursor to the *Datum* toolbar on the *Ribbon* toolbar and click on the **Datum Plane** tool icon as shown.

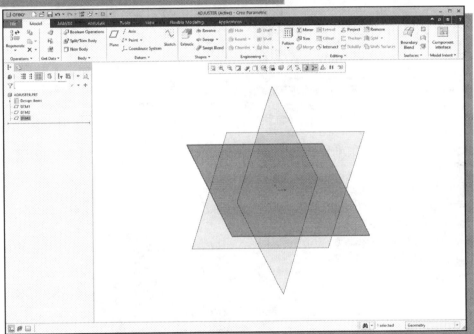

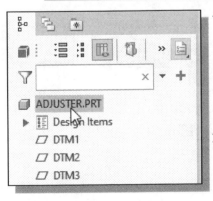

❖ In the *Navigator Model Tree* window and the display area, three datum planes represented by three rectangles are displayed. Datum planes are infinite planes and they are perpendicular to each other. We can consider these planes as XY, YZ, and ZX planes of a Cartesian coordinate system.

2. Click the model name, *Adjuster.prt*, in the *Navigator* window to **deselect** the last created feature.

➤ Switching On/Off the Plane Tag Display

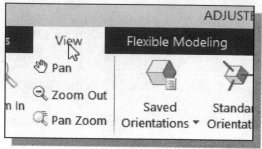

1. Click on the **View** tab in the Ribbon toolbar to show the view related commands in Creo.

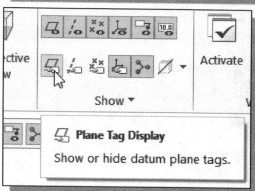

2. Click on the **Plane Tag Display** icon to toggle on/off the display of the plane tag.

➤ Note the other options available to display the **Axis Tag** and/or **Point Tag**.

➤ On your own, experiment with turning on/off the Plane Tag; set the datum planes to display with the associated names as shown before proceeding to the next section.

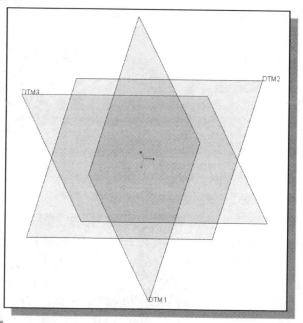

3. Click on the **Model** tab in the *Ribbon toolbar* to return to the Model toolbar as shown.

Step 2: Determine/Set up the Base Solid Feature

- For the *Adjuster* design, we will create an extruded solid as the base feature.

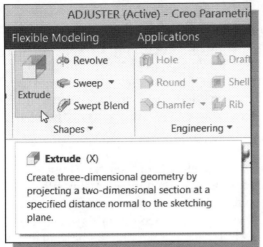

1. In the **Shapes** toolbar (the fourth group in the *Ribbon* toolbar), click on the **Extrude** tool icon as shown.

- The **Feature Option Dashboard**, which contains applicable construction options, is displayed in the *Ribbon* toolbar of the *Creo Parametric* main window.

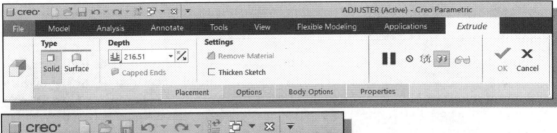

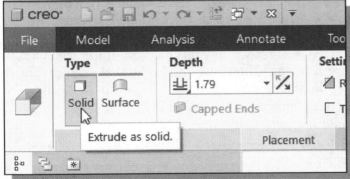

2. On your own, move the cursor over the icons and read the descriptions of the different options available. Note that the default extrude option is set to **Extrude as solid**.

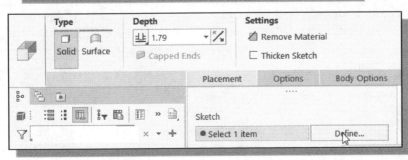

3. Click the **Placement** option and choose **Define** to begin creating a new *internal sketch*.

Sketching plane – It is an XY CRT, but an XYZ World

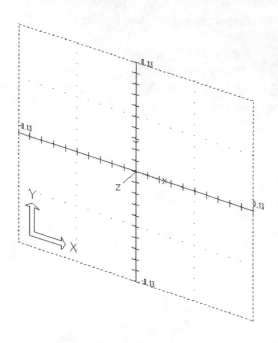

Design modeling software is becoming more powerful and user friendly, yet the system still does only what the user tells it to do. When using a geometric modeler, we therefore need to have a good understanding of what its inherent limitations are. We should also have a good understanding of what we want to do and what to expect, as the results are based on what is available.

In most 3D geometric modelers, 3D objects are located and defined in what is usually called **world space** or **global space**. Although a number of different coordinate systems can be used to create and manipulate objects in a 3D modeling system, the objects are typically defined and stored using the world space. The world space is usually a **3D Cartesian coordinate system** that the user cannot change or manipulate.

In most engineering designs, models can be very complex, and it would be tedious and confusing if only the world coordinate system were available. Practical 3D modeling systems allow the user to define **Local Coordinate Systems (LCS)** or **User Coordinate Systems (UCS)** relative to the world coordinate system. Once a local coordinate system is defined, we can then create geometry in terms of this more convenient system.

Although objects are created and stored in 3D space coordinates, most of the geometric entities can be referenced using 2D Cartesian coordinate systems. Typical input devices such as a mouse or digitizer are two-dimensional by nature; the movement of the input device is interpreted by the system in a planar sense. The same limitation is true of common output devices, such as CRT displays and plotters. The modeling software performs a series of three-dimensional to two-dimensional transformations to correctly project 3D objects onto the 2D display plane.

The *Creo Parametric **sketching plane*** is a special construction approach that enables the planar nature of the 2D input devices to be directly mapped into the 3D coordinate system. The *sketching plane* is a local coordinate system that can be aligned to an existing face of a part or a reference plane.

Think of the sketching plane as the surface on which we can sketch the 2D sections of the parts. It is similar to a piece of paper, a white board, or a chalkboard that can be attached to any planar surface. The first sketch we create is usually drawn on one of the established datum planes. Subsequent sketches/features can then be created on sketching planes that are aligned to existing **planar faces of the solid part** or **datum planes.**

Defining the Sketching Plane

The *sketching plane* is a reference location where two-dimensional sketches are created. The *sketching plane* can be any planar part surface or datum plane. Note that *Creo Parametric* uses a two-step approach in setting up the selection and alignment of the sketching plane.

❖ In the *Section Placement* window, the selection of the sketch plane and the orientation of the sketching plane are organized into two groups as shown in the figure. The **Sketch Plane** can be set to any surfaces, including datum planes. The **Sketch Orientation** is set based on the selection of the Sketch plane.

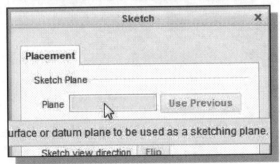

1. Notice the **Plane** option box in the *Sketch* window is activated, and the message *"Select a plane or surface to define sketch plane."* is displayed in the quick help tip and also in the message area.

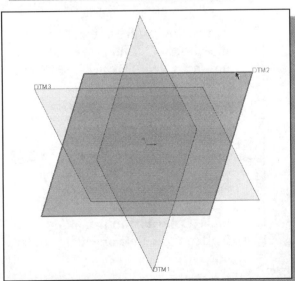

2. In the *graphic area*, select **DTM2** by clicking on any edge of the plane as shown.

❖ Notice an arrow appears on the left edge of DTM2. The arrow direction indicates the viewing aligned direction of the sketch plane. The viewing direction can be reversed by clicking on the **Flip** button in the *Sketch Orientation* section of the pop-up window.

Defining the Orientation of the Sketching Plane

Although we have selected the sketching plane, *Creo Parametric* still needs additional information to define the orientation of the sketch plane. *Creo Parametric* expects us to choose a reference plane (any plane that is perpendicular to the selected sketch plane) and the orientation of the reference plane is relative to the computer screen.

❑ **To define the orientation of the sketching plane, select the facing direction of the reference plane with respect to the computer screen.**

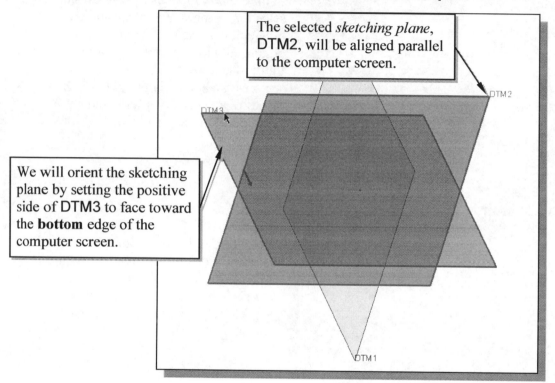

The selected *sketching plane*, DTM2, will be aligned parallel to the computer screen.

We will orient the sketching plane by setting the positive side of DTM3 to face toward the **bottom** edge of the computer screen.

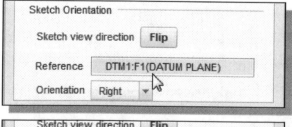

1. Notice the **Reference** option box in the Sketch Orientation window is now activated. The message "*Select a reference, such as surface, plane or edge to define view orientation.*" is displayed in the message area.

2. In the *graphic area*, select **DTM3** by clicking on one of the datum plane edges as shown in the above figure.

3. In the **Orientation** list, pick **Bottom** to set the orientation of the reference plane.

4. Pick **Sketch** to exit the *Section Placement* window and proceed to enter the *Creo Parametric Sketcher* mode.

5. To orient the sketching plane parallel to the screen, the **Sketch View** icon in the *Display View* toolbar is available as shown. Note the orientation of the sketching plane is adjusted based on the setup on the previous page.

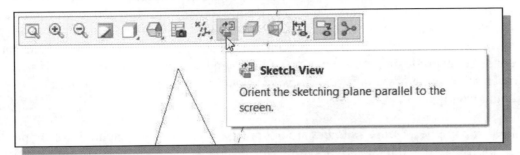

- *Creo Parametric* will now rotate the three *datum planes*: DTM2 is now aligned to the screen and the positive side of DTM3 is facing toward the bottom edge of the computer screen.

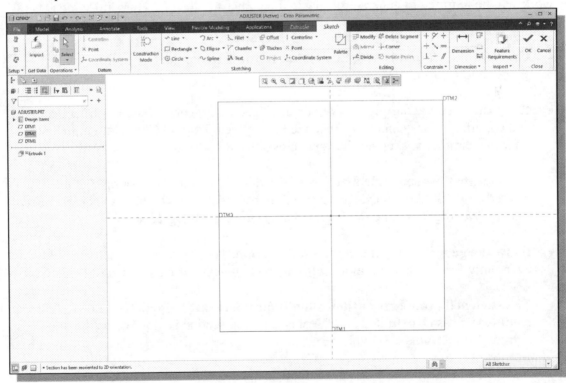

❖ Note the orientation of the *sketching plane* can be very confusing to new users at first. It is suggested that you read through this section carefully a few times to make sure you understand the steps involved.

Step 3: Creating 2D Rough Sketches

♦ Shape Before Size – Creating Rough Sketches

Quite often during the early design stage, the shape of a design may not have any precise dimensions. Most conventional CAD systems require the user to input the precise lengths and location dimensions of all geometric entities defining the design, and some of the values may not be available during the early design stage. With *parametric modeling*, we can use the computer to elaborate and formulate the design idea further during the initial design stage. With *Creo Parametric*, we can use the computer as an electronic sketchpad to help us concentrate on the formulation of forms and shapes for the design. This approach is the main advantage of *parametric modeling* over conventional solid-modeling techniques.

As the name implies, *rough sketches* are not precise at all. When sketching, we simply sketch the geometry so it closely resembles the desired shape. Precise scale or dimensions are not needed. *Creo Parametric* provides us with many tools to assist in finalizing sketches, known as *sections*. For example, geometric entities such as horizontal and vertical lines are set automatically. However, if the rough sketches are poor, much more work will be required to generate the desired parametric sketches. Here are some general guidelines for creating sketches in *Creo Parametric*:

- **Create a sketch that is proportional to the desired shape.** Concentrate on the shapes and forms of the design.

- **Keep the sketches simple.** Leave out small geometry features such as fillets, rounds, and chamfers. They can easily be placed using the Fillet and Chamfer commands after the parametric sketches have been established.

- **Exaggerate the geometric features of the desired shape.** For example, if the desired angle is 85 degrees, create an angle that is 50 or 60 degrees. Otherwise, *Creo Parametric* might assume the intended angle to be a 90-degree angle.

- **Draw the geometry so that it does not overlap.** The sketched geometry should eventually form a closed region. *Self-intersecting* geometric shapes are not allowed.

- **The sketched geometric entities should form a closed region.** To create a solid feature, such as an extruded solid, a closed region section is required so that the extruded solid forms a 3D volume.

➢ **Note:** The concepts and principles involved in *parametric modeling* are very different, and sometimes they are totally opposite, to those of the conventional computer aided drafting systems. In order to understand and fully utilize *Creo Parametric's* functionality, it will be helpful to take a *Zen* approach to learning the topics presented in this text: **Temporarily forget your knowledge and experiences using conventional computer aided drafting systems.**

◆ The Creo Parametric SKETCHER and INTENT MANAGER

In previous generation CAD programs, construction of models relies on exact dimensional values, and adjustments to dimensional values are quite difficult once the model is built. With *Creo Parametric*, we can now treat the sketch as if it is being done on a napkin, and it is the general shape of the design that we are more interested in defining. The *Creo Parametric* part model contains more than just the final geometry. It also contains the *design intent* that governs what will happen when geometry changes. The design philosophy of "**shape before size**" is implemented through the use of the *Creo Parametric Sketcher*. This allows the designer to construct solid models in a higher level and leave all the geometric details to *Creo Parametric*.

In *Creo Parametric,* previously known as Pro/ENGINEER, one of the more important functionalities is the ***Intent Manager*** in the *2D Sketcher*.

The ***Intent Manager*** enables us to do:

- Dynamic dimensioning and constraints
- Add or delete constraints explicitly
- Undo any *Sketcher* operation

The first thing that *Creo Parametric Sketcher* expects us to do, which is displayed in the *References* window, is to specify ***sketching references***. In the previous sections, we created the three datum planes to help orient the model in 3D space. Now we need to orient the 2D sketch with respect to the three datum planes. At least two references are required to orient the sketch in the horizontal direction and in the vertical direction. By default, the two planes (in our example, DTM1 and DTM3) that are perpendicular to the sketching plane (DTM2) are automatically selected.

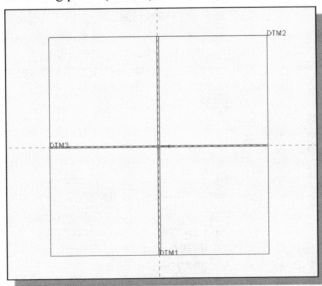

> Note that DTM1 and DTM3 are pre-selected as the sketching references. In the graphics area, the two references are displayed with two **dashed lines**.

- In *Creo Parametric*, a 2D sketch needs to be Fully Placed with respect to at least two references. In this case, DTM1 is used to control the horizontal placement of geometry, where DTM3 is used to control the vertical placements.

❖ Next, we will create a rough sketch by using some of the visual aids available, and then update the design through the associated control parameters.

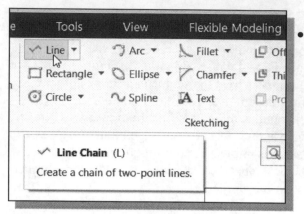

- Move the graphics cursor to the **Line** icon in the *Sketching* toolbar. A *help-tip* box appears next to the cursor to provide a brief description of the command.

❖ The *Sketching* toolbar provides tools for creating the basic 2D geometry that can be used to create features and parts.

Graphics Cursors

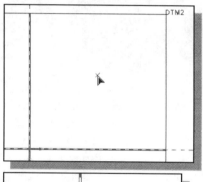

❖ Notice the cursor changes from an arrow to an arrow with a small crosshair when graphical input is expected.

1. As you move the graphics cursor, you will see different symbols appear at different locations.

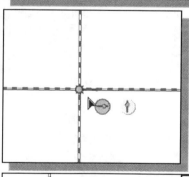

2. Move the cursor near the intersection of the two references, and notice that the small crosshair attached to the cursor will automatically snap to the intersection point. **Left-click** once to place the starting point as shown. Notice the small geometric constraint symbol next to the cursor indicating the first line endpoint is Coincident with the intersection point.

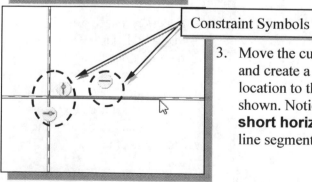

Constraint Symbols

3. Move the cursor along the horizontal reference and create a short horizontal line by clicking at a location to the right side of the starting point as shown. Notice the geometric constraint symbol, a **short horizontal line**, indicating the created line segment is Horizontal.

Geometric Constraint Symbols

Creo Parametric displays different visual clues, or symbols, to show you alignments, perpendicularities, tangencies, etc. These constraints are used to capture the *design intent* by creating constraints where they are recognized. *Creo Parametric* displays the governing geometric rules as models are built.

|	Vertical	indicates a line segment is vertical
—	Horizontal	indicates a line segment is horizontal
=	Equal Length	indicates two line segments are of equal length
	or	
	Equal Radii	indicates two curves are of equal radii
♂	Tangent	indicates two entities are tangent to each other
****	Parallel	indicates a segment is parallel to other entities
⋎	Perpendicular	indicates a segment is perpendicular to other entities
→ ←	Symmetry	indicates two points are symmetrical
○	Point on Entity	indicates the point is on another entity

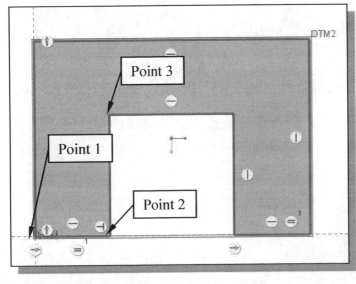

1. Complete the sketch as shown, a closed region ending at the starting point (*Point 1*). Watch the displayed constraint symbols while sketching, especially the **Equal Length** constraint, **=**, applied to the bottom two horizontal edges.

➤ Note that all segments are either *vertical* or *horizontal*.

2. Inside the *graphics area*, click **twice** with the **middle-mouse-button** to end the current line sketch.

❖ *Creo Parametric's* **Intent Manager** automatically places dimensions and constraints on the sketched geometry. This is known as the **Dynamic Dimensioning and Constraints** feature. Constraints and dimensions are added "on the fly." Do not be concerned with the size of the sketched geometry or the displayed dimensional values; we will modify the sketched geometry in the following sections.

Dynamic Viewing Functions

Creo Parametric provides a special user interface, **Dynamic Viewing**, which enables convenient viewing of the entities in the display area at any time. The **Dynamic Viewing** functions are controlled with the combinations of the middle mouse button, the **[Ctrl]** key and the **[Shift]** key on the keyboard.

Zooming – Turn the **mouse wheel** or **[Ctrl]** key and **[middle-mouse-button]**

Use the **mouse wheel** to perform the zooming option; turning the wheel forward will reduce the scale of display. Hold down the **[Ctrl]** key and press down the middle-mouse-button in the display area. Drag the mouse vertically on the screen to adjust the scale of the display. Moving upward will reduce the scale of the display, making the entities display smaller on the screen. Moving downward will magnify the scale of the display.

Zoom   Ctrl   +   Middle mouse button

Panning – **[Shift]** key and **[middle-mouse-button]**

Hold down the **[Shift]** key and press down the middle-mouse-button in the display area. Drag the mouse to pan the display. This allows you to reposition the display while maintaining the same scale factor of the display. This function acts as if you are using a video camera. You control the display by moving the mouse.

Pan   Shift   +   Middle mouse button

➢ On your own, use the *Dynamic Viewing* functions to reposition and magnify the scale of the 2D sketch to the center of the screen so that it is easier to work with.

Step 4: Apply/Modify Constraints and Dimensions

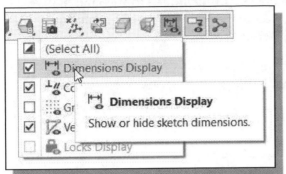

1. Switch **on** the Dimensions Display option by clicking the display option icon as shown.

➢ As the sketch is made, *Creo Parametric* automatically applies geometric constraints (such as **Horizontal**, **Vertical** and **Equal Length**) and dimensions to the sketched geometry. We can continue to modify the geometry, apply additional constraints and/or dimensions, or define/modify the size and location of the existing geometry. It is more than likely that some of the automatically applied dimensions may not match with the design intent we have in mind. For example, we might want to have dimensions identifying the overall-height, overall-width, and the width of the inside-cut of the design, as shown in the figures below.

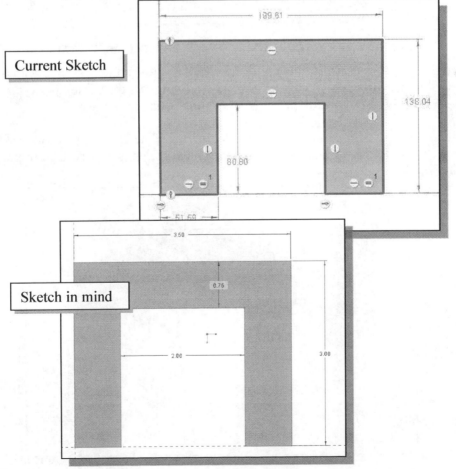

Current Sketch

Sketch in mind

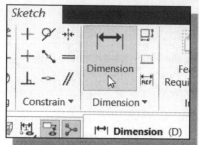

2. Click on the **Normal Dimension** icon in the *Sketching* toolbar as shown. This command allows us to create defining dimensions.

3. Select the **Right Vertical line** by left-clicking once on the line as shown.

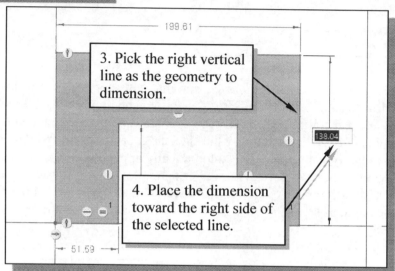

3. Pick the right vertical line as the geometry to dimension.

4. Place the dimension toward the right side of the selected line.

4. Move the graphics cursor to the right of the selected line and click once with the **middle-mouse-button** to place the dimension. (Note that the value displayed on your screen might be different than what is shown in the above figure.)

5. In the *dimension value* box, the current length of the line is displayed. Enter **3.0** as the new value for the dimension. Notice the entire sketch is now resized in proportion to the original shape.

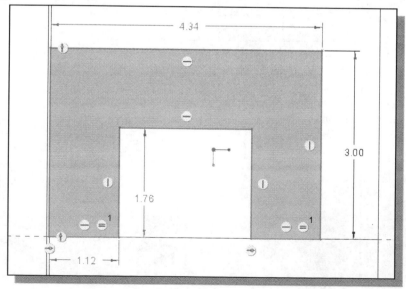

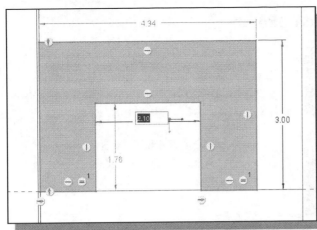

6. Select the *inside horizontal line*.

7. Place the dimension, by clicking once with the **middle-mouse-button**, at a location below the line.

8. Click again with the **middle-mouse-button** to exit the *Edit Dimension* mode.

❖ The Dimension command will create a length dimension if a single line is selected. Also notice the lower-left size dimension applied automatically by the *Intent Manager* is removed as the new dimension is defined. Note that the dimensions we just created are displayed with a different color than those that are applied automatically. The dimensions created by the *Intent Manager* are called **weak dimensions**, which can be replaced/deleted as we create specific defining dimensions to satisfy our design intent.

9. Select the **top horizontal line** as shown below.

10. Select the **inside horizontal line** as shown below.

11. Place the dimension, by clicking once with the **middle-mouse-button,** at a location in between the selected lines as shown below.

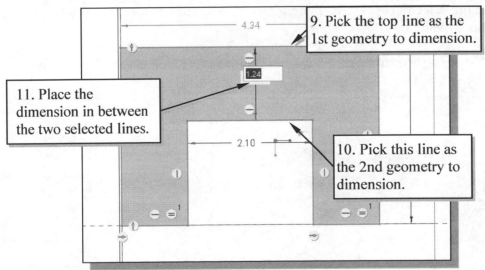

9. Pick the top line as the 1st geometry to dimension.

11. Place the dimension in between the two selected lines.

10. Pick this line as the 2nd geometry to dimension.

12. Click again with the **middle-mouse-button** to exit the *Edit Dimension* mode.

❖ When two parallel lines are selected, the Dimension command will create a dimension measuring the distance in between.

Modifying the Dimensions in a Sketch

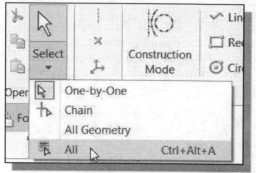

1. In the pull-down menu area, click on the down-arrow on **Select** to display the option list and select the **All** option as shown.

   (Note that using the hotkey combination **Crtl+Alt+A** can also activate this option.)

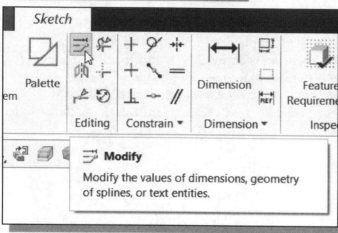

2. In the *Editing* toolbar, click on the **Modify** icon as shown.

- With the pre-selection option, all dimensions are selected and listed in the *Modify Dimensions* dialog box.

3. Turn *off* the **Regenerate** option by unchecking the option in the *Modify Dimensions* dialog box as shown.

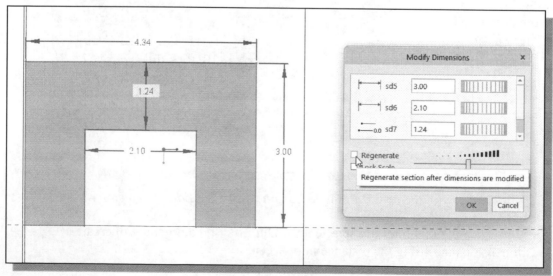

4. On your own, adjust the dimensions as shown below. Note that the dimension selected in the *Modify Dimensions* dialog box is identified with an enclosed box in the display area.

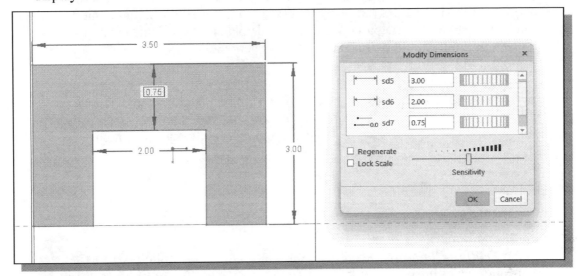

5. Inside the *Modify Dimensions* dialog box, click on the **OK** button to regenerate the sketched geometry and exit the Modify Dimensions command.

6. On your own, use the dynamic zooming function and adjust the display roughly as shown in the above figure.

Repositioning Dimensions

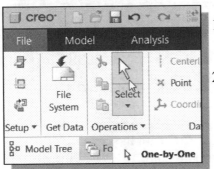

1. Confirm the **Select** icon, in the *Operations toolbar*, is activated as shown.

2. Press and hold down the **left-mouse-button** on any dimension text, then drag the dimension to a new location in the display area. (Note the cursor is changed to a moving arrow icon during this operation.)

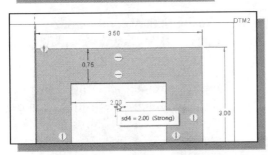

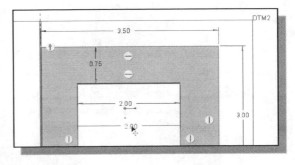

Step 5: Completing the Base Solid Feature

➢ Now that the 2D sketch is completed, we will proceed to the next step: creating a 3D part from the 2D section. Extruding a 2D section is one of the common methods that can be used to create 3D parts. We can extrude planar faces along a path. In *Creo Parametric*, the default extrusion direction is perpendicular to the selected sketching plane, DTM2.

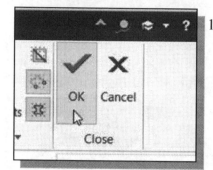

1. In the *Ribbon* toolbar, click **OK** to exit the *Creo Parametric 2D Sketcher*. The 2D sketch is the first element of the *Extrude* feature definition.

2. In the *Feature Option Dashboard*, confirm the *depth value* option is set as shown. This option sets the extrusion of the section by **Extrude from sketch plane by a specified depth value**.

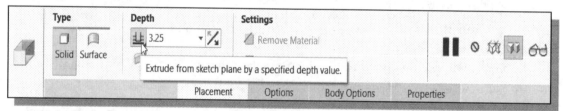

3. In the *depth value* box, enter **2.5** as the extrusion depth.

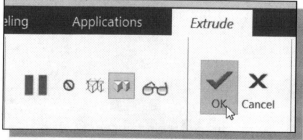

4. In the *Dashboard* area, click **OK** to proceed with the creation of the solid feature.

5. Use the hotkey combination **CTRL+D** to reset the display to the default 3D orientation.

➢ Note that all dimensions disappeared from the screen. All parametric definitions are stored in the ***Creo Parametric* database**, and any of the parametric definitions can be displayed and edited at any time.

The Third Dynamic Viewing Function

3D Dynamic Rotation – [middle mouse button]

Press down the middle-mouse-button in the display area. Drag the mouse on the screen to rotate the model about the screen.

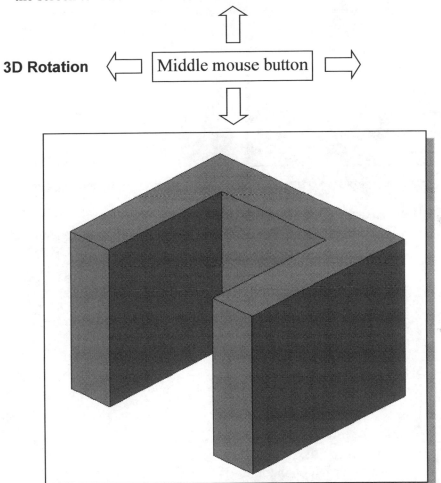

❖ On your own, practice the use of the *Dynamic Viewing functions*; note that these are convenient viewing functions at any time.

Display Modes: Wireframe, Shaded, Hidden Edge, No Hidden

The display in the graphics window has six display modes: Shading with Edges, Shading with Reflections, Shading, No Hidden lines, Hidden Line, and Wireframe image. To change the display mode in the active window, click on the display mode button in the *Display* toolbar to display the list as shown.

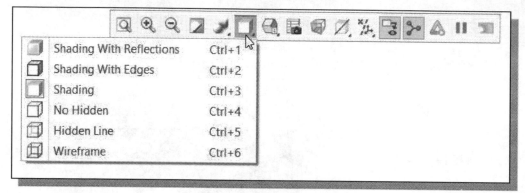

Shading With Reflections	Ctrl+1	
Shading With Edges	Ctrl+2	
Shading	Ctrl+3	
No Hidden	Ctrl+4	
Hidden Line	Ctrl+5	
Wireframe	Ctrl+6	

❖ **Shading With Reflections:**
The second icon in the display mode button group generates a more enhanced shaded image of the 3D object.

❖ **Shading With Edges:**
The first icon in the display mode button group generates a shaded image of the 3D object with edges highlighted.

❖ **Shading:**
The third icon in the display mode button group generates a shaded image of the 3D object.

❖ **No Hidden:**
The fourth icon in the display mode button group can be used to generate a wireframe image of the 3D object with all the back lines removed.

❖ **Hidden Line:**
The fifth icon in the display mode button group can be used to generate a wireframe image of the 3D object with all the back lines shown as hidden lines.

❖ **Wireframe:**
The sixth icon in the display mode button group allows the display of 3D objects using the basic wireframe representation scheme.

Step 6: Adding Additional Features

> Next, we will create another extrusion feature that will be added to the existing solid object.

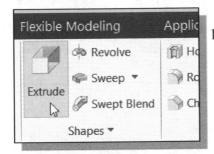

1. In the *Shapes* toolbar (the fourth toolbar in the *Ribbon* toolbar), select the **Extrude** tool option as shown.

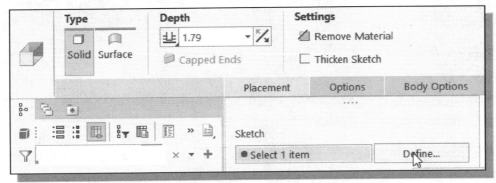

2. Click the **Placement** option and choose **Define** to begin creating a new *internal sketch*.

3. Pick the right vertical face of the solid model as the sketching plane as shown in the figure below.

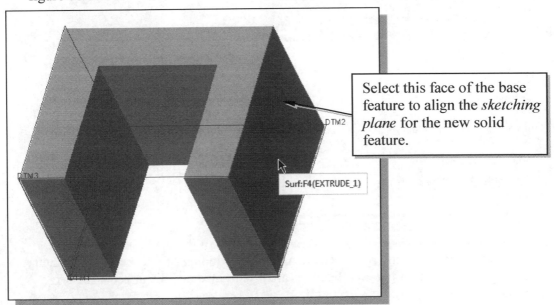

Select this face of the base feature to align the *sketching plane* for the new solid feature.

4. In the display area, pick the **top face** of the base feature as the reference plane as shown.

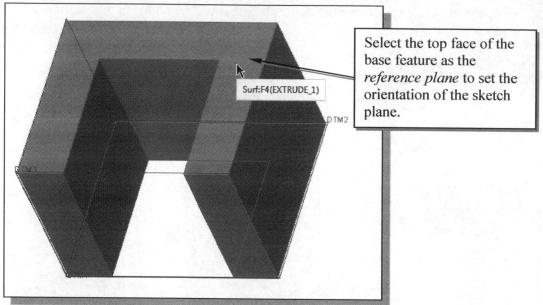

Select the top face of the base feature as the *reference plane* to set the orientation of the sketch plane.

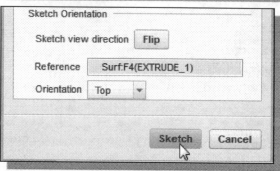

5. In the Sketch Orientation menu, pick **Top** to set the reference plane Orientation.

6. Pick **Sketch** to exit the *Section Placement* window and proceed to enter the *Creo Parametric Sketcher* mode.

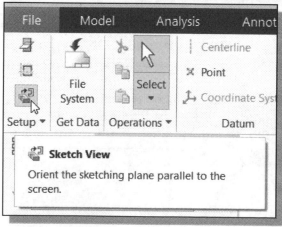

7. On your own, use the **dynamic rotation** option and change the display to a random orientation.

8. To orient the sketching plane parallel to the screen, click on the **Sketch View** icon in the *Setup* toolbar as shown.

9. Note that the top surface of the solid model and one of the datum planes are pre-selected as the sketching references to aid the positioning of the sketched geometry. In the graphics area, the two references are highlighted and displayed with two dashed lines.

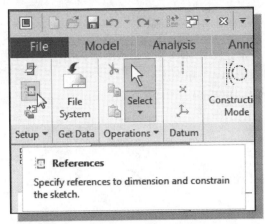

10. In the *Setup* toolbar, click on **References** to display the option list and select the References option.

➢ This will bring up the *References* dialog box.

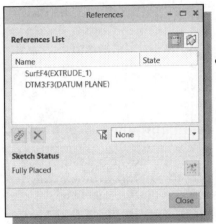

• Note that, in the *References* dialog box, the top surface of the solid model and DTM3 are pre-selected as the sketching references as shown.

11. Select the **right edge** and the **bottom edge** of the base feature so that the four sides of the selected sketching plane, or corresponding datum planes, are used as references as shown.

12. Click **Update** to apply the changes.

13. In the *References* dialog box, click on the **Close** button to accept the selections.

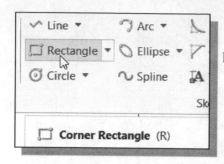

14. In the *Sketching* toolbar, click on the **Corner Rectangle** icon as shown to activate the Rectangle command.

15. Create a rectangle by clicking on the **lower left corner** of the solid model as shown below.

16. Move the cursor upward and place the opposite corner of the rectangle along the **right edge** of the base solid as shown below.

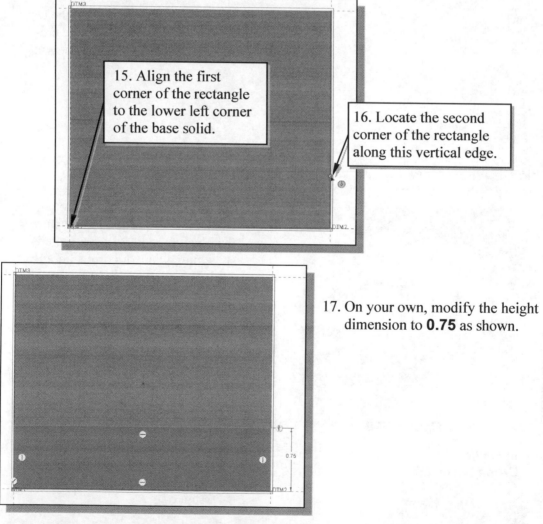

15. Align the first corner of the rectangle to the lower left corner of the base solid.

16. Locate the second corner of the rectangle along this vertical edge.

17. On your own, modify the height dimension to **0.75** as shown.

0.75

- Note that only one dimension, the height dimension, is applied to the 2D sketch; the width of the rectangle is defined by the references.

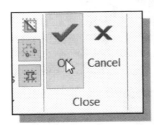

18. In the *Ribbon* toolbar, click on the **OK** icon to end the *Creo Parametric 2D Sketcher* and proceed to the next element of the feature definition.

19. In the *Feature Option Dashboard*, confirm the **depth value** option is set and enter **2.5** as the extrusion depth as shown.

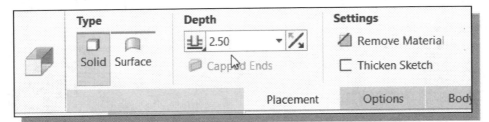

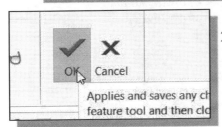

20. In the *dashboard* area, click **OK** to proceed with the creation of the solid feature.

Creating a CUT Feature

We will create a circular cut as the next solid feature of the design. Note that the procedure in creating a *cut feature* is almost the same as creating a *protrusion feature*.

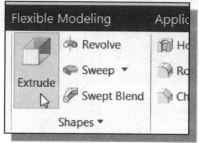

1. In the *Shapes* toolbar (the fourth toolbar in the *Ribbon* toolbar area), select the **Extrude** tool option as shown.

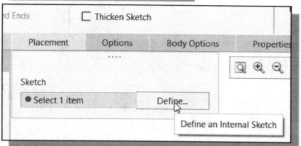

2. Click the **Placement** option and choose **Define** to begin creating a new *internal sketch*.

3. We will use the top surface of the last feature as the sketching plane. Click once, with the **left-mouse-button**, inside the top surface of the rectangular solid feature as shown in the figure below.

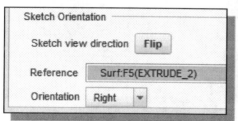

4. In the Sketch Orientation menu, confirm the reference plane Orientation is set to **Right**.

5. Pick the **right vertical face** of the second solid feature as the reference plane, which will be oriented toward the right edge of the computer screen.

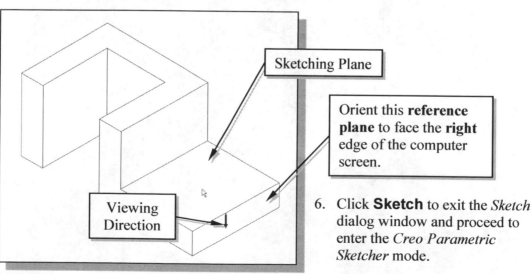

Sketching Plane

Orient this **reference plane** to face the **right** edge of the computer screen.

Viewing Direction

6. Click **Sketch** to exit the *Sketch* dialog window and proceed to enter the *Creo Parametric Sketcher* mode.

Creating the 2D Section of the CUT Feature

1. Note that the **right vertical plane** is pre-selected as a reference for the new sketch.

    *   Note that at least one horizontal reference and one vertical reference are required to position a 2D sketch. We will need at least one more vertical reference for this sketch.

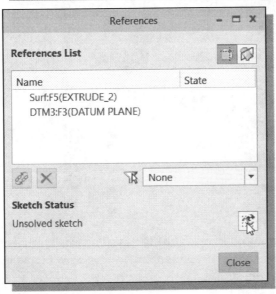

2. Select **DTM3** as the vertical sketching reference as shown. In the graphics area, the two references are highlighted and displayed with two dashed lines.

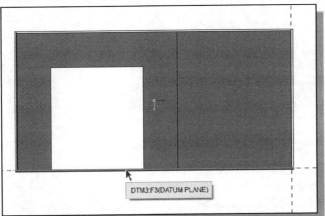

3. Click **Update** to apply the changes.

4. Click on the **Close** button to accept the selected references and proceed to entering the *Creo Parametric Sketcher* module.

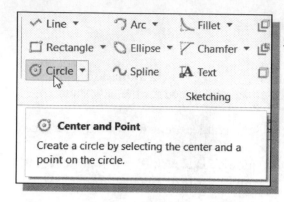

5. In the *Sketching* toolbar, select **Circle** as shown. The default option is to create a circle by specifying the center point and a point through which the circle will pass. The message "*Select the center of a circle*" is displayed in the message area.

6. On your own, create a circle of arbitrary size on the sketching plane as shown.

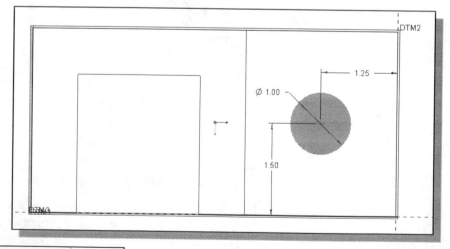

7. On your own, edit/modify the dimensions as shown.

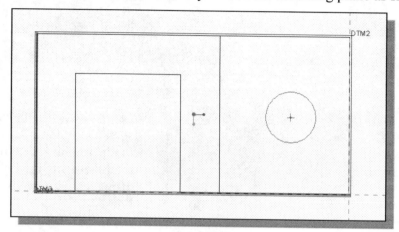

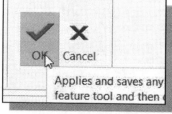

8. In the *Ribbon* toolbar, click OK to exit the *Creo Parametric 2D Sketcher* and proceed to the next element of the feature definition.

9. Switch on the **Remove Material** option as shown in the figure below.

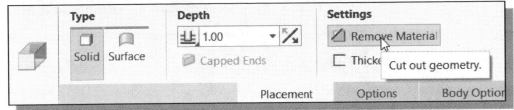

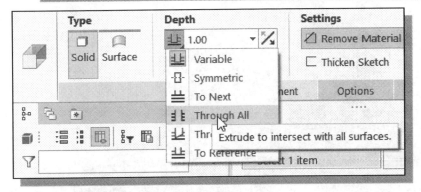

10. In the *Feature Option Dashboard*, select the **Through All** option as shown.

- Note that this Through All option does not require us to enter a value to define the depth of the extrusion; *Creo Parametric* will calculate the required value to assure the extrusion is through the entire solid model.

11. On your own, use the *Dynamic Rotate* function to view the feature.

12. Click on the **Flip direction** icon as shown in the figure below to set the cut direction.

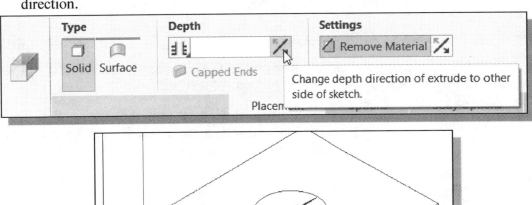

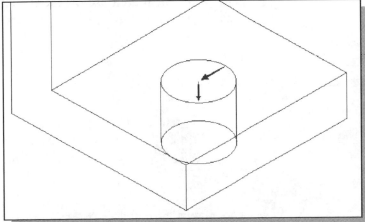

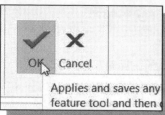

13. Click on **OK** to proceed with the extrusion option.

Creating another CUT Feature

We will next create a triangular cut as the next solid feature of the design.

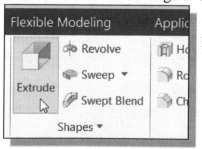

1. In the *Shapes* toolbar (the fourth toolbar in the *Ribbon* toolbar area), select the **Extrude** tool option as shown.

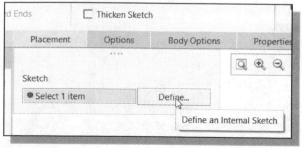

2. Click the **Placement** option and choose **Define** to begin creating a new *internal sketch*.

3. We will use the right vertical surface of the first solid feature as the sketching plane. Click once, with the **left-mouse-button**, inside the top surface of the rectangular solid feature as shown in the figure below.

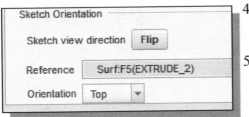

4. In the Sketch Orientation menu, set the reference plane Orientation to **top**.

5. Pick the **top horizontal face** of the second solid feature as the reference plane, which will be oriented toward the top edge of the computer screen.

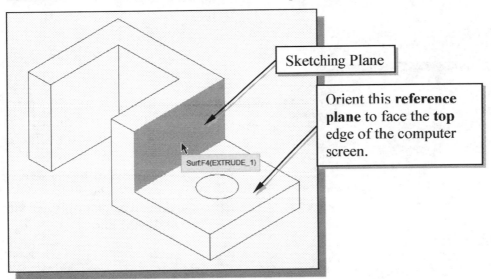

Sketching Plane

Orient this **reference plane** to face the **top** edge of the computer screen.

Surf:F4(EXTRUDE_1)

6. Click **Sketch** to exit the *Sketch* dialog window and proceed to enter the *Creo Parametric Sketcher* mode.

Delete/Select the Sketching References

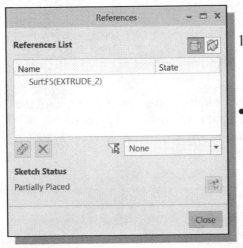

1. Note that currently the **top horizontal plane** of the second solid feature is pre-selected as a reference for the new sketch.

- Note that at least one horizontal reference and one vertical reference are required to position a 2D sketch. We will need at least one more vertical reference for this sketch.

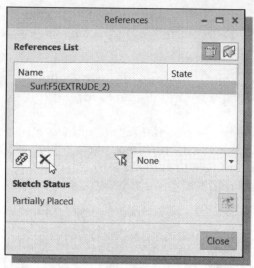

2. Select the **referenced surface** in the item list.

3. Click **Delete** to remove the item as a reference plane.

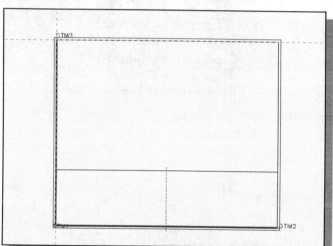

4. Select the left vertical edge, **DTM3**, and the **top surface** as sketching references as shown. In the graphics area, the two references are highlighted and displayed with two dashed lines.

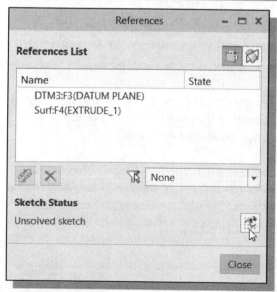

5. Click **Update** to apply the changes.

6. Click on the **Close** button to accept the selected references and proceed to entering the *Creo Parametric Sketcher* module.

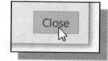

Create the 2D Section

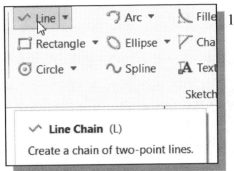

1. In the *Sketching* toolbar, select **Line** as shown. The default option is to create a chain of lines by specifying the endpoint of the line segments.

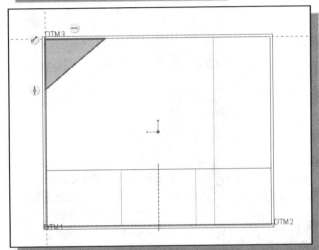

2. On your own, start at the top left corner and create a triangle of arbitrary size on the sketching plane as shown.

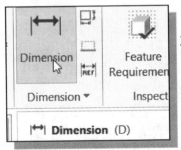

3. On your own, create and modify the length dimension of the top edge of the triangle as shown.

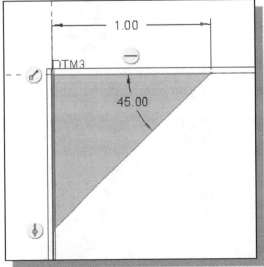

4. Create the angle dimension by selecting the two adjacent lines and place the angular dimension *inside* the *desired quadrant*.

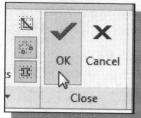

5. In the *Ribbon* toolbar, click **OK** to exit the *Creo Parametric 2D Sketcher* and proceed to the next element of the feature definition.

6. Switch on the **Remove Material** and **Flip direction** option as shown in the figure below.

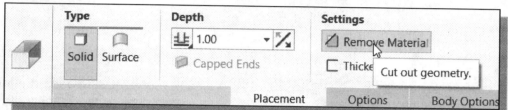

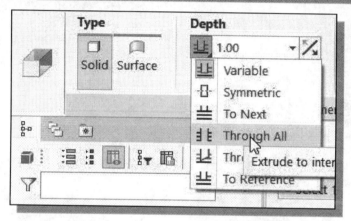

7. In the *Feature Option Dashboard*, select the **To Next** option as shown.

*   Note that with the **To Next** option, *Creo Parametric* will calculate the required value to assure the extrusion is to the next surface.

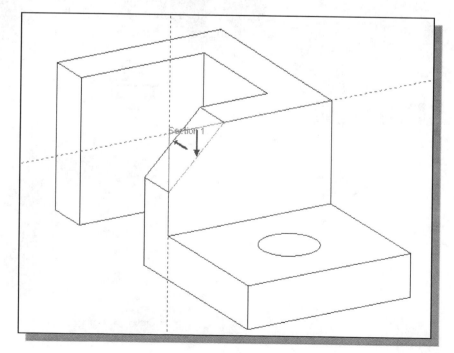

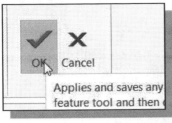

8. Click on **OK** to proceed with the extrusion option.

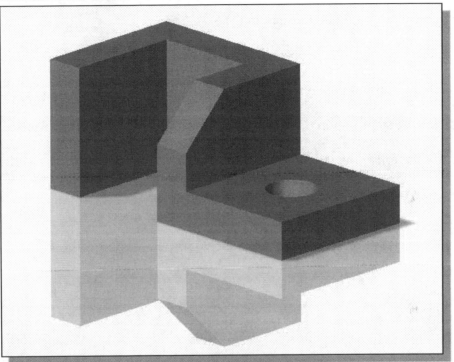

Save the Part

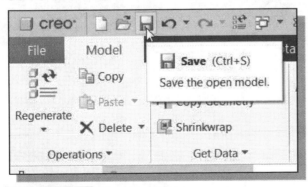

1. Select **Save** in the *Standard* toolbar, or you can also use the "**Ctrl-S**" combination (press down the **[Ctrl]** key and hit the **[S]** key once) to save the part.

2. In the *message area*, the part name is displayed. Click on the **OK** button to save the file.

❖ It is a good habit to save your model periodically. In general, you should save your work onto the disk at an interval of every 15 to 20 minutes.

Review Questions:

1. What is the first thing we should set up in *Creo Parametric* when creating a new model?

2. How do we modify more than one dimension in the *Sketcher*?

3. How do we reposition dimensions in the *Sketcher*?

4. List three of the geometric constraint symbols used by the *Creo Parametric Sketcher*.

5. Describe two different ways to modify dimensions in the *Sketcher*.

6. Describe the steps required to define the orientation of the sketching plane.

7. Identify the following quick-keys commands:

(a)

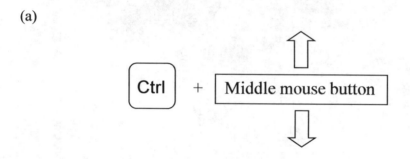

(b)

Exercises: (All dimensions are in inches.)

1. **Angle Spacer** (Plate Thickness: 0.25)

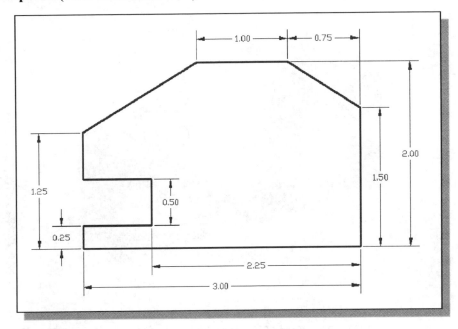

2. **Spacer Plate** (Plate Thickness: 0.125)

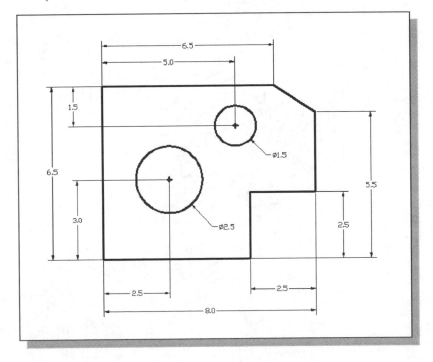

3. Positioning Stop

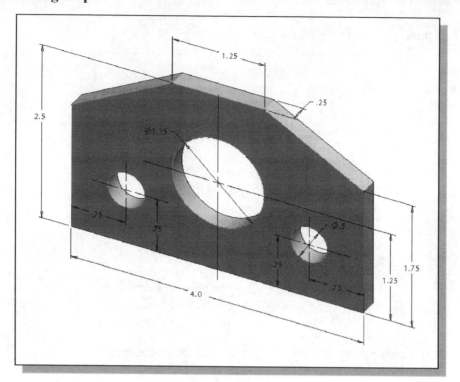

4. Guide Block

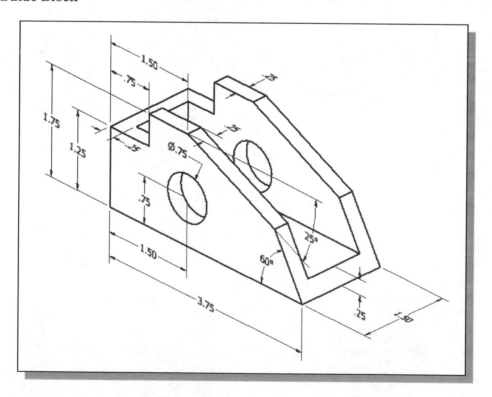

5. Coupler

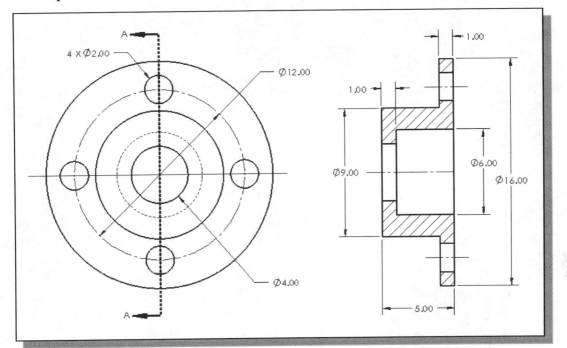

Notes:

Chapter 2
Constructive Solid Geometry Concepts

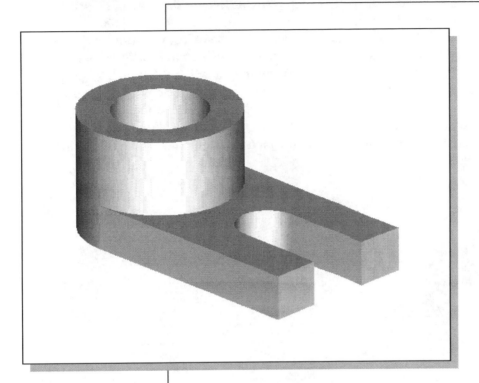

Learning Objectives

- ♦ **Understand the Constructive Solid Geometry Concepts**
- ♦ **Create a Binary Tree**
- ♦ **Understand the basic Boolean Operations**
- ♦ **Understand the importance of Order of Features**
- ♦ **Create Placed Features**
- ♦ **Use the different Extrusion options**

Introduction

In the 1980s, one of the main advancements in **solid modeling** was the development of the **Constructive Solid Geometry** (CSG) method. CSG describes the solid model as combinations of basic three-dimensional shapes (**primitive solids**). The basic primitive solid set typically includes Rectangular-prism (Block), Cylinder, Cone, Sphere, and Torus (Tube). Two solid objects can be combined into one object in various ways, and these operations are known as **Boolean operations**. There are three basic Boolean operations: **JOIN (Union)**, **CUT (Difference)**, and **INTERSECT**. The JOIN operation combines the two volumes included in the different solids into a single solid. The CUT operation subtracts the volume of one solid object from the other solid object. The INTERSECT operation keeps only the volume common to both solid objects. The CSG method is also known as the **Machinist's Approach**, as the method is parallel to machine shop practices.

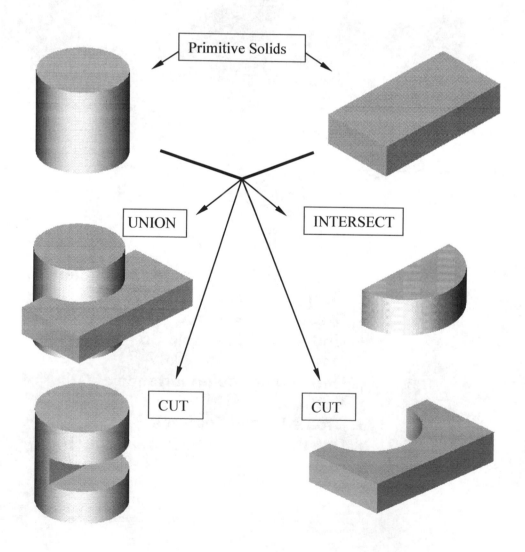

Binary Tree

Constructive Solid Geometry (CSG) is also referred to as the method used to store a solid model in the database. The resulting solid can be easily represented by what is called a **binary tree**. In a binary tree, the terminal branches (leaves) are the various primitives that are linked together to make the final solid object (the root). The binary tree is an effective way to keep track of the *history* of the resulting solid. By keeping track of the history, the solid model can be rebuilt by re-linking through the binary tree. This provides a convenient way to modify the model. We can make modifications at the appropriate links in the binary tree and re-link the rest of the history tree without building a new model.

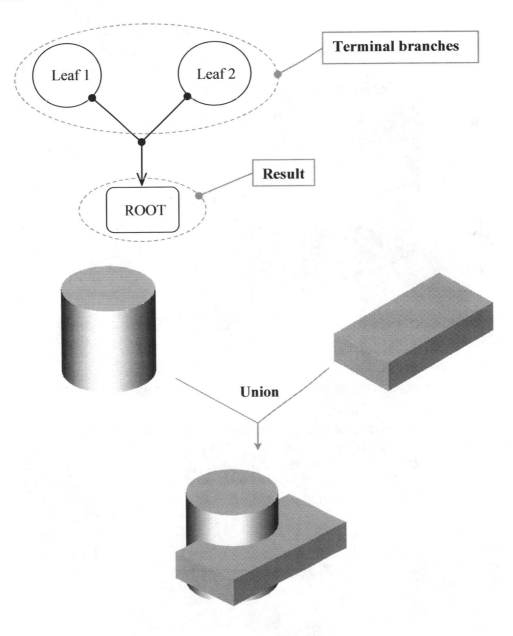

The *Locator* Design

The CSG concept is one of the important building blocks for feature-based modeling. In *Creo Parametric*, the CSG concept can be used as a planning tool to determine the number of features that are needed to construct the model. It is also a good practice to create features that parallel the manufacturing process required for the design. With parametric modeling, we are no longer limited to using only the predefined basic solid shapes. In fact, any solid features we create in *Creo Parametric* are used as primitive solids; parametric modeling allows us to maintain full control of the design variables that are used to describe the features. In this lesson, a more in-depth look at the parametric modeling procedure is presented. The equivalent CSG operation for each feature is also illustrated.

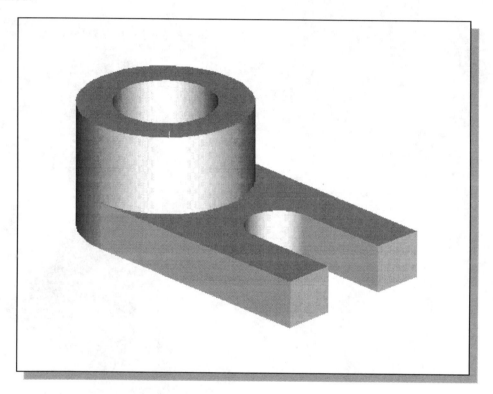

➢ Before going through the tutorial, on your own make a sketch of a CSG binary tree of the *Locator* design using only two basic types of primitive solids: cylinder and rectangular prism. In your sketch, how many *Boolean operations* will be required to create the model? What is your choice for the first primitive solid to use, and why? Take a few minutes to consider these questions and do the preliminary planning by sketching on a piece of paper. Compare the sketch you make to the CSG binary tree steps shown on the next page. Note that there are many different possibilities in combining the basic primitive solids to form the desired solid model. Even for the simplest design, it is possible to take several different approaches to creating the same solid model.

Modeling Strategy – CSG Binary Tree

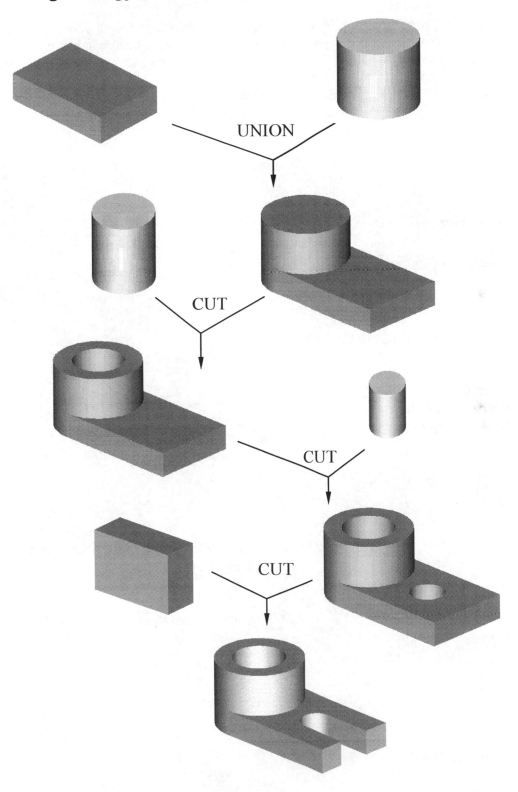

Starting *Creo Parametric*

1. Select the **Creo Parametric** icon on the desktop or type [*Creo*] at your system search area to start *Creo Parametric*. The *Creo Parametric* main window will appear on the screen.

2. Click on the **New** icon as shown. Notice that we can also use the key combination **Ctrl-N** to start a new object.

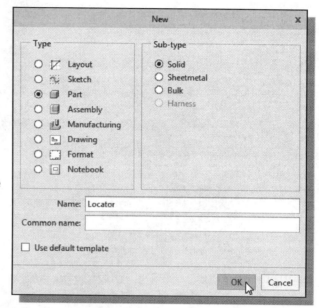

3. In the *New* form, enter **Locator** as the solid part file **Name**.

4. Turn *off* the Use default template option.

5. Click **OK** to continue. The *Model Tree* window and the *Feature* toolbars appear on the screen.

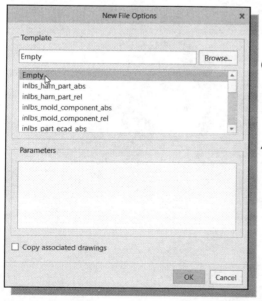

6. In the *New File Options* dialog box, select **Empty** in the option list to not use any template file.

7. Click on the **OK** button to accept the settings and enter the *Creo Parametric Part Modeling* mode.

Units Setup

When starting a new model, the first thing we should do is to choose the set of units we want to use.

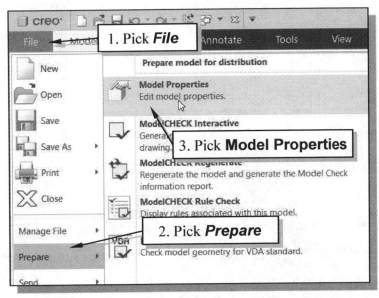

1. Use the left-mouse-button and select **File** pull-down menu.

2. Use the left-mouse-button and select **Prepare** in the **pull-down list** as shown.

3. Select **Model Properties** in the expanded list as shown.

➢ Note that the *Creo Parametric* menu system is context-sensitive, which means that the menu items and icons of the non-applicable options are grayed out (temporarily disabled).

4. Select the **Change** option that is to the right of the **Units** option in the *Model Properties* window.

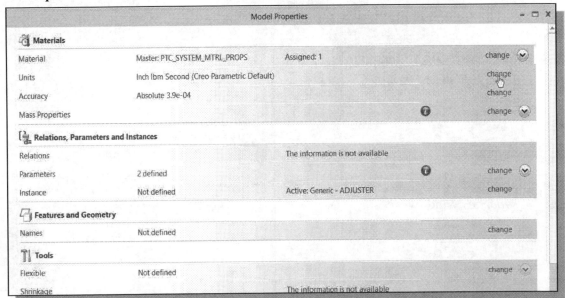

5. In the **Units Manager – Systems of Units** form, the *Creo Parametric* default setting Inch lbm Second is displayed. The set of units is stored with the model file when you save. Pick **millimeter Newton Second (mmNs***)* by clicking in the list window as shown.

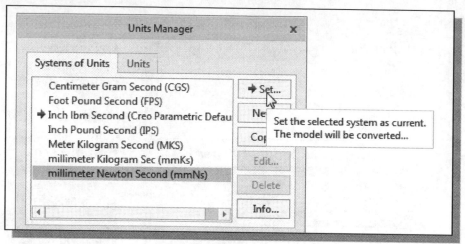

6. Click on the **Set** button to accept the selection.

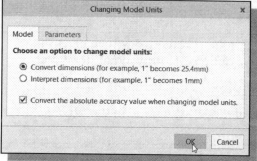

7. In the *Changing Model Units* dialog box, click on the **OK** button to accept the change of the units.

➢ Note that *Creo Parametric* allows us to change model units even after the model has been constructed.

8. Click on the **Close** button to exit the *Units Manager* dialog box.

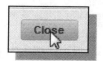

9. Pick **Close** to exit the *Model Properties* window.

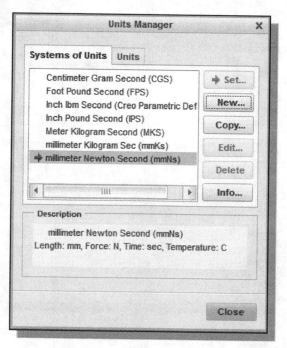

Adding the First Part Features – Datum Planes

Creo Parametric provides many powerful tools for model creation. In doing feature-based parametric modeling, it is a good practice to establish three reference planes to locate the part in space. The reference planes can also be used as location references in feature constructions.

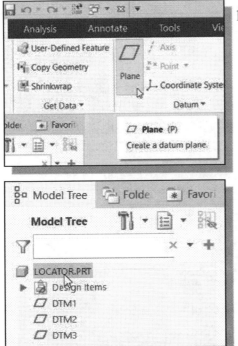

1. In the *Datum* toolbar, click on the **Datum Plane** tool icon to create three datum planes as shown.

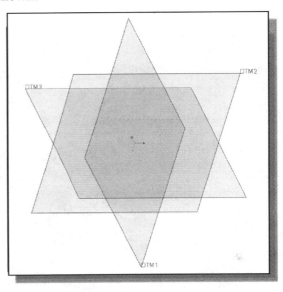

2. Click the model name, *Locator.prt*, in the *Navigator* window to deselect DTM3.

Set up Grid Display

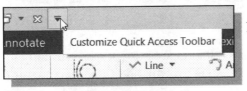

1. To set up the Grid display, click on the **Customize Quick Access toolbar** icon as shown.

2. Choose **More Commands** in the pull-down menu list as shown.

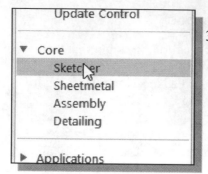

3. In the *Creo Parametric Options* dialog box, select **Core → Sketcher** as shown.

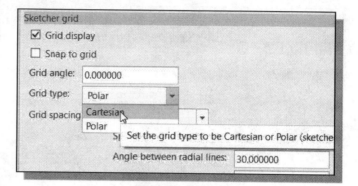

4. In the *Sketcher Grid* option, switch on the **Grid display** and choose **Cartesian** as shown.

 5. Click on the **OK** button to accept the settings.

6. Click **Yes** to save the configuration settings when the warning window appears.

7. On your own, save the configuration file at the default file location.

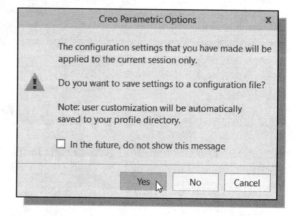

Base Feature

In *parametric modeling*, the first solid feature is called the **base feature**, which usually is the primary shape of the model. Depending upon the design intent, additional features are added to the base feature.

Some of the considerations involved in selecting the base feature:

- **Design Intent** – Determine the functionality of the design; identify the feature that is central to the design.

- **Order of features** – Choose the feature that is the logical base in terms of the order of features in the design.

- **Ease of making modifications** – Select the feature that is more stable and is less likely to be changed.

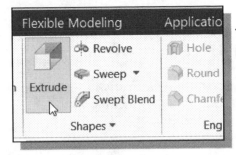

❖ We will create a rectangular block as the base feature of the *Locator* design.

1. In the *Shapes* toolbar, select the **Extrude** tool option as shown.

• The *Feature Option Dashboard*, which contains applicable construction options, is displayed above the message area near the bottom of the *Creo Parametric* main window.

2. Click the **Placement** option and choose **Define** to begin creating a new *internal sketch*.

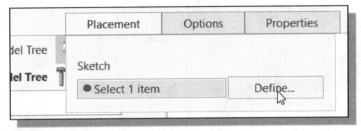

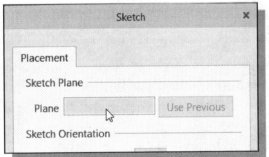

3. Move the cursor inside the **Plane** option box in the *Sketch* window as shown. The message "*Select a plane surface or datum plane to be used as a sketching plane.*" is displayed in the message area.

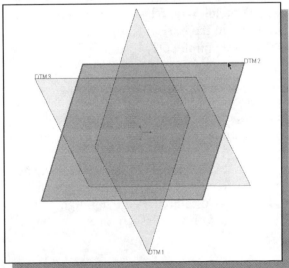

4. In the graphic area, select **DTM2** by clicking on the text DTM2 as shown.

❖ Notice an arrow appears on the edge of DTM2. The arrow direction indicates the viewing direction of the sketch plane. The viewing direction can be reversed by clicking on the Flip button in the Sketch Orientation section of the pop-up window.

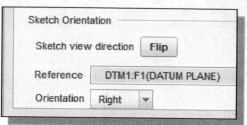

5. Confirm the Sketch Orientation options are set to **DTM1** and **Right**, as shown.

6. Pick **Sketch** to exit the *Section Placement* window and proceed to enter the *Creo Parametric Sketcher* mode.

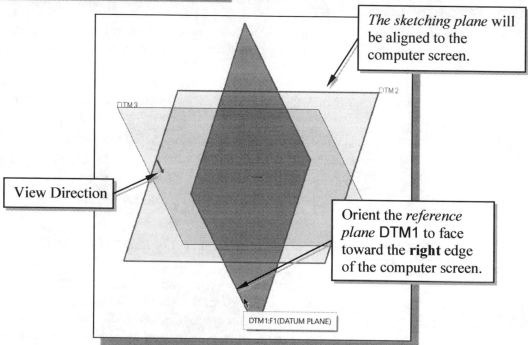

The sketching plane will be aligned to the computer screen.

View Direction

Orient the *reference plane* DTM1 to face toward the **right** edge of the computer screen.

Create a 2D Parametric Section

The first thing that *Creo Parametric Sketcher* expects us to do is specify **sketching references**. In the previous sections, we created the three datum planes to help orient the model in 3D space. Now we need to orient the 2D sketch with respect to the three datum planes. At least two references are required to orient in the horizontal direction and in the vertical direction. By default, the two planes (in our example, DTM1 and DTM3) that are perpendicular to the sketching plane (DTM2) are automatically selected.

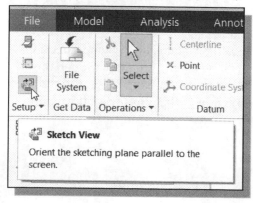

1. To orient the sketching plane parallel to the screen, click on the **Sketch View** icon in the *Setup* toolbar as shown.

❖ Note that DTM1 and DTM3 are pre-selected as the sketching references. In the graphics area, the two references are highlighted and displayed with two dashed lines.

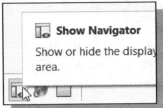

2. Near the bottom of the graphics window, click on the **Navigator** icon to toggle *off* the display of the *Model Tree* window.

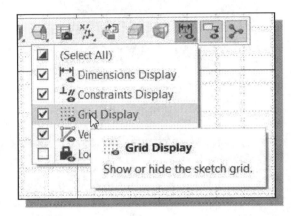

3. In the ***Display Control*** toolbar, switch *on* the *Grid Display*.

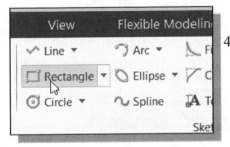

4. In the *Sketching* toolbar, click on the **Rectangle** icon as shown to activate the Create Rectangle command.

5. Create a rectangle as shown, with one corner on the vertical axis (DTM1). (Do not be concerned if the dimensional values on your screen are different than what are shown here.)

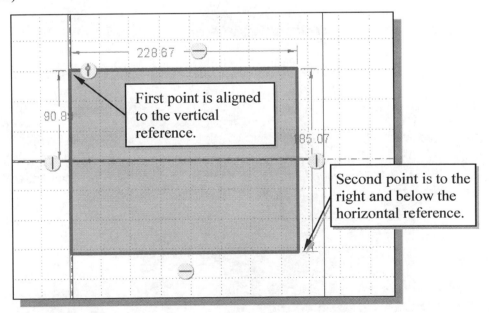

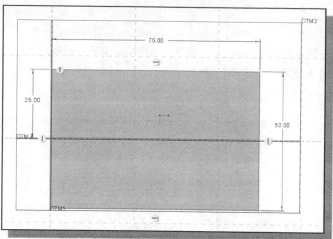

6. On your own, modify the dimensional values to **75mm x 50mm** and centered vertically.

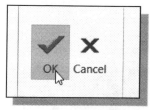

7. In the *Ribbon* toolbar, click on the **OK** icon to exit the *Creo Parametric 2D Sketcher* and proceed to the next element of the feature definition.

8. In the *Feature Options Dashboard*, confirm the *depth value* option is set and enter **15** as the *extrusion depth* as shown. (Hit the **ENTER** key once.)

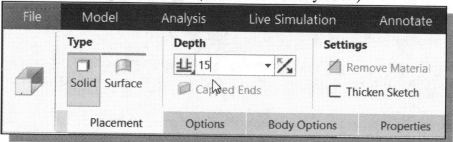

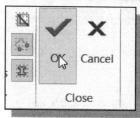

9. In the message area, click OK to proceed with the creation of the solid feature.

➢ Note that all dimensions disappeared from the screen. All parametric definitions are stored in the *Creo Parametric* **database** and can be displayed and edited at anytime.

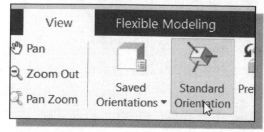

10. In the **View** tab, pick the **Standard Orientation** option [quick-key: **Ctrl-D**] in the *Ribbon* toolbar.

Create the Second Solid Feature

We will add a cylinder to the part as the second solid feature of the *Locator*.

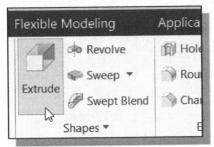

1. Switch to the **Model** tab, and in the *Shapes* toolbar (the fourth toolbar in the *Ribbon* toolbar), select the **Extrude** tool option as shown.

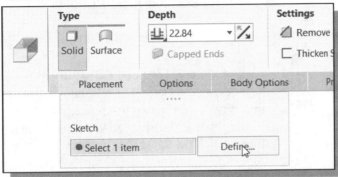

2. Click the **Placement** option and choose **Define** to begin creating a new *internal sketch*.

3. We will use datum plane **DTM2** as the *sketching plane*. Pick **DTM2**.

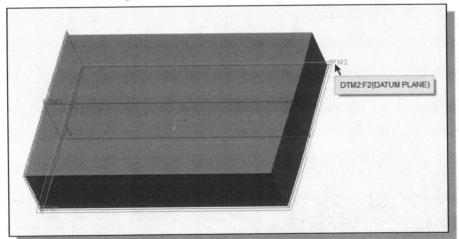

4. An arrow appears on one of the edges of DTM2 to indicate the viewing direction of the sketching plane.

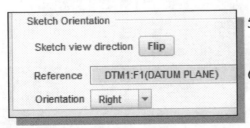

5. Confirm the Orientation option is set to reference **DTM1** and **Right** as shown.

6. Pick **Sketch** to exit the *Section Placement* window and proceed to enter the *Creo Parametric Sketcher* mode.

Create the 2D Sketch

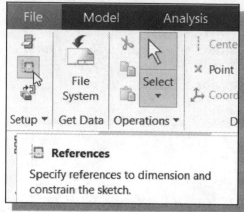

1. In the *Setup* toolbar, click on **References** to show the *References* options.

    ➢ This will bring up the *References* dialog box.

2. On your own, orient the sketching plane parallel to the screen by clicking on the **Sketch View** icon in the *Setup* toolbar.

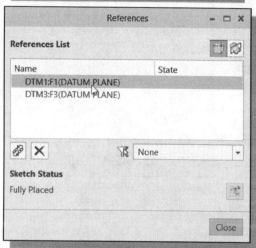

3. Note that the **DTM1** and **DTM3** are pre-selected as the sketching references. In the graphics area, the two references are highlighted and displayed with two dashed lines. Click on the **DTM1** reference in the *References* dialog box and notice it is highlighted in the graphics window.

4. References can be easily added/deleted in *Creo Parametric Sketcher*. Click on the **top edge** of the base feature as shown.

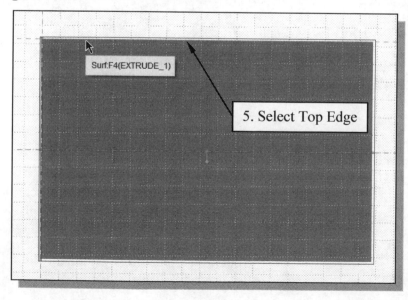

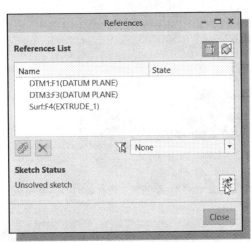

5. In the *References* dialog box, the two edges are added to the reference list. Click **Update** to apply the changes.

6. Click on the **Close** button to exit the References command.

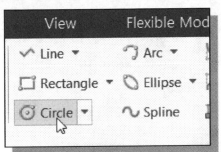

7. In the *Sketching* toolbar, select **Circle** as shown. The default option is to create a circle by specifying the center point and a point through which the circle will pass. The message "*Select the center of a circle*" is displayed in the message area.

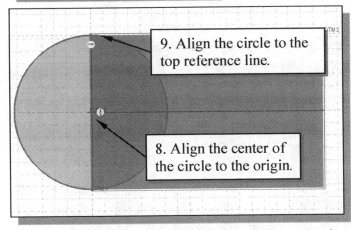

9. Align the circle to the top reference line.

8. Align the center of the circle to the origin.

8. Move the cursor along the axes and watch for the alignment symbol. Pick the **origin** as the center of the circle.

9. Click the **top reference line** when the Tangent symbol is shown.

- The circle is fully defined and no dimension is needed. The selected references, which are aligned to the base solid, control the size of the circle.

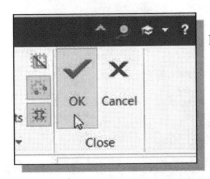

10. In the *Ribbon* toolbar, click **OK** to exit the *Creo Parametric 2D Sketcher* and proceed to the next element of the feature definition.

11. In the *depth* value box, enter **40** as the extrusion depth.

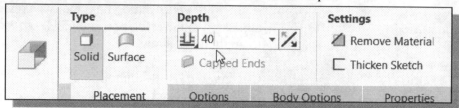

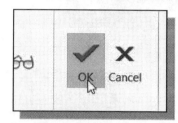

12. In the *message area*, click on the **OK** button and proceed with the feature definition.

13. Pick the **Standard Orientation** option in the *Display Control* toolbar or use the quick-key [**Ctrl-D**] option to reorient the display.

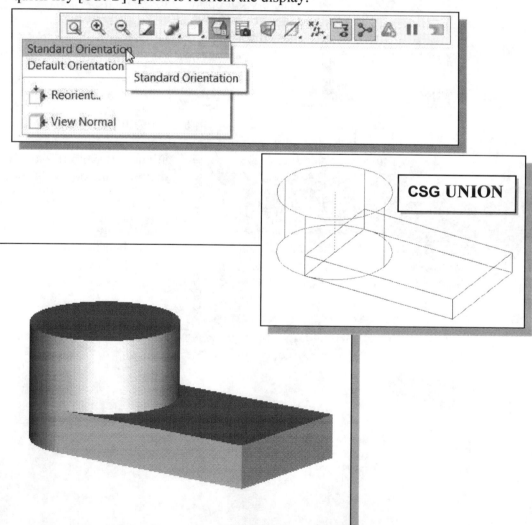

Create a CUT Feature

We will create a circular cut as the next solid feature of the *Locator*.

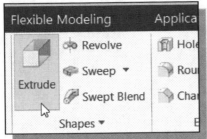

1. In the *Shapes* toolbar (the fourth toolbar in the *Ribbon* toolbar area), select the **Extrude** tool option as shown.

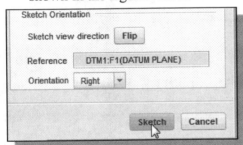

2. Click the **Placement** option and choose **Define** to begin creating a new *internal sketch*.

3. We will use the **top surface** of the last feature as the sketching plane. Click once, with the left-mouse-button, inside the top surface of the circular solid feature as shown in the figure below.

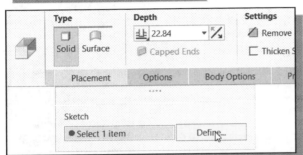

4. In the Sketch Orientation menu, confirm the reference plane Orientation is set to reference **DTM1** and orientation to **Right**.

5. Click **Sketch** to exit the *Section Placement* window and proceed to enter the *Creo Parametric Sketcher* mode.

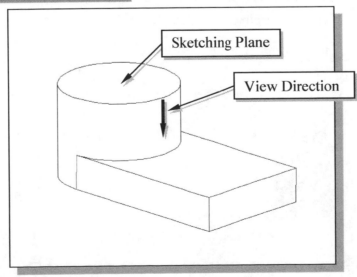

Create a 2D Section

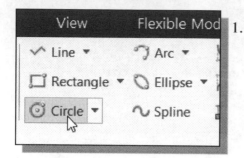

1. In the *Sketching* toolbar, select **Circle** as shown. The default option is to create a circle by specifying the center point and a point through which the circle will pass. The message "*Select the center of a circle*" is displayed in the message area.

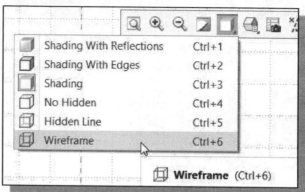

2. Switch to the **Wireframe** display using the *Display Control* toolbar.

3. Move the cursor along the axes and watch for the alignment symbol. Pick the origin as the center of the circle.

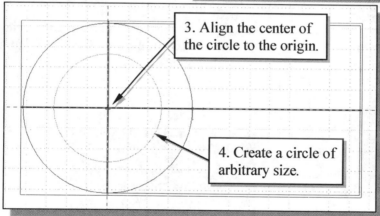

3. Align the center of the circle to the origin.

4. Create a circle of arbitrary size.

4. On your own, create a circle of arbitrary size.

5. On your own, modify the diameter to **40mm**.

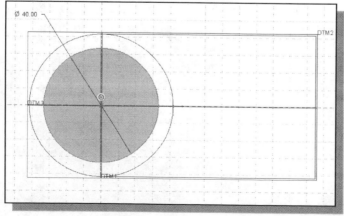

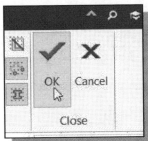

6. In the *Close* toolbar, click on **OK** to exit the *2D Sketcher* and proceed to the next element of the extrude feature definition.

7. Pick the **Standard Orientation** option in the *Display Control* toolbar or use the quick-key [**Ctrl-D**] option to reorient the display.

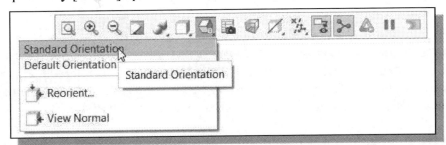

❖ Note that the default setting of the **Extrude** tool command is set to **add material**. The arrow indicates the extrusion direction.

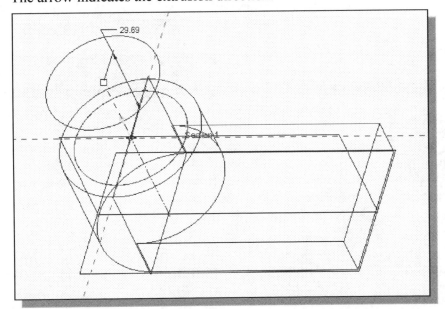

8. Click on the **Remove Material** icon as shown in the figure below.

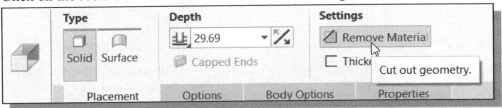

9. In the *Feature Option Dashboard*, select the **Through All** option as shown.

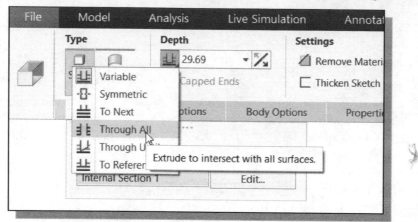

*   Note that the **Extrude to intersect with all surface** option does not require us to enter a value to define the depth of the extrusion; *Creo Parametric* will calculate the required value to assure the extrusion is through the entire solid model.

10. Click on the **Flip direction** icon, as shown in the figure below, to set the cut direction.

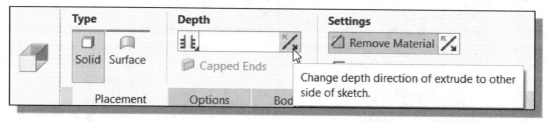

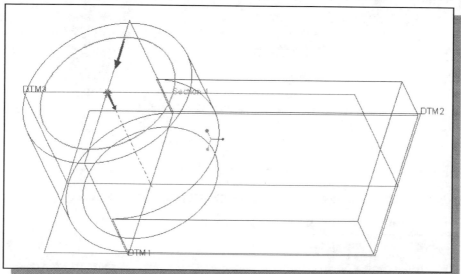

❖   We are now returned to the feature dialog box with all the required elements defined. Prior to creating the feature, we can preview and/or modify any of the elements that are currently defined.

11. Click on the **Preview** button to examine the current solid feature.

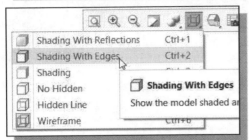

12. Switch the display to **Shading with edges** to redisplay the model in the shaded format.

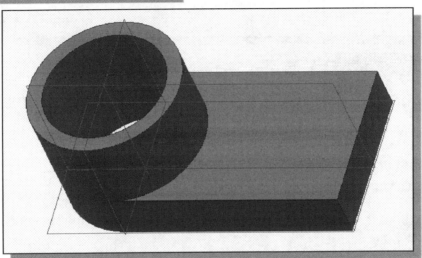

Redefine a Feature Element

Quite often during the design stage, minor modifications become necessary as the model is being formed. We will illustrate the flexibility and functionality of the feature-based modeling concept by redefining one of the elements of the feature.

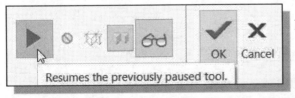

1. Click on the **Resume** button to exit the *Preview* mode.

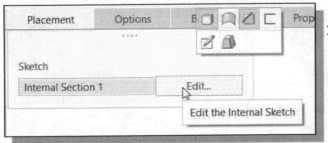

2. Click the **Placement** option and choose **Edit** to edit the current *internal sketch*.

3. On your own, modify the diameter dimension to **30mm** as shown in the figure.

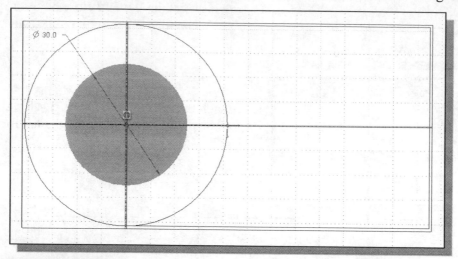

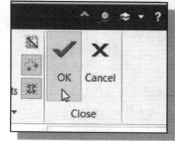

4. In the *Ribbon* toolbar, click on the **OK** icon to end the *Creo Parametric 2D Sketcher*.

5. Click on **OK** to proceed with the extrusion option.

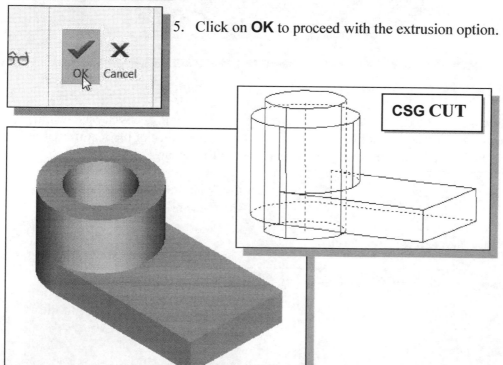

Create a Placed HOLE Feature

In *parametric modeling*, there are two types of geometric features: **placed features** and **sketched features**. The last cut feature we created is a *sketched feature*, where we created a 2D section and performed an extrusion operation. We can also create a hole feature, which is a *placed feature*. A *placed feature* is a feature that does not need a sketch and can be created automatically. Holes, fillets, chamfers, and shells are all placed features.

1. In the *Engineering* toolbar, select the **Hole** tool option as shown.

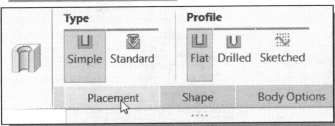

2. In the *Feature Option Dashboard*, click on the **Placement** option as shown.

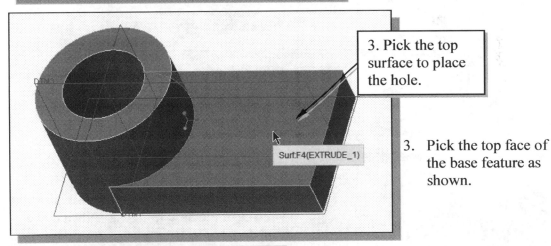

3. Pick the top surface to place the hole.

Surf:F4(EXTRUDE_1)

3. Pick the top face of the base feature as shown.

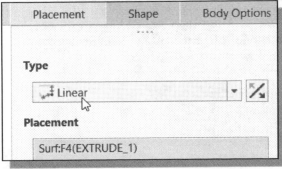

4. Confirm the placement type is set to **Linear** as shown. Note the Linear option requires the setup of two linear dimensions.

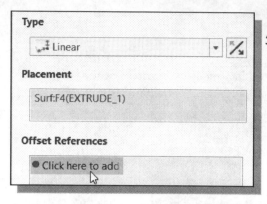

5. In the *Feature Option Dashboard*, click once with the left-mouse-button inside the **Offset References** option box as shown.

6. In the graphics area, select **DTM3** as the first secondary reference as shown in the figure below.

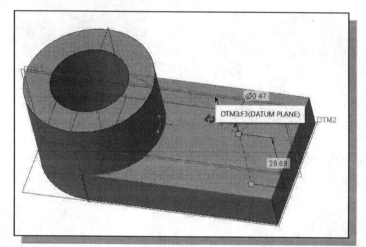

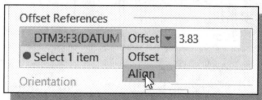

7. In the **Secondary references** section, set the alignment option to **Align** as shown.

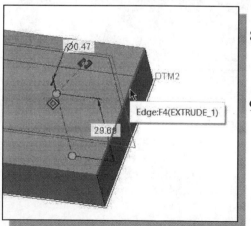

8. Hold down the **[Ctrl]** key and select the **right surface** of the base feature as shown in the figure to the left.

9. Enter **30** for the **Offset** distance of the second reference as shown in the figure below.

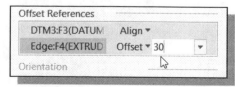

10. In the *Feature Option Dashboard*, set the *feature diameter* to **20mm** as shown.

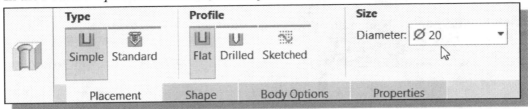

11. In the *Feature Option Dashboard*, select the **Drill to intersect with all surfaces** option as shown.

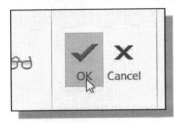

12. Click on the **OK** icon and proceed to create the hole feature.

➢ The last two features we have created produced the same result even though they were accomplished by using two different approaches. The Extrude tool command is more flexible in the shapes that can be used, but it also requires the definition of more elements.

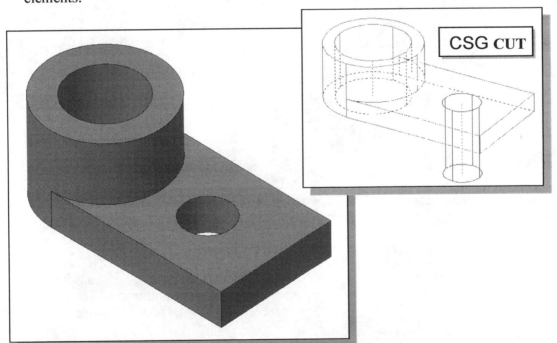

CSG CUT

Create the Final Feature

We will create a rectangular cut as the final solid feature of the *Locator*.

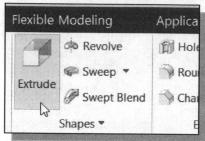

1. In the *Shapes* toolbar, select the **Extrude** tool option as shown.

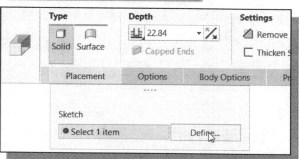

2. Click the Placement option and choose **Define** to begin creating a new *internal sketch*.

3. We will use the right vertical surface of the base feature as the sketching plane. Click once, with the left-mouse-button, inside the right vertical surface of the base feature as shown in the figure below.

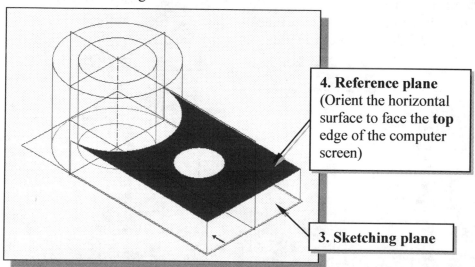

4. Reference plane (Orient the horizontal surface to face the **top** edge of the computer screen)

3. Sketching plane

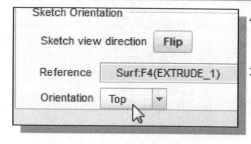

4. Select the **top surface** of the base feature as the orientation reference as shown.

5. In the Sketch Orientation menu, set the reference plane Orientation to **Top**, and proceed to enter the *Sketcher* mode.

Create a vertical 2D Section

1. On your own, set up to use DTM2, DTM3, and the top surface of the base feature as the references for the 2D sketch. *Creo Parametric* will use these references to assure the 2D sketch is created at the correct location.

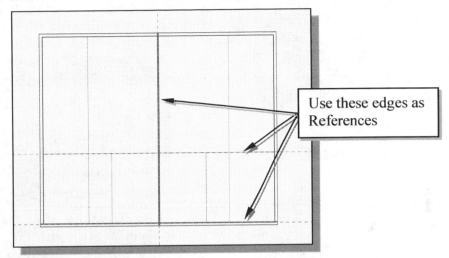

Use these edges as References

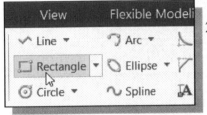

2. In the *Sketching* toolbar, click on the **Rectangle** icon as shown to activate the Rectangle command.

3. Create a rectangle as shown, with one corner on the horizontal axis (DTM2) and the other on the top surface of the base feature.

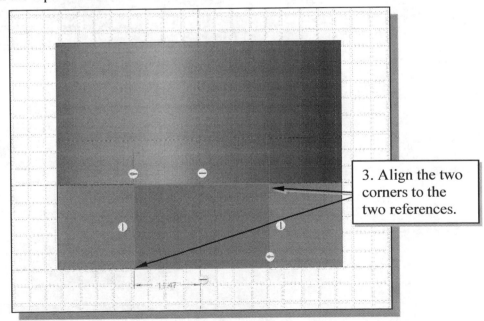

3. Align the two corners to the two references.

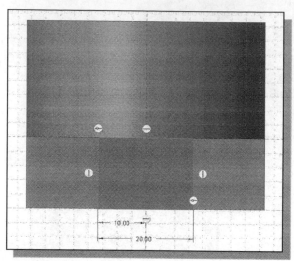

4. On your own, use the **Modify** command and adjust the rectangle to **20mm** wide and centered as shown.

5. In the *Ribbon* toolbar, click on the **OK** button to end the *Creo Parametric 2D Sketcher*.

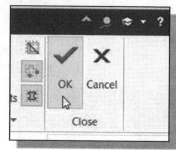

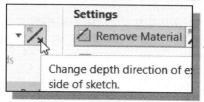

6. In the *Feature Option Dashboard*, select the **Remove Material** option as shown.

7. In the *Feature Option Dashboard*, click the **Flip Direction** icon to change the extrusion direction.

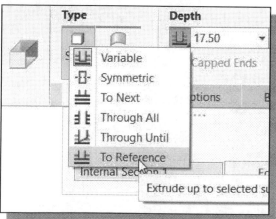

8. In the *Feature Option Dashboard*, select the **Extrude to selected point, curve, plane or surface** option as shown.

9. Choose the **center axis** of the hole feature as shown.

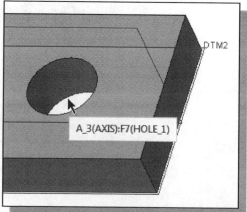

10. On your own, use the **Preview** option to examine the current feature.

11. On your own, use the *Dynamic Viewing* functions to reorient the 3D model and confirm the cut stops at the center axis of the smaller hole.

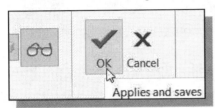

12. Click on the **OK** icon and proceed to create the hole feature.

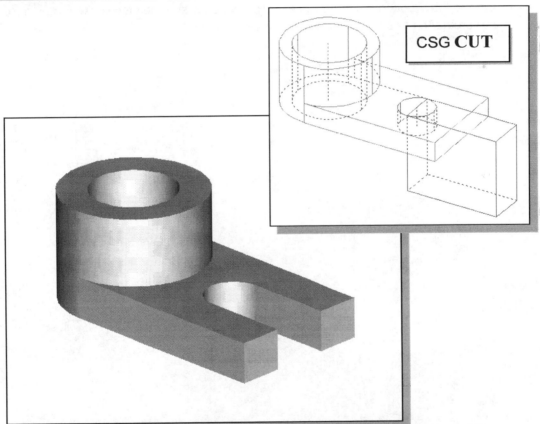

CSG **CUT**

➤ Note that the above *Locator design* was constructed using rectangular blocks and cylinders to illustrate the basic CSG procedure. The CSG procedure is embedded inside the parametric modeling system. It is also important to point out that the modern parametric modeling software is much more flexible, and all parametric features created in *Creo Parametric* can be treated as primitive solids.

Review Questions:

1. What are the three basic *Boolean operations* commonly used in computer geometric modeling software?

2. What is a *primitive solid*?

3. What does *CSG* stand for?

4. Which *Boolean operation* keeps only the volume common to the two solid objects?

5. What are the differences in creating a CUT feature and creating a HOLE feature in *Creo Parametric*?

6. Using the CSG concept, create Binary Tree sketches showing the steps you plan to use to create the two models shown on the next page:

Ex.1)

Ex.2)

Exercises: (All dimensions are in inches.)

1. **L-Bracket**

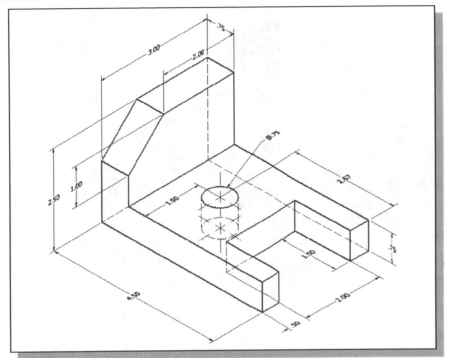

2. **Tube Mount**

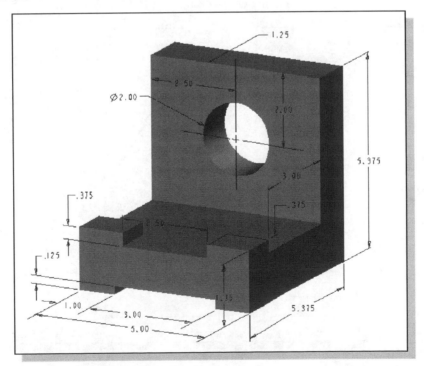

3. Shaft Support

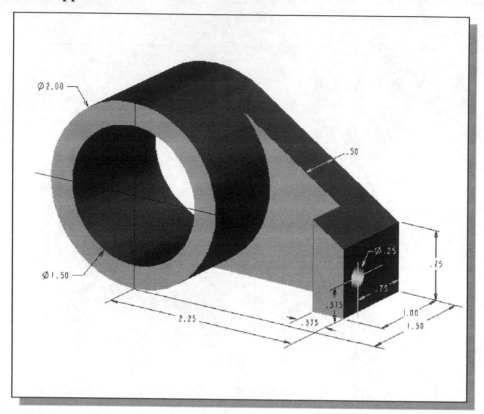

4. Guide Plate (Thickness: **0.25** inches. Boss height: **0.125** inches.)

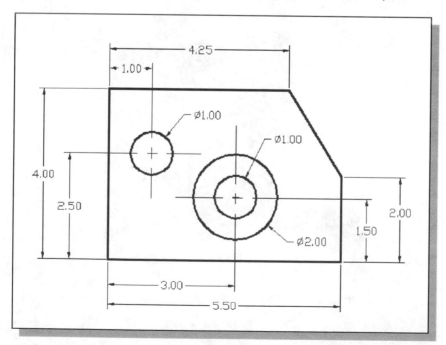

5. Support Base

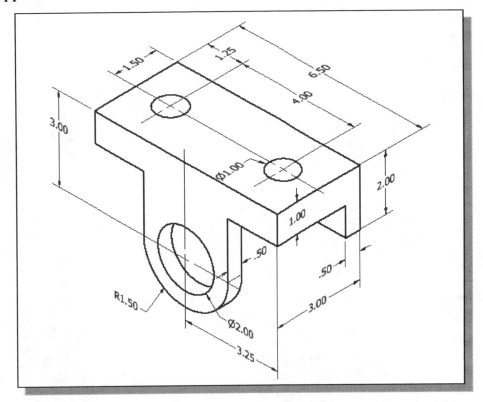

Notes:

Chapter 3
Model History Tree

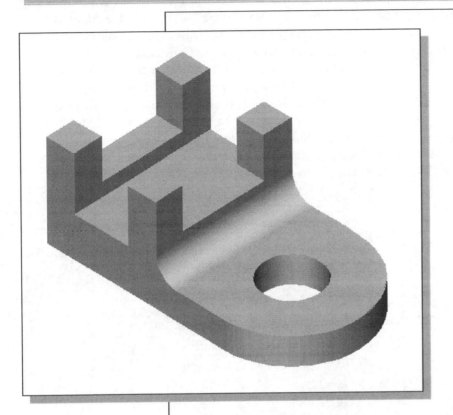

Learning Objectives

- ♦ **Understand Feature Interactions**
- ♦ **Use the Part Browser**
- ♦ **Modify and Update Feature Dimensions**
- ♦ **Perform History-Based Part Modifications**
- ♦ **Change the Names of Created Features**
- ♦ **Perform Basic Design Changes**

Introduction

In *Creo Parametric*, the *design intents* are stored as features in the **history tree**. The structure of the model history tree resembles that of a **CSG binary tree**. A CSG binary tree contains only Boolean relations, while the *Creo Parametric* **history tree** contains all features, including Boolean relations. A history tree is a sequential record of the features used to create the part. This history tree contains the construction steps, plus the rules defining the design intent at each construction operation. In a history tree, each time a new modeling event is created, previously defined features can be used to define information such as size, location and orientation. It is therefore important to think about your modeling strategy before you start creating anything. It is important, but also difficult, to plan ahead for all possible design changes that might occur. This approach in modeling is a major difference of **feature-based cad software**, such as *Creo Parametric*, from previous generation CAD systems.

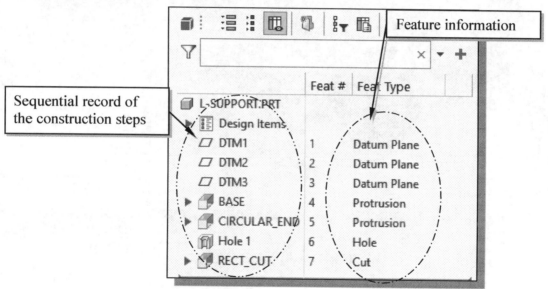

Feature based parametric modeling is a cumulative process. Every time a new feature is added, a new result is created and the feature is also added to the history tree. The database also includes parameters of features that were used to define them. All of this happens automatically as features are created and manipulated. At this point, it is important to understand that all of this information is retained, and modifications are done based on the same input information.

In *Creo Parametric*, the *Model Tree* gives information about modeling order and other information about the feature. Part modifications can be done through accessing the features in the history tree. It is important to understand the concept of the history tree when you modify parts. *Creo Parametric* remembers the history of a part, including all the rules that were used to create it, so that changes can be made to any operation that was performed to create the part. To modify a feature in *Creo Parametric*, you may select the feature by picking it in the display area or selecting the name of the feature in the *Model Tree* window.

The *L-Support* Design

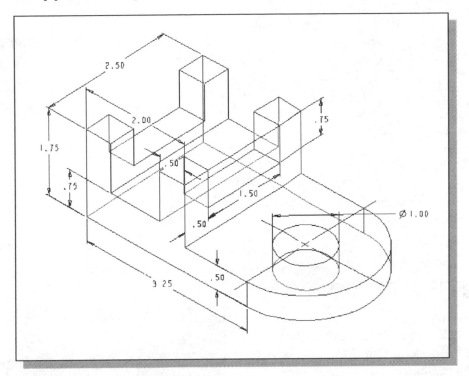

❖ Based on your knowledge of *Creo Parametric* so far, how many features would you use to create the design? Which feature would you choose as the **base feature**, the first feature, of the model? What is your choice in arranging the order of the features? Would you organize the features differently if additional fillets were to be added in the design? Take a few minutes to consider these questions and do preliminary planning by sketching on a piece of paper. You are also encouraged to create the model on your own prior to following through the tutorial.

Starting *Creo Parametric*

1. Select the **Creo Parametric** option on the *Start* menu or select the **Creo Parametric** icon on the desktop to start *Creo Parametric*. The *Creo Parametric* main window will appear on the screen.

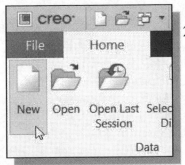

2. Click on the **New** icon, located in the *Ribbon toolbar* as shown.

3. In the *New* dialog box, confirm the model **Type** is set to **Part** (**Solid** Sub-type).

4. Enter **L-Support** as the part **Name** as shown in the figure.

5. Turn *off* the **Use default template** option.

6. Click on the **OK** button to accept the settings.

7. In the *New File Options* dialog box, select **Empty** in the option list to not use any template file.

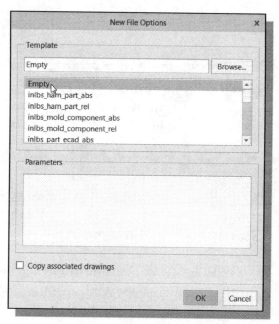

8. Click on the **OK** button to accept the settings and enter the *Creo Parametric Part Modeling* mode.

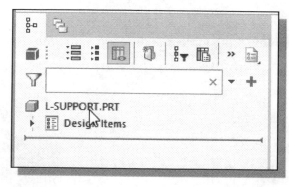

❖ The *Model Tree* window and the *Menu Manager* window appear on the screen. Notice in the *Model Tree* window, the part name is displayed. The *Creo Parametric Model Tree* window presents the model structure, feature by feature, in the order in which the features are created.

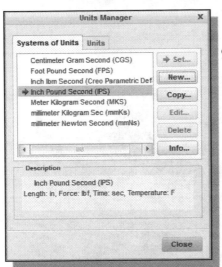

9. On your own, set up the **Units** to **Inch-Pound-Second (IPS)**.

10. Click on the **Datum Plane** icon and create the basic set of three datum planes.

❖ In the *Model Tree* window, the datum planes are new features added into the model structure.

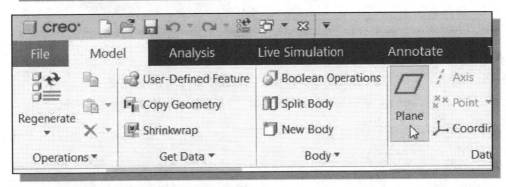

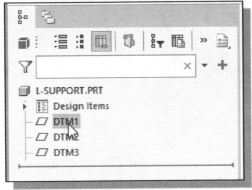

11. Click on **DTM1** in the *Model Tree* window and notice the selected feature is highlighted in the display area.

12. Click on **DTM2**; the selected feature is highlighted in the display area.

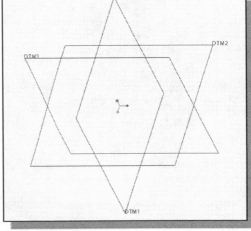

13. Select **DTM3** as the pre-selected datum plane before moving on to the next section.

❖ **Any selection of features can also be done through the *Model Tree* window.**

Modeling Strategy

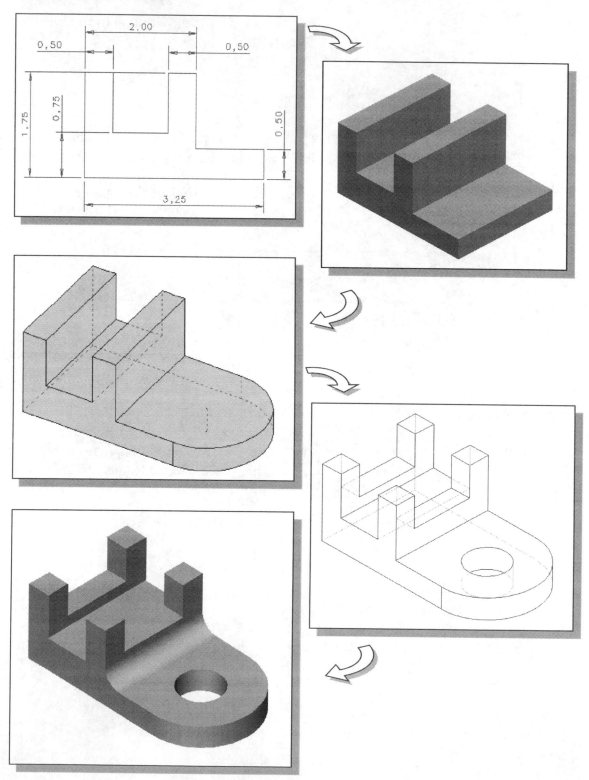

Create the Base Feature

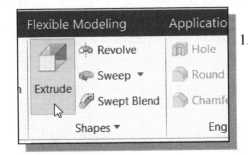

1. In the *Shapes* toolbar, select the **Extrude** tool option as shown.

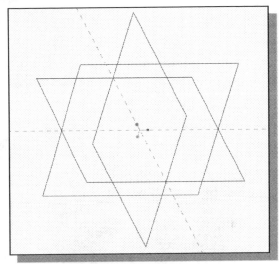

2. In the graphics area, notice that **DTM3** is set as the Sketch Plane and the Sketch Orientation is set to **DTM1-Right** orientation as shown.

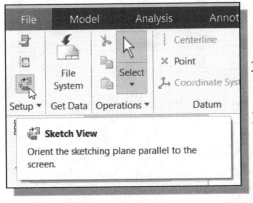

3. Click **Sketch View** to align the *Sketching* plane to the screen if necessary.

➤ With DTM3 preselected, *Creo Parametric* will align the sketching plane to the pre-selected plane and DTM1 is used as the orientation reference.

2D Sketch of the Base Feature

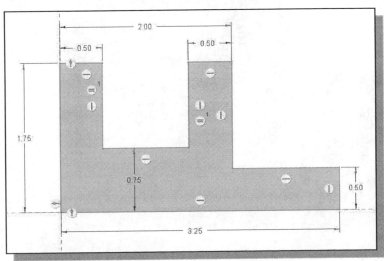

1. On your own, create the 2D section as shown. (Hint: Start at the lower left corner, aligned to the origin, and go clockwise to create the connected line segments.)

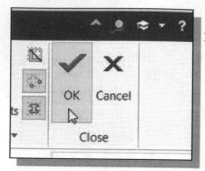

2. Now that the 2D sketch is completed, we will proceed to the next *element*. In the *Ribbon* toolbar, click on the **OK** button to end the *Creo Parametric Sketcher*.

3. In the *Feature Option Dashboard*, choose the **Symmetric** option as shown. This option sets the extrusion of the section to **Extrude on both sides of sketch plane**.

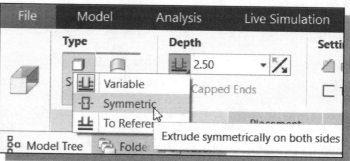

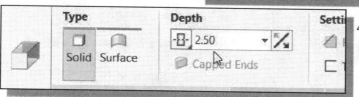

4. In the *depth* value box, enter **2.5** as the extrusion depth.

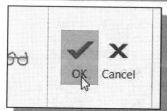

5. In the *Extrude Option Dashboard*, click **OK** to proceed with the solid feature creation.

➤ Note that **DTM3** passes through the center of the base feature and the extrusion feature is added as the last item in the *Model Tree* window.

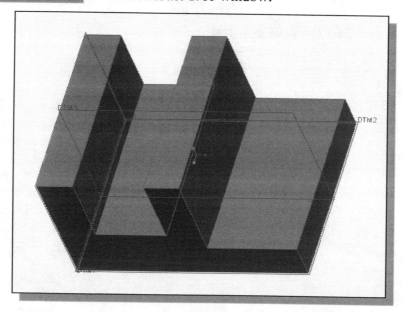

Set up the Default Model View and Sketcher View

1. Click on the **triangle** icon in the *Quick Access* toolbar. The *Customize Quick Access Toolbar* option list is displayed. Click on the **More Commands** option; the *Creo Parametric Options* dialog box appears, which allows us to adjust various settings.

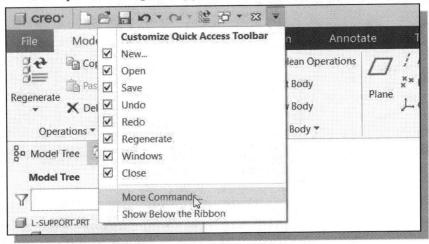

2. In the **Model Display** section, choose **Isometric** as the Default model orientation as shown.

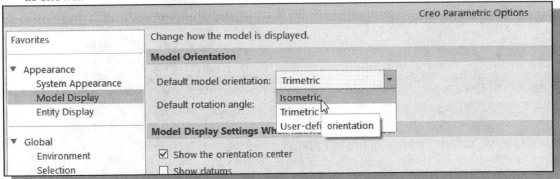

3. Click **Sketcher** and switch on the **Sketcher startup** option as shown.

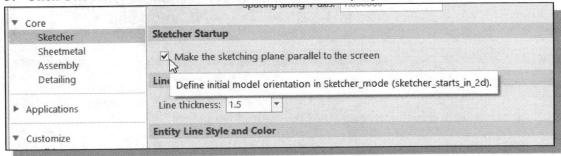

4. Click **OK** to exit the form and choose **NO** to not save the settings to a file.

5. On your own, rotate the model and then use the quick-key combination [Ctrl-D] to view an *Isometric* view orientation in the display area.

Add the Second Solid Feature

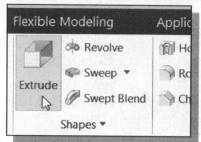

1. In the *Shapes* toolbar, select the **Extrude** tool option as shown.

2. Click the **Placement** option and choose **Define** to begin creating a new *internal sketch*.

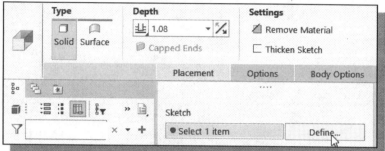

3. In the *Model Tree* window, pick **DTM2** as the sketching plane as shown.

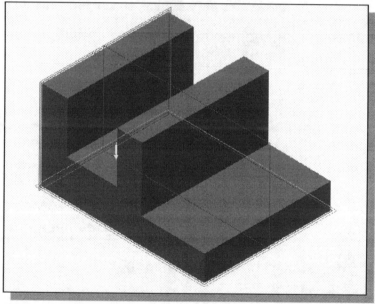

❖ As more features are created, selecting through the *Model Tree* is an extremely convenient option. As can be seen, *Creo Parametric* provides us a variety of tools to aid the selection and construction of the desired features.

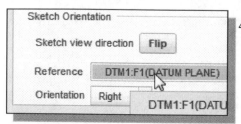

4. In the Sketch Orientation options, select the **Reference** option box (currently DTM1 is listed) by clicking once with the left-mouse button. Note that the Orientation option is set to **Right**.

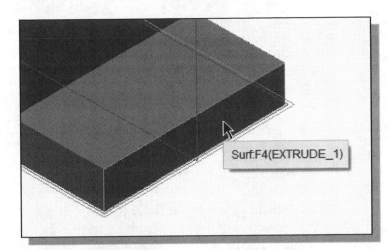

5. Pick the right vertical face of the base feature as shown.

6. Pick **Sketch** to enter the *Creo Parametric Sketcher* mode.

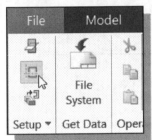

7. On your own, activate the *Sketch Reference* dialog box.

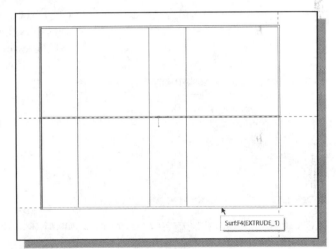

8. Select the front surface of the base feature as an additional sketching reference as shown in the figure.

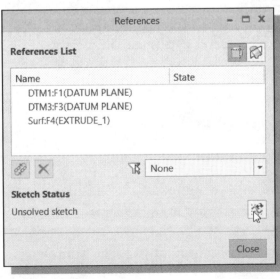

9. In the *References* dialog box, three references are listed as shown. Click on the **Update** button to accept the selections.

10. Click on the **Close** button to exit the *References* option settings.

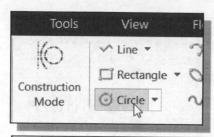

11. In the *Sketching* toolbar, select **Circle** as shown. The default option is to create a circle by specifying the center point and a point through which the circle will pass. The message "*Select the center of a circle*" is displayed in the message area.

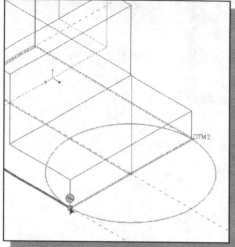

12. Press [**Ctrl+D**] to reset the display to isometric view as shown.

13. Align the center of the circle to the intersection of the two sketching references and create the circle as shown. (Note the **coincident** constraint symbol.)

- The circle is fully defined, and no dimension is needed. The base solid controls the size of the circle.

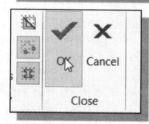

14. Now that the 2D sketch is completed, we will proceed to the next *element.* In the *Ribbon* toolbar, click on the **OK** button to end the *Creo Parametric Sketcher* mode.

15. In the *Feature Option Dashboard*, select the **To Reference** (*Extrude to selected point, curve, plane or surface*) option as shown.

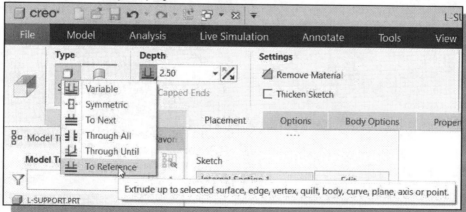

➤ Note that the Until Selected option does not require us to enter a value to define the depth of the extrusion; the extrusion distance is calculated based on the termination entity.

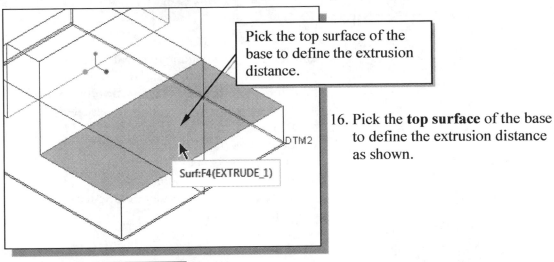

Pick the top surface of the base to define the extrusion distance.

Surf:F4(EXTRUDE_1)

16. Pick the **top surface** of the base to define the extrusion distance as shown.

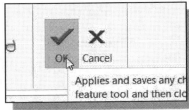

OK   Cancel

Applies and saves any ch
feature tool and then clc

17. Click on **OK** to proceed with the extrusion option.

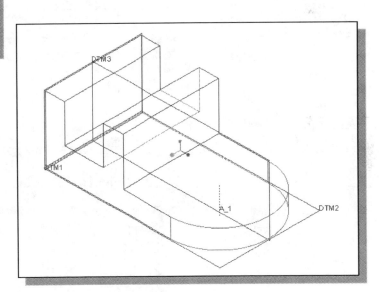

L-SUPPORT.PRT
- Design Items
  - DTM1
  - DTM2
  - DTM3
  - Extrude 1
  - Extrude 2

➤ Notice the newly created feature is now part of the *Model History* with the feature shown in the *Model Tree* window.

Rename the Part Features

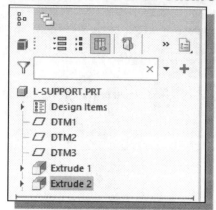

Our model now contains five features: three datum planes and two protrusions. The last feature is highlighted in the display area as well as in the *Model Tree* window. Each time a new feature is created, the feature is also added in the *Model Tree* window. By default, *Creo Parametric* will use generic names for part features, but when we begin to deal with parts with a large number of features, it will be much easier to identify the features using more meaningful names.

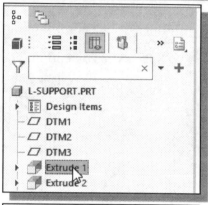

1. Inside the *Model Tree* window, click twice with the **left-mouse-button** on the first protrusion feature, **Extrude 1**, to edit the name of the feature.

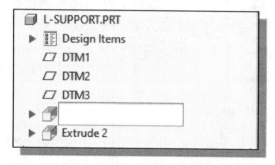

2. Enter **BASE** as the new feature name as shown.

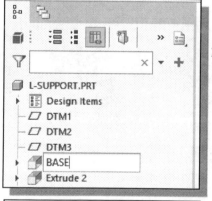

3. On your own, rename the second protrusion to **CIRCULAR_END** as shown. (Note that *Creo Parametric* does not allow spaces to be used as part of the feature names.)

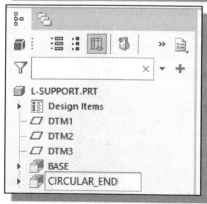

- In the *Model Tree* window, the new feature names are displayed as shown.

Expand the Model Tree Listing

1. In the *Model Tree* window, click on the **Tree Columns** icon to set the **column display options**.

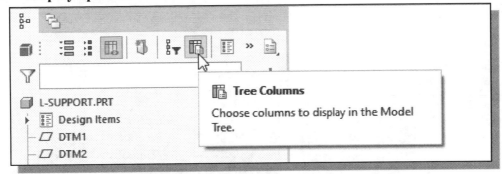

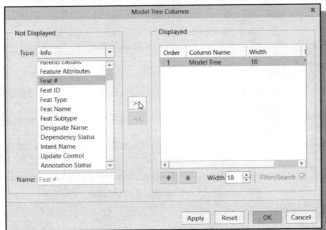

2. In the *Model Tree Columns* form, select

[Feat #] → [>>]

3. In the *Model Tree Columns* form, select

[Feat Type] → [>>]

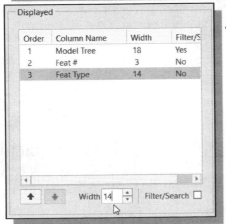

4. In the **Displayed** section, set the column **Width** of **Feat #** to **3** and **Feat Type** to **14** as shown in the figure.

5. Click on the **OK** button to apply the settings and exit the form.

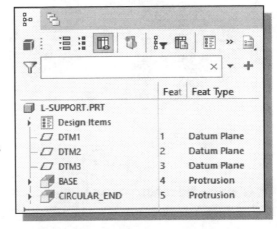

*   In the Model Tree window, notice the additional feature information is shown.

Create a Placed HOLE Feature

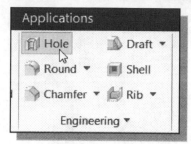

1. In the *Engineering* toolbar, select the **Hole** tool option as shown.

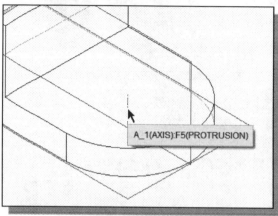

2. Pick the **center axis** of the Circular_End feature as shown.

❖ Note that selecting an axis will create a coaxial hole aligned to the selected axis.

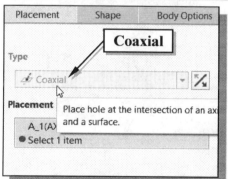

3. In the *Feature Option Dashboard*, click the **Placement** option and examine the placement settings. Note the **Coaxial** option is being used.

➢ The **Coaxial** option requires the selection of an existing axis and a placement plane. The red color of the placement option indicates additional placement element is required.

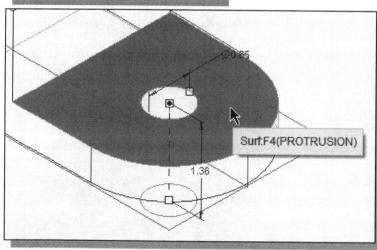

4. Hold down the **[Ctrl]** key and select the top plane of the Circular_End feature as shown.

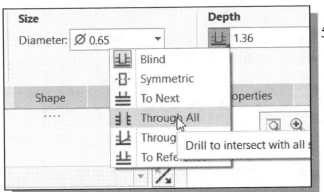

5. In the *Feature Option Dashboard*, select the **Drill to intersect with all surfaces** option as shown.

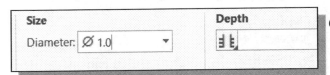

6. In the *Feature Option Dashboard*, set the feature *diameter* to **1.0** as shown.

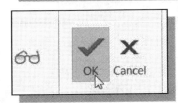

7. Click on the **OK** icon and proceed to create the hole feature.

Create a Rectangular Cutout

We will create a rectangular cutout as the final solid feature of the *Locator*.

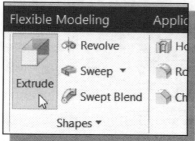

1. In the *Shapes* toolbar, select the **Extrude** tool option as shown.

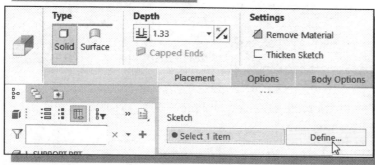

2. Click the **Placement** option and choose **Define** to begin creating a new *internal sketch*.

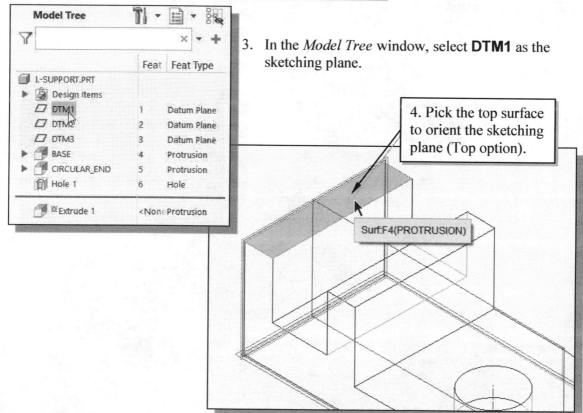

3. In the *Model Tree* window, select **DTM1** as the sketching plane.

4. Pick the top surface to orient the sketching plane (Top option).

4. Select the **top surface** of the base feature as the orientation reference as shown in the above figure.

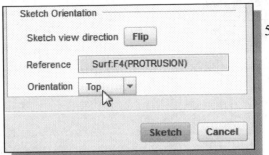

5. In the Sketch Orientation menu, confirm the reference plane Orientation is set to **Top**.

6. Pick **Sketch** to exit the *Section Placement* window and proceed to enter the *Creo Parametric Sketcher* mode.

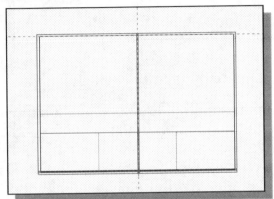

7. Note that **DTM3** and the **top horizontal plane** are set as the *sketching references* as shown.

8. Create a **rectangle** of arbitrary size, with the top edge aligned to the top surface of the solid model, as shown in the figure below.

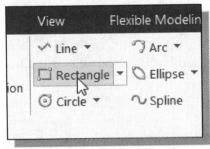

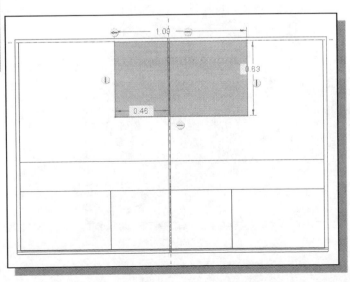

9. Press [Ctrl]+[Alt]+[A] to select all the entities created so far.

10. In the *Editing* toolbar, click on the **Modify** command.

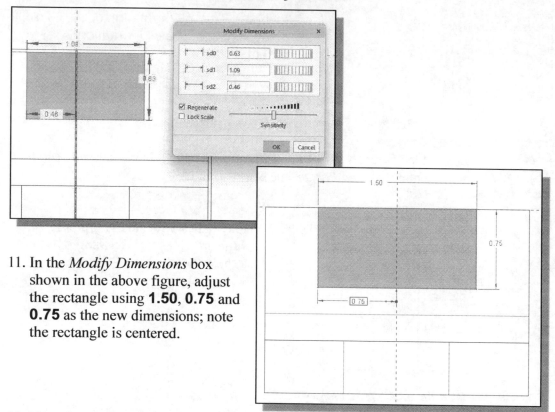

11. In the *Modify Dimensions* box shown in the above figure, adjust the rectangle using **1.50**, **0.75** and **0.75** as the new dimensions; note the rectangle is centered.

12. Now that the 2D sketch is completed, we will proceed to the next element. In the *Ribbon* toolbar, click on the **OK** button to end the *Creo Parametric Sketcher* mode.

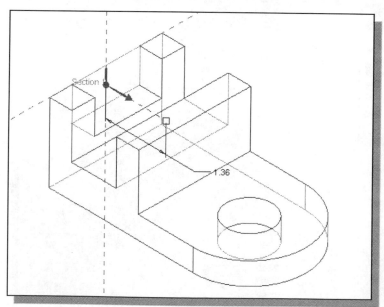

13. Reset the display to the default orientation [quick-key: **Ctrl-D**].

❖ Note that Creo automatically changes the default setting of the **Extrude** tool command to **Remove material**. The displayed arrow indicates the extrusion direction.

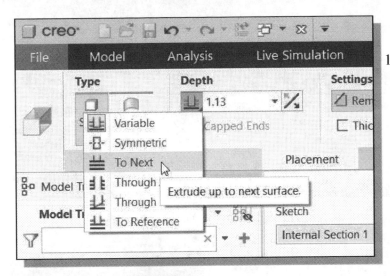

14. In the *Feature Option Dashboard*, select the **Extrude up to next surface** option as shown.

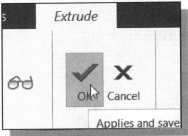

15. Click **OK** to proceed with creating the cut feature.

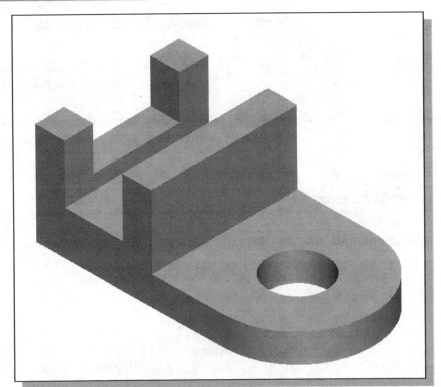

Examine the Model History

1. On your own, change the name of the last feature to **Rect_CUT**.

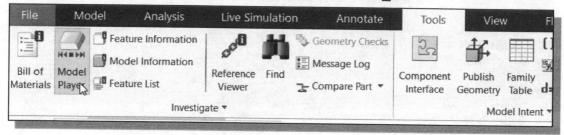

2. Select **Tools** tab in the *Ribbon* toolbars area, then pick the **Model Player** option.

3. Click on the **Go to the beginning of the model** button in the *Model Player* window as shown.

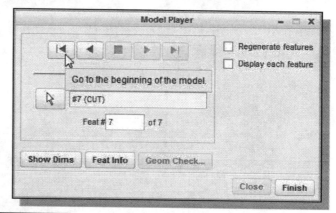

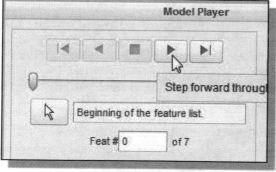

4. Click on the **Step Forward** button to look at the first feature created.

♦ Notice that we have literally gone back in time. We are at the first feature of the model.

5. On your own, click the **Step Forward** button and review all of the solid features used to create the model.

6. Click **Finish** to exit the *Model Player*.

❖ Note that the *Model Player* simply redisplays the features recorded in the *Model Tree*. The *Creo Parametric Model Tree* is a sequential record of the features used to create the part. We can review and make modifications to assure the accuracy of our model at any time.

History-based Part Modifications

Creo Parametric uses the *history-based part modification* approach, which enables us to make modifications to the appropriate features in the *Model History Tree* and re-link the rest of the *History Tree*. We can think of it as going back in time and modifying some aspects of the modeling steps used to create the part. We can modify any feature that we have created. As an example, we will adjust the depth of the rectangular cutout.

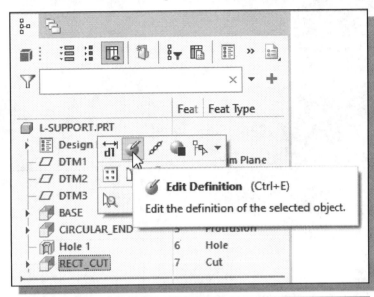

1. In the *Model Tree* window, move the cursor on top of **RECT_CUT**, then click once with the left-mouse-button. A pop-up menu is displayed.

2. Select **Edit Definition** in the pop-up menu. Notice the feature dashboard appears near the top of the main window.

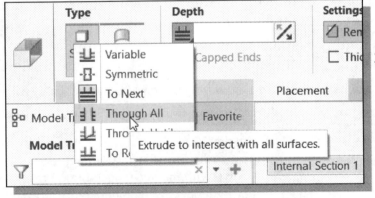

3. Change the *depth* option to **Through all - extrude to intersect with all surfaces** as shown.

4. Click on the **Preview** button to examine the effect of the modification.

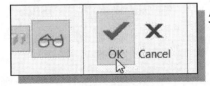

5. Click on the **OK** icon and proceed to create the hole feature.

❖ As can be seen, the *Creo Parametric* history-based modification approach is very straightforward, and it only took a few seconds to do this modification.

Modify Dimensional Values

Once a feature has been created, it is very easy to modify its dimensional values in *Creo Parametric*.

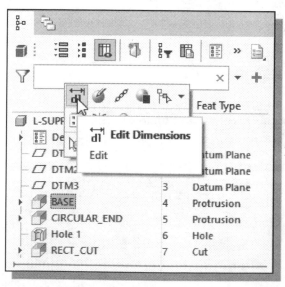

1. In the *Model Tree* window, move the cursor on top of **BASE**, then press down with the left-mouse-button to bring up the option menu. Note more options are available with the single right-mouse-click.

2. Select **Edit Dimensions** in the *option menu*. Notice the 2D section, along with all the dimensions, appears in the graphics area.

3. Pick the width dimension of the upper section of the part (**2.0**) by double-clicking with the left-mouse-button.

4. Enter **2.25** as the new dimension value.

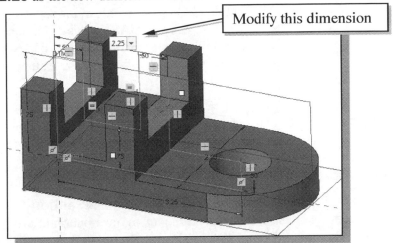

Modify this dimension

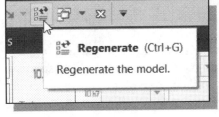

5. Click **Regenerate** in the *Quick Access* toolbar to apply the modification.

Modify the 2D Section of the Base Feature

We can also make changes to the original 2D sections we have created.

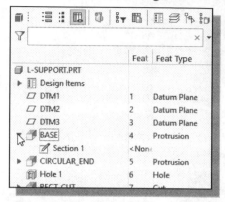

1. In the *Model Tree* window, move the cursor to the arrow in front of **BASE**, then click once with the left-mouse-button to expand the list. Note that one 2D section is used for the base feature.

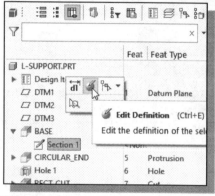

2. Move the cursor on top of **Section 1**, then click once with the left-mouse-button to bring up the option menu.

3. Pick **Edit Definition** in the *option menu*.

❖ The *Creo Parametric Sketcher* toolbar appears on the screen and the original 2D section is displayed in the display area. We have literally gone back in time to the point where the 2D section was being created.

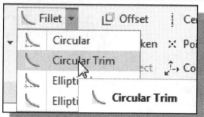

4. In the *Sketching* toolbar, select the **Fillet-Circular Trim** command.

5. Pick the two lines on the right side, as shown in the figure below, to create a fillet.

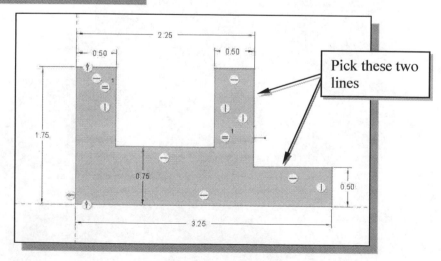

Pick these two lines

6. On your own, modify the radius of the fillet to **0.25** inches.

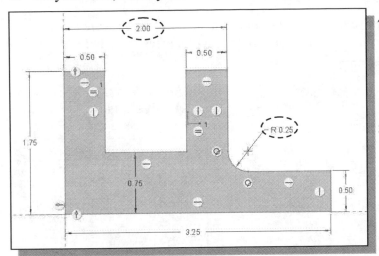

7. On your own, adjust the width dimension of the upper section of the part from **2.25** to **2.0** as shown in the figure.

8. In the *Ribbon* toolbar, click on the **OK** button to accept the modification and exit the *Creo Parametric Sketcher* mode.

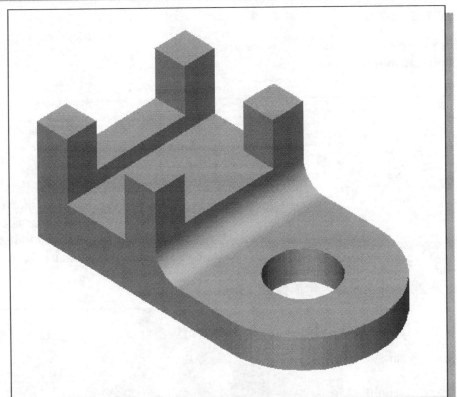

❖ In a typical design process, the initial design will undergo many analyses, testing and reviews. The *history-based part modification* approach is an extremely powerful tool that enables us to quickly update the design. At the same time, it is quite clear that PLANNING AHEAD becomes more important in doing feature-based modeling.

Use the Measure Tools

Creo Parametric also provides several measure tools that allow us to measure area, perimeter and additional information of the constructed 2D sketches.

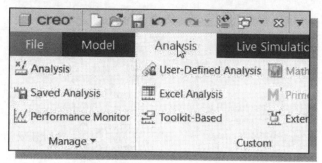

1. In the *Ribbon* toolbar area, switch to the **Analysis** tab.

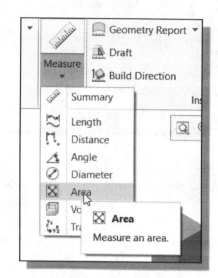

2. In the *Measure* toolbar, click the **Area** option as shown.

- Note that **other** measurement options are also available in the pull-down list.

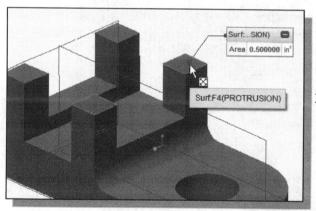

3. Click on one of the small flat surfaces of the model as shown, and notice the adjacent surface is also highlighted; the highlighted surface area is calculated as 0.5.

4. Click on the **cylindrical surface**, on the inside of the hole; the surface area is calculated as shown.

5. Click **Close** to end the Measure: Area analysis option.

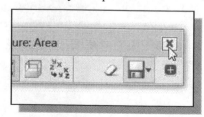

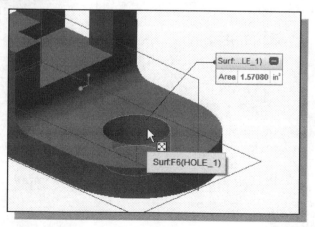

Calculate the Associated Physical Properties

Material properties can be assigned to *Creo Parametric* models. The *Material properties* can be used to create assembly bills of materials, drawing parts lists, and can also be used in stress analyses. With *Material properties*, we can also set and calculate physical properties for a part or assembly using the material library. This will allow us to examine the physical properties, such as weight or center of gravity, of the created model.

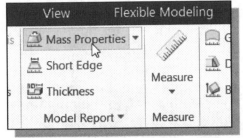

1. In the **Analysis** tab, click on the **Mass Properties** icon in the *Model Report* toolbar as shown.

➢ The **Mass Properties** command can be used to get a quick sense of the volume and weight of the created model.

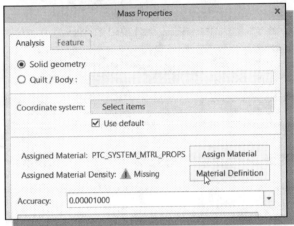

2. In the *Mass Properties* dialog box, notice the material density is **missing** as shown.

3. Click on the **Material Definition** icon to adjust the Material Density.

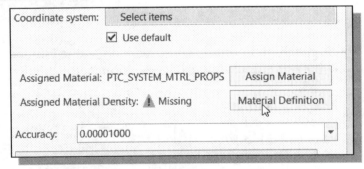

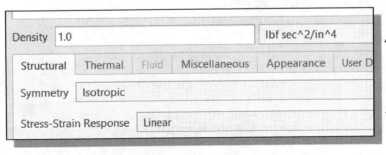

4. Enter **1.0** as the new Material Density as shown.

5. Click **OK** to accept the settings.

6. Click on the **Preview** icon to perform the calculation based on the current settings.

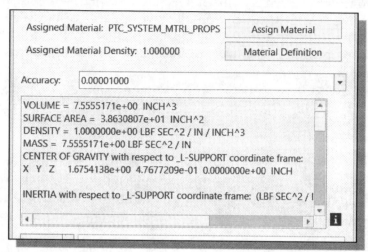

7. Click on the **Material Definition** icon to adjust the Material Density.

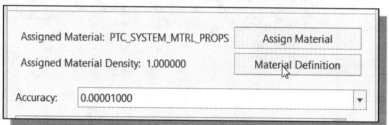

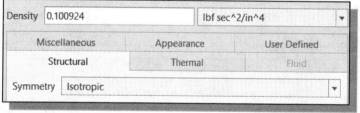

8. In the *Mass Properties* dialog box, enter **0.100924** (this is the density for AL2014) as the density of the material as shown.

9. Click on the **Preview** icon to perform the calculation.

➢ Notice the physical properties, such as volume, surface area, mass, center of gravity, and mass moment inertia, are all calculated and displayed.

10. Click **OK** to end the Mass Properties analysis option.

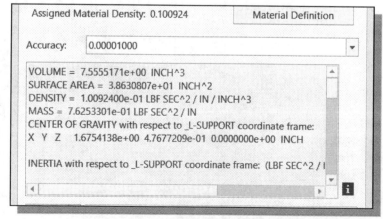

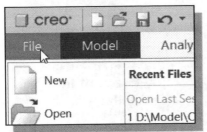

11. Click on the **File** *pull-down menu* as shown.

12. Select **Prepare** in the **pull-down list** and **Model Properties** as shown.

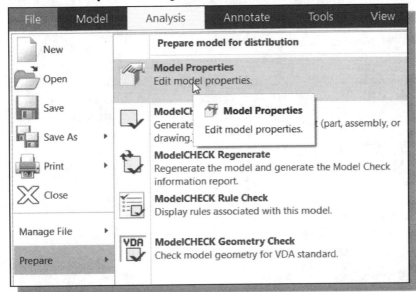

13. Select the **Change** icon that is associated to the **Material** option in the *Model Properties* window.

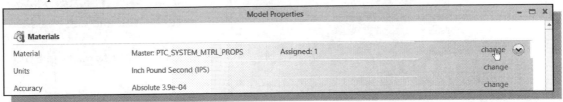

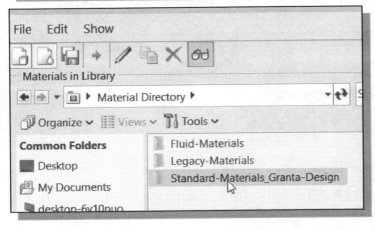

14. Double-click the *Standard-Materials_Granta-Design* library, which comes with many of the commonly used materials. Note that new material properties can also be added and all material properties can be edited.

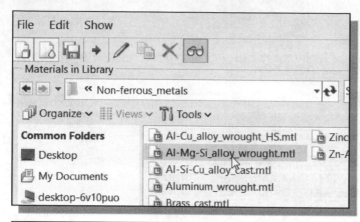

15. Choose **Non-ferrous-metals → Al-Mg-Si-alloy-wrought** from the material library as shown.

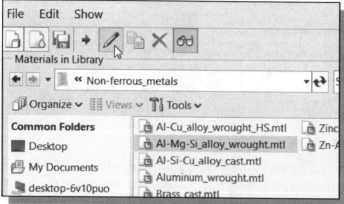

16. In the *Materials* dialog box, click the **Edit** icon as shown.

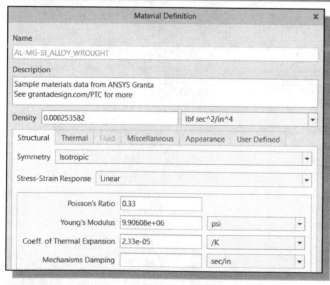

17. In the *Material Definition* dialog box, the material properties of **Al-Mg-Si-alloy** are displayed and can be edited.

18. Click **Cancel** to exit the Edit option.

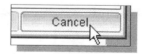

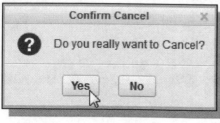

19. Click **Yes** to confirm the canceling of the Edit command.

20. On your own, select **Steel_Cast** and move it to the **Materials in Model** box by double-clicking on the material as shown.

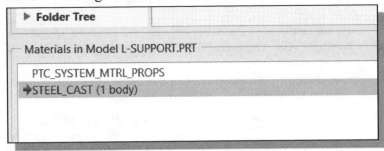

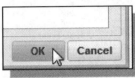

21. Click **OK** to accept the selection and assignment.

22. Click **Close** to exit the *Model Properties* dialog box as shown.

23. On your own, activate the **Mass Properties** analysis option.

24. Compute and display the **mass properties** of the model, which has been assigned the material properties of **STEEL_CAST**.

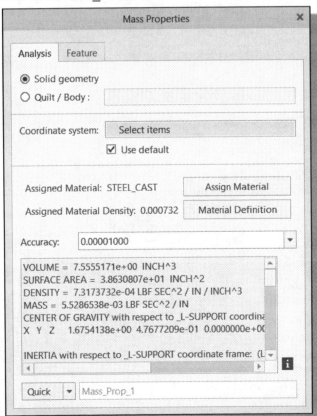

Review Questions:

1. What are stored in the *Creo Parametric History Tree*?

2. When extruding, what is the difference between *Depth Value* and *Thru All*?

3. What is the *history-based part modification* approach?

4. What determines how a model reacts when other features in the model change?

5. Describe two methods in *Creo Parametric* to change existing dimensions.

6. Create *History Tree* sketches showing the steps you plan to use to create the two models shown on the next page:

Ex.1)

Ex.2)

Exercises: (All dimensions are in inches.)

1. **Rod Slide** (The overall height is 2.5)

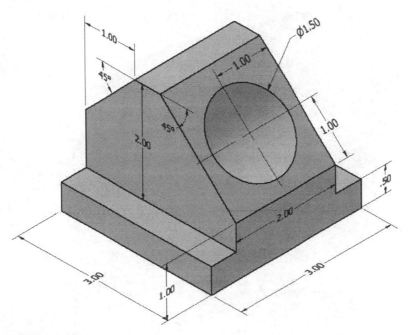

2. **Coupling Base**

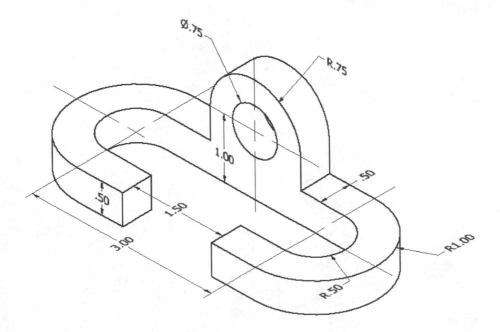

3. Latch Plate

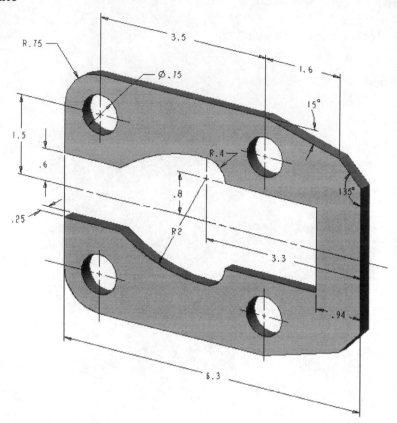

4. Transfer Fork

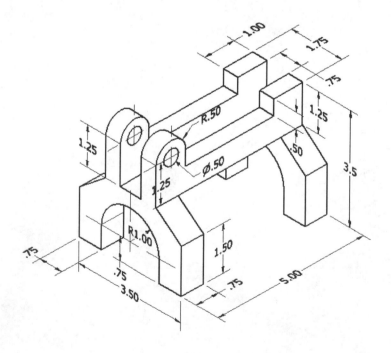

5. Pivot Lock (The center points of all of the circular features in the design are aligned to the two centers at the base.)

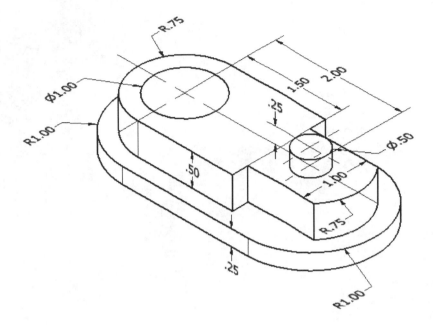

Notes:

Chapter 4
Constraints and Parametric Relations

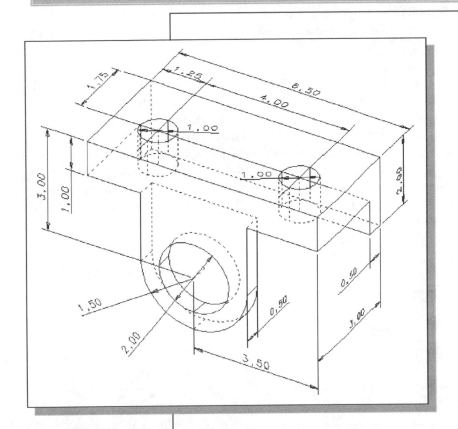

Learning Objectives

- ♦ **Understand the importance of Fully Constrained Sketches**
- ♦ **Apply Constraints manually**
- ♦ **Create Parametric Relations**
- ♦ **Create Sketch Files**
- ♦ **Edit by Drag-and-Drop**
- ♦ **Display and Modify Parametric Relations**
- ♦ **Use the Basic Editing commands**
- ♦ **Create Offset Features**

Geometric Construction and Constraints

The main characteristics of solid modeling are the accuracy and completeness of the geometric database of the three-dimensional objects. However, working in three-dimensional space using input and output devices that are largely two-dimensional in nature is potentially tedious and confusing. The **Creo Parametric 2D Sketcher** provides an assortment of two-dimensional construction tools to make the creation of 2D geometry easier and more efficient.

In doing geometric constructions, dimensional values are necessary to describe the **SIZE** and **LOCATION** of constructed geometric entities. Besides using dimensions to define the geometry, we can also apply geometric rules to control geometric entities. This set of geometric rules is known as **geometric constraints** in parametric modeling. The geometric constraints are **geometric restrictions** that can be applied to geometric entities, such as tangent, parallel, perpendicular, etc. In *parametric modeling*, **parametric relations** and **geometric constraints** are commonly used in 2D sketches to assist the capture of individual design intent. In this next example, the use of *Creo Parametric Sketcher's* geometric construction tools, including the application of **geometric constraints**, is illustrated. The parametric relation is also illustrated in a later section of this chapter.

The *Gasket* Design

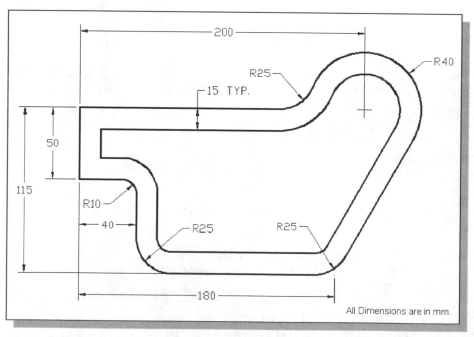

❖ Based on your knowledge of *Creo Parametric* so far, how would you create this design? What is the more difficult geometry involved in the design? Take a few minutes to consider a modeling strategy and do preliminary planning by sketching on a piece of paper. You are also encouraged to create the design on your own prior to following through the tutorial.

Modeling Strategy

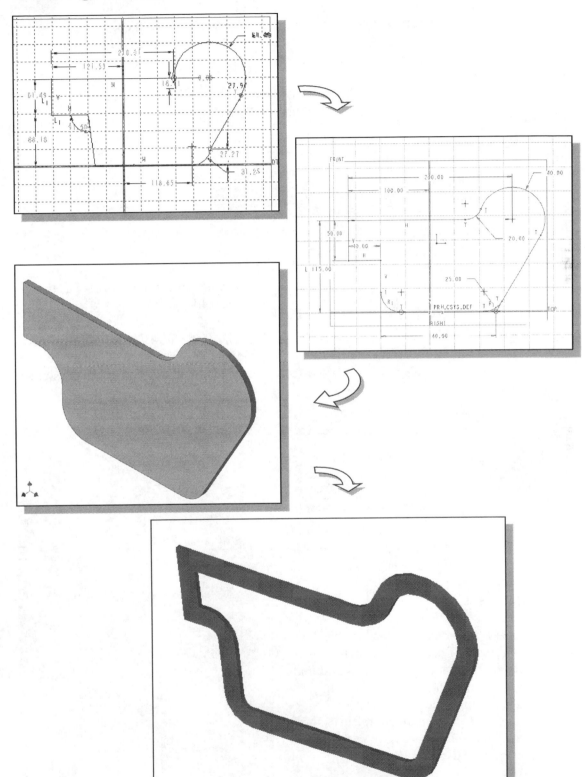

Starting *Creo Parametric* and Using a Metric Template

1. Select the **Creo Parametric** option on the *Start* menu or select the **Creo Parametric** icon on the desktop to start *Creo Parametric*. The *Creo Parametric* main window will appear on the screen.

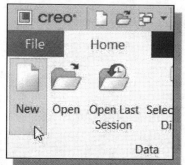

2. Pick the **New** icon in the toolbar. (You can also use the key combination **Ctrl-N** to start a new object.)

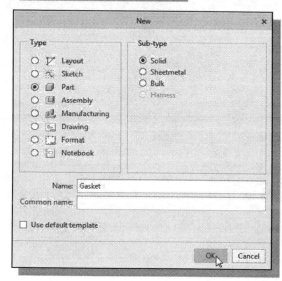

3. In the *New* form, enter **Gasket** as the solid part file **Name**.

4. Switch *off* the **Use Default Template** option as shown. (We will select the metric template for the *Gasket* design.)

5. Pick **OK** to continue. The *Model Tree* window and the *Menu Manager* window appear on the screen.

6. In the *New File Options* window, select **mmns_part_solid_abs** as the template to use. This is the template for using with the metric system (length in **millimeters**).

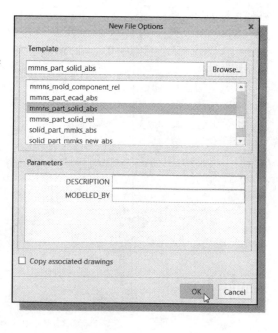

❖ In *Creo Parametric*, a **template** file can contain a set of predefined settings, such as the datum planes, datum axes and units setup. The template files allow us to reduce startup repetitions. The **part_solid_abs** template refers to the use of the *relative accuracy setting* as described on the next page.

Model Accuracy

Model Properties		
Materials		
Material	Master: PTC_SYSTEM_MTRL_PROPS	Assigned: 1
Units	millimeter Newton Second (mmNs)	
Accuracy	Relative 0.0012	

In Creo Parametric, *model accuracy* depends on the granularity, or precision, with which the software creates the geometry. For example, model accuracy determines such characteristics as how finely the software tessellates the model's curves. A finely tessellated curve is a smoother curve, whereas a coarsely tessellated curve would give the appearance of a series of straight lines that approximate the shape of the curve. Model accuracy in assemblies can determine whether geometry is treated as merged or separate as well as how successfully you can mesh two components.

The two types of model accuracy available are:
• **Absolute accuracy**—The absolute accuracy of a model defines the smallest allowable size of the unit that Creo Parametric can display or interpret without any error. For example, if the absolute accuracy of the model is 0.001 inches, Creo Parametric can accurately display the edges with a length greater than or equal to 0.001 inches but may not display edges with a length less than 0.001 inches.

• **Relative accuracy**— The relative accuracy of a model specifies the comparative ratio of the smallest model dimension to the part size. For example, if the relative accuracy of a model is 0.001 and the part size of the model is 10 inches, Creo Parametric can accurately display the edges with a length greater than or equal to 0.01 inches but may not display edges with a length less than 0.01 inches. Creo Parametric applies a relative accuracy by default when it creates geometry. The default relative value is the same regardless of the part size.

Creo Parametric determines the overall accuracy of the geometry by multiplying the relative accuracy value by the part size to determine the absolute accuracy of the geometry. The relation between absolute accuracy and relative accuracy is defined as
Absolute accuracy = Relative accuracy * Part Size

Thus, using the default relative accuracy, the software would create a small part with a greater degree of geometric refinement than a large part. For example, if the size of a small part is 100 and the relative accuracy is 0.012, then the absolute accuracy of the geometry would be 1.2. If the size of the larger part is 10,000 and the relative accuracy is 0.012, the absolute accuracy of the geometry would be 120.

The significant difference between the geometric accuracy of these two parts may be a cause for concern during the meshing of the assembly of the two parts.

Adjusting the Default Model Orientation and Sketcher Startup

1. Click on the **triangle** icon in the *Quick Access* toolbar. The *Customize Quick Access Toolbar* option list is displayed. Click on the **More Commands** option; the *Creo Parametric Options* dialog box appears, which allows us to adjust various settings.

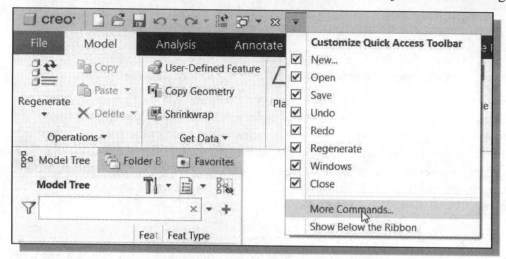

2. In the *Sketcher Grid* option, switch *on* the **Sketcher startup** option as shown.

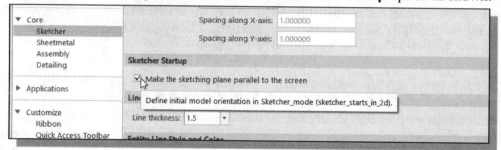

3. Scroll down to the bottom, click **Sketcher** and switch on the **Grid display** and choose **Cartesian** as shown.

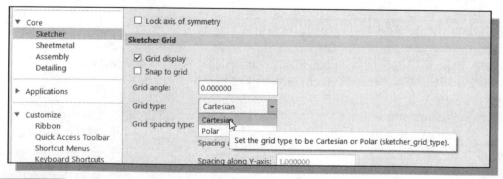

4. Click **OK** to exit the *Options* form.

5. On your own, save the settings to a configuration file.

Create the Base Feature

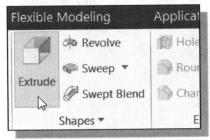

1. In the *Shapes* toolbar, select the **Extrude** tool option as shown.

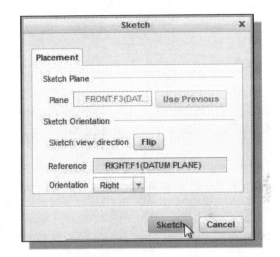

2. Click the **Placement** option and choose **Define** to begin creating a new *internal sketch*.

3. On your own, set up the **FRONT** datum plane as the sketch plane with the **RIGHT** datum plane facing the **right** edge of the computer screen. (Hint: Use the *Model Tree* to aid the selection.)

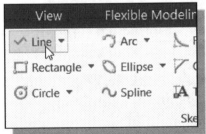

4. Choose the **Line** command in the *Sketching* toolbar.

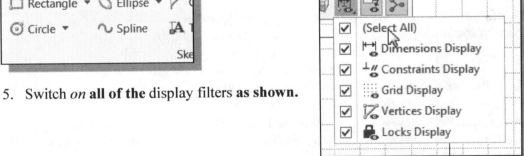

5. Switch *on* **all of the** display filters **as shown.**

6. Create the geometry, with the geometric constraints shown, by selecting points near the grid-points indicated.

❖ Do not end the Line command. We will continue to sketch after the following discussion of the implicit geometric relationships.

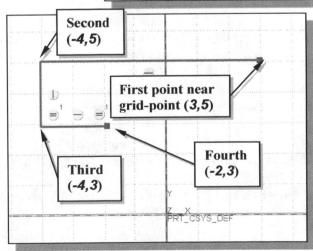

Note that the *Intent Manager* displays different constraint symbols to show perpendicularities, verticals, equal length, etc. The three geometric constraint symbols shown on the screen represent three types of geometric constraints: Vertical, Horizontal and Equal Length.

| Vertical | indicates a segment is vertical |

—   Horizontal   indicates a segment is horizontal

=   Equal Length   indicates two segments are of equal length

Implicit Geometric Relationships

In parametric modeling software, such as *Creo Parametric*, **geometric constraints** can be applied to geometric entities to ensure the shape of a part behaves predictably as changes are made. By default, *Creo Parametric* automatically adds constraints to geometry as they are constructed. The following is a list of assumptions about geometric relationships in *Creo Parametric* sketches.

- **If the sketched lines are approximately horizontal or vertical,** *Creo Parametric* **makes them exactly horizontal or vertical.**

- **If the sketched lines are approximately parallel or perpendicular,** *Creo Parametric* **makes them exactly parallel or perpendicular.**

- **If the sketched lines are approximately equal to the length of an existing line,** *Creo Parametric* **makes them equal to the length of the existing line.**

- **If the sketched arcs or circles are approximately equal to the radius of an existing arc or circle,** *Creo Parametric* **makes them equal to the radius of the existing arc or circle.**

- **If the sketched entities are approximately symmetrical about a centerline,** *Creo Parametric* **makes them symmetrical about the centerline.**

- **If the sketched entities are approximately tangent to each other,** *Creo Parametric* **makes them tangent to each other.**

- **If the sketched entities are approximately collinear to each other,** *Creo Parametric* **makes them collinear to each other.**

- **If the sketched endpoints are approximately on other entities,** *Creo Parametric* **makes them exactly on the entities.**

Note that we can override these implicit rules by adding appropriate dimensions and/or geometric constraints. We can also disable the implicit constraints during geometric constructions. A single click of the right-mouse-button (in the display area) will toggle on/off the displayed constraint. We can always modify the constraints to ensure the design intent.

A good rule of thumb to follow in using an implicit-geometric-constraint-based modeler, such as *Creo Parametric*, is to exaggerate the feature during the initial sketch. For example, if we want to create an angle between two lines, we should exaggerate when sketching (that is, sketch an eighty-five degree angle at sixty degrees) so that *Creo Parametric* will not make the lines perpendicular.

Lock/Disable/Enable Constraints

Geometric constraints can be modified while sketching; the rules to **lock, disable** or **enable** constraints are as follows:
- The current constraint is highlighted in red.
- Lock the current constraint by pressing the **right-mouse-button**. Small red circles on the symbols signify the constraint is activated and locked.
- Disable the current constraint by pressing the **right-mouse-button** again.
- Pressing the **right-mouse-button** again will enable the constraint.
- When more than one constraint is present, you can walk through them all by using the [**TAB**] key. Each one will be made current in turn and can then be disabled or locked.

➤ We will next demonstrate the usage of the rules in the current sketch.

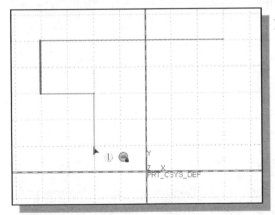

1. Move the cursor below the last endpoint we created, near grid point (**-2,1**); note that both the **Vertical** and **Equal Length** constraint symbols are displayed. Click once with the **right-mouse-button** to **lock** the Equal Length constraint. (The small lock appeared next to the symbol.)

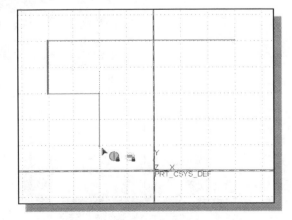

2. Hit the [**TAB**] key once to shift to the next active constraint.

3. Click the **right-mouse-button** once again and the current constraint, the Vertical constraint in this case, is also locked.

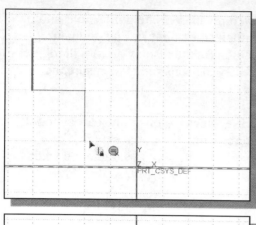

4. Hit the **[TAB]** key once to shift to the next constraint.

5. Click the **right-mouse-button** again to disable the active constraint, which is the Equal Length constraint.

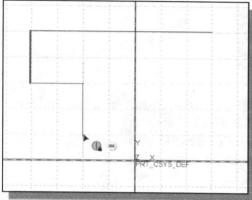

6. Click the **right-mouse-button** again to enable the Equal Length constraint; this is the default state of the constraint with no circles or lines attached to the symbols.

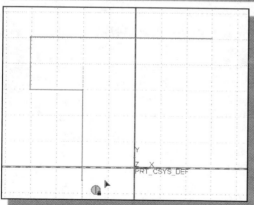

7. Note that the Vertical constraint is still locked, which prevents us from moving the cursor to create an inclined line segment.

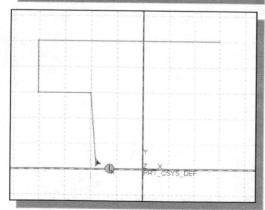

8. Click the **right-mouse-button** again and the current constraint, the Vertical constraint, is disabled.

9. Click the right-mouse-button again to remove the disabled constraint.

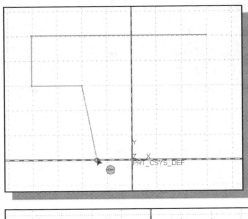

10. Move the cursor near the horizontal axis (datum plane **TOP**). Notice the cursor will *SNAP* onto the horizontal axis.

11. Pick near grid point (**-1.5,0**) and create an inclined line as shown.

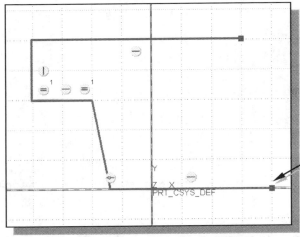

12. Pick near grid point (**4,0**) to create a horizontal line on the horizontal axis.

> **Sixth point: near grid point (*4,0*)**

13. In the *display area*, click the **middle-mouse-button** twice to end the Line command.

❖ Notice that dimensions are added automatically to describe the length and positions of the created lines. Note the values shown here may be very different than what is shown on your screen.

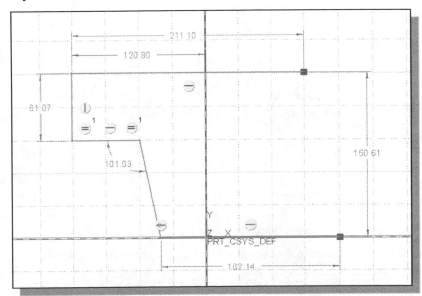

Create and Divide a Circle

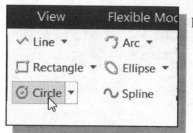

1. In the *Sketching* toolbar, select **Circle** as shown. The default option is to create a circle by specifying the center point and a point through which the circle will pass.

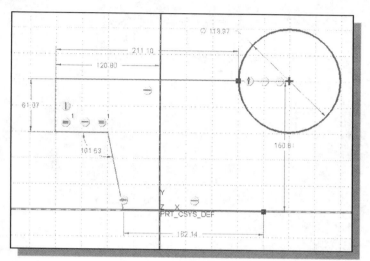

2. Place the center of the circle toward the right side of the sketch, approximately at grid point (**5,5**).

3. Pick near the right endpoint (**3,5**) of the top horizontal line; an implicit alignment constraint will be applied automatically.

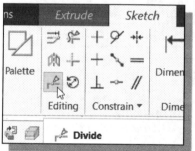

4. In the *Editing* toolbar, activate the **Divide** command by selecting the icon in the icon stack.

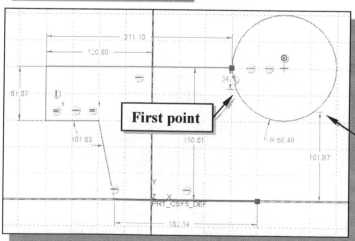

5. Click on the lower portion of the circle at two locations to divide the circle into two arcs, as shown in the figure.

First point

Second point

6. Click once with the middle-mouse-button to end the command.

❖ Note the *Intent Manager* automatically added additional dimensions as the circle is divided into two arcs.

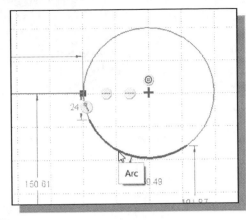

7. Select the lower arc by clicking with the **left-mouse-button**. Notice the selected entity is highlighted.

8. Inside the *graphics area*, press down and hold the right-mouse-button to display the option menu and select **Delete**.

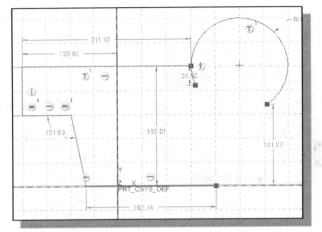

Create a Closed Region Sketch

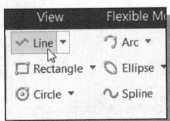

1. Pick **Line** in the *Sketching* toolbar.

2. Pick the right endpoint of the lower horizontal line to attach the first endpoint of the new line.

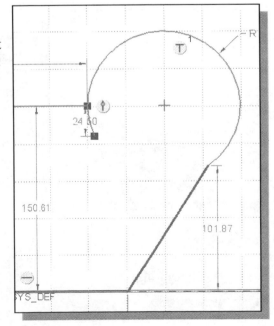

3. Move the cursor along the arc and note that different types of constraints are feasible at different locations. Attach the end of the line to the end of the arc. (Note that a Tangent constraint may be applied automatically depending upon the size of the arc you have created.)

4. Inside the display area, click the **middle-mouse-button** once to end the Line command.

❖ Note the *Intent Manager* automatically adds and removes additional dimensions and/or constraints to maintain a fully described sketch.

Modify Geometry by Drag-and-Drop

In *Creo Parametric*, geometric entities can also be modified by *drag-and-drop* with the left-mouse-button. Note that the applied geometric constraints, such as Vertical or Horizontal, are maintained by the system during the editing.

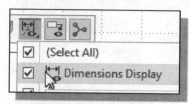

1. Click on the **Display dimensions** icon to switch *off* the display of the dimensions. Note the other display icons available in the *Standard* toolbar area.

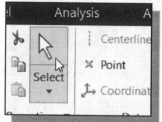

2. Click on the **Select** icon in the *Operations* toolbar.

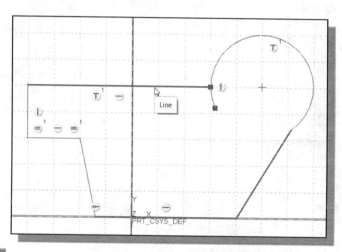

3. Use the left-mouse-button to drag the top-horizontal line upward and notice that the geometry is adjusted while maintaining the applied geometric constraints.

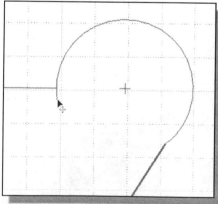

- We can dynamically modify geometric entities using **drag-and-drop**. The drag-and-drop capabilities provide a more flexible modeling environment, especially when exact dimension values are not known.

4. On your own, experiment with adjusting the locations of the endpoint of the arc, the center point of the arc, etc.

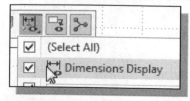

5. Click on the **Display dimensions** icon to switch *on* the display of the dimensions.

Controlling Geometric Constraints

In *Creo Parametric*, a set of rules can be applied to geometric entities to ensure the capture of the design intent. This set of rules is called the **geometric constraints**. As we create geometric entities such as lines and circles in the *Sketcher* mode, the *Intent Manager* automatically adds dimensions and/or geometric constraints. The application of different constraints will affect the geometry differently, and the design intent can be embedded into the model file through the proper use of constraints.

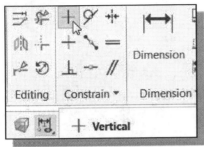

❖ The **Constrain** toolbar contains nine different icons in the *Ribbon* toolbars area. Note that the nine constraint icons actually provide more than a dozen of the applicable constraint options.

Vertical	Make an entity vertical
Line Up Vertical	Line up two points vertically
Horizontal	Make an entity horizontal
Line Up Horizontal	Line up two points horizontally
Perpendicular	Make entities perpendicular
Tangent	Make entities tangent to each other
Midpoint	Place point on the middle of the line
Collinear	Make two lines collinear
Point On Entity	Move a point on entity
Same Points	Make two points coincident
Symmetric	Make two points symmetric about a centerline

Equal Radii   Make arcs or circles equal radii
Equal Lengths Make lines equal length

Parallel   Make entities parallel

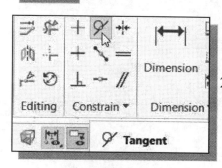

1. Pick **Tangent** in the *Constrain* list.

2. Pick the arc and the line as shown.

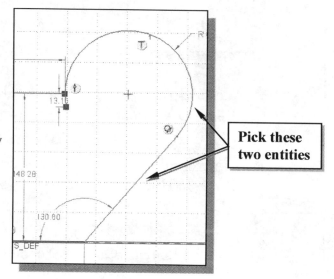

Pick these two entities

❖ Note that *Creo Parametric* adjusts the two entities to satisfy the applied **Tangent** constraint.

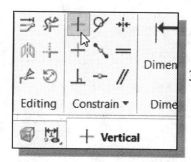

3. Pick **Vertical** in the *Constrain* list.

4. Pick the inclined line on the left. Notice the **Vertical** symbol is displayed as the constraint is applied.

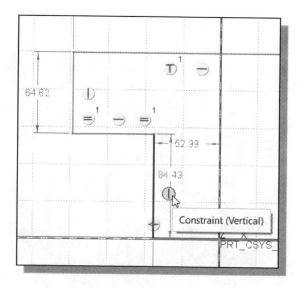

Add Rounded Corners

1. Pick **Undo** in the *Standard* toolbar.

❖ *Creo Parametric* records all the steps used while we are in *Sketcher* mode. All *Sketcher* operations can be undone with the **Undo** command. We can use **Undo** and **Redo** to experiment with different design options we have in mind.

2. Pick **Redo** to reapply the Vertical constraint.

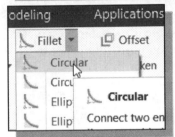

3. In the *Sketching* toolbar, select **Fillet-Circular**.

4. The Fillet command allows us to create a fillet by picking two intersecting entities. Create the three fillets as shown.

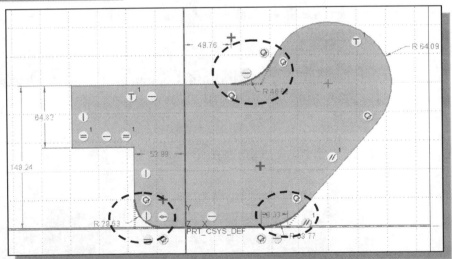

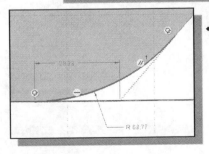

❖ Note that in *Creo Parametric*, the default Fillet-Circular command automatically trims two lines and added construction points of the original corners. To create a fillet with trimmed edges, use the *Fillet - Circular Trim* command.

Adjusting the Constructed Geometry

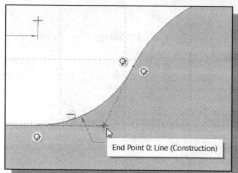

1. Select the construction point on the top as shown.

2. Inside the graphics area, press down and hold the right-mouse-button to display the option menu and select **Delete** as shown.

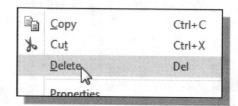

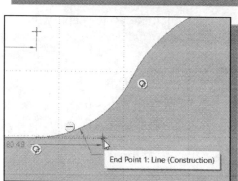

3. On your own, repeat the above steps to delete the second construction point aligned at the same location.

4. On your own, use the **Equal Radii** option to make the two lower arcs equal radii.

5. On your own, use the **Perpendicular** option to align the top horizontal line and the center point of the top arc if necessary.

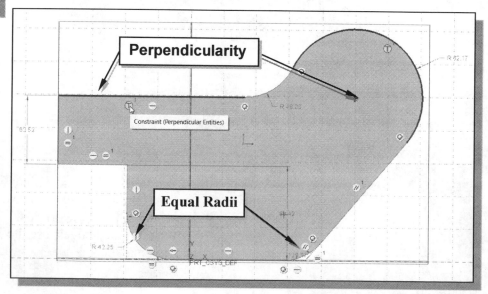

6. Use the **Dimension** command and add the two dimensions as shown. (Accept the default values and do not adjust any of the dimensions at this point.)

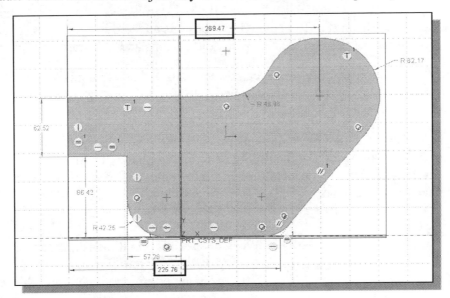

Deleting Geometric Constraints

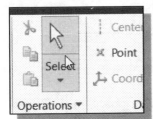

1. Choose **Select** in the *Operations* toolbar.

2. Pick the equal length symbol (=₁) on the left vertical edge as shown in the below figure.

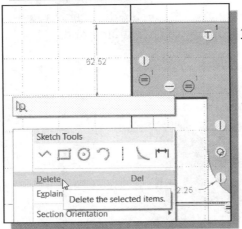

3. Press and hold down the right-mouse-button to display the option menu and select **Delete** to remove the Equal Length constraint.

❖ Notice *Creo Parametric* removed the Equal Length constraint and a horizontal dimension is added to the sketched section.

❖ In *Creo Parametric*, completed 2D sketches (**sections**) are always fully constrained. *Creo Parametric's Intent Manager* incorporates the *Dynamic Dimensioning* feature, which means that *Sketcher* adds or removes dimensions and constraints *on the fly* to keep sections fully constrained, even in the middle of sketching. 2D sections are never under-dimensioned or over-dimensioned.

Weak Dimensions and Constraints

Dimensions and constraints are called "**weak**" if *Creo Parametric Sketcher* can remove them without any confirmation from us. All dimensions added dynamically by *Sketcher* are *weak*, as are many of the constraints. Weak dimensions and constraints appear blue.

There is no need to remove undesired weak dimensions and constraints explicitly. Instead, simply add the desired dimensions and constraints. The weak ones will be removed automatically as appropriate.

If you attempt to add a constraint or dimension that conflicts with other *strong* dimensions or constraints, the conflicting items will immediately be highlighted, and you will be prompted to select one or more of them to be removed. This keeps the section from ever becoming over-constrained or over-dimensioned.

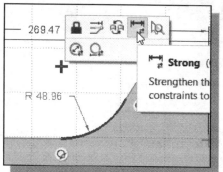

1. Weak constraints and dimensions can be "*strengthened*" by left-clicking and choosing **Strong** in the pop-up option menu. This will ensure that they are never removed without confirmation. Modifying weak dimensions also strengthens them.

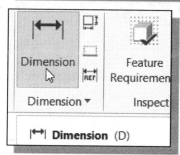

2. On your own, create/strengthen additional dimensions so that the sketch appears as shown in the figure below. (Do not be overly concerned with the dimension values; concentrate on applying proper constraints and dimensions.)

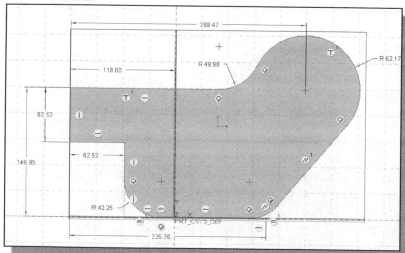

Complete the Base Feature

1. On your own, adjust the dimensions as shown in the figure below.

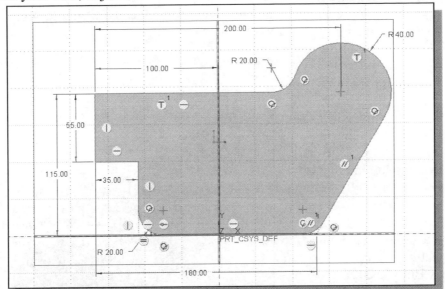

2. In the *Ribbon* toolbar, click the **OK** icon to end the *Creo Parametric 2D Sketcher* and proceed to the next element of the feature definition.

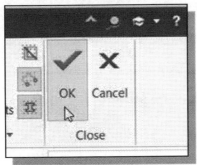

3. In the *Extrude Options Dashboard*, confirm the *depth* option is set and enter **5** as the *extrusion depth* as shown in the figure.

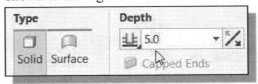

4. Pick **OK** in the feature dialog box to create the solid.

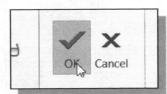

➤ On your own, use the *Dynamic Viewing* functions to view the completed 3D solid model.

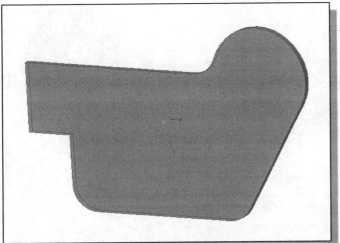

Using the Offset Loop Option to Create a CUT Feature

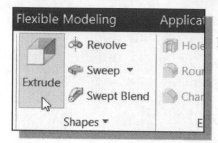

1. In the *Shapes* toolbar select the **Extrude** tool option as shown.

2. Pick the **front face** of the solid model as the sketching plane and set the view direction into the solid model.

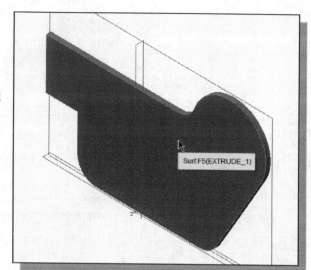

➢ Note that we are using the auto-selection feature of *Creo Parametric*. Both **RIGHT** and **TOP** datum planes are selected as sketching references.

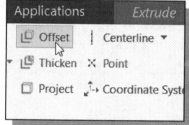

3. In the *Sketching* toolbar, select **Offset**.

4. Pick any edge/arc of the base feature as shown.

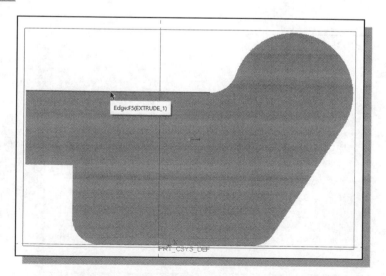

5. Hold down the **Shift** key and click on the inside of the front surface of the base feature. The contour of the front surface is automatically selected.

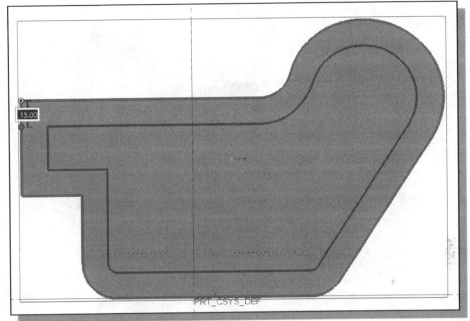

6. On your own, enter **15** to set the offset curve on the inside of the base feature.

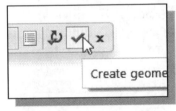

7. Click the **OK** button to accept the entered value and create offset geometry.

➤ Notice there is only the offset dimension needed to define the geometry. The **Offset Edge** command creates parent/child relationships with the referenced feature. The offset entities are updated automatically as the base geometry is modified.

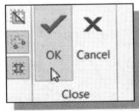

8. In the *Ribbon* toolbar, click the **OK** icon to end the *Creo Parametric 2D Sketcher* and proceed to the next element of the feature definition.

9. On your own, **flip** the extrusion direction and set extrude options to **Remove Material** and **Thru All** and complete the offset cut feature.

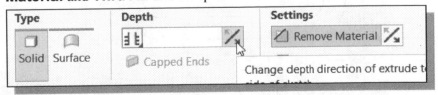

➤ Use the *Dynamic Viewing* functions to view the completed 3D model.

➤ On your own, modify the design by creating one additional (Radius: **10**) fillet on the left side as shown, adjust some of the dimensions and notice the offset curve is maintained and updated automatically. **Save** the model before proceeding to the next section.

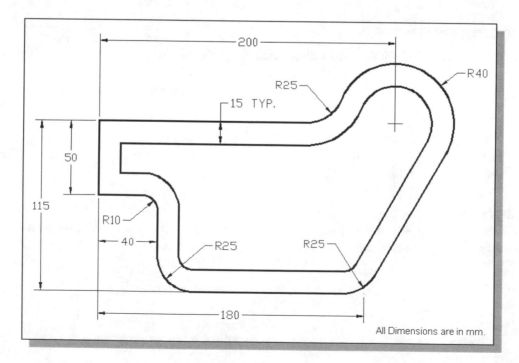

Parametric Relations and Geometric Constraints

A primary and essential difference between parametric modeling and previous generation computer modeling is that parametric modeling captures the *design intent*. In the previous lessons, we have seen that the design philosophy of *"shape before size"* is implemented through the use of *Creo Parametric's **Intent Manager*** and ***Sketcher***. In performing geometric constructions, dimensional values are necessary to describe the **size** and **location** of constructed geometric entities. Besides using dimensions to define the geometry, we can also apply geometric rules to control geometric entities. More importantly, *Creo Parametric* can capture design intent through the use of **geometric constraints**, **dimensional constraints**, and **parametric relations.** In *Creo Parametric*, there are two types of constraints: **geometric constraints** and **dimensional constraints**. For part modeling in *Creo Parametric*, constraints are applied to *2D sketches*, as they were illustrated in the previous section. **Geometric constraints** are **geometric restrictions** that can be applied to geometric entities; for example, Horizontal, Parallel, Perpendicular, and Tangent are commonly used *geometric constraints* in parametric modeling. **Dimensional constraints** are used to describe the SIZE and LOCATION of individual geometric shapes. In *Creo Parametric*, **parametric relations** are user-defined mathematical equations composed of *dimensional variables* and/or *design variables*. In parametric modeling, features are made of geometric entities with both relations and constraints describing individual design intent. In this section, we will discuss the use of parametric relations.

Create a Simple *PLATE* Design

In parametric modeling, dimensions are design parameters that are used to control the sizes and locations of geometric features. Dimensions are more than just values; they can also be used as feature control variables. This concept is illustrated by the following example.

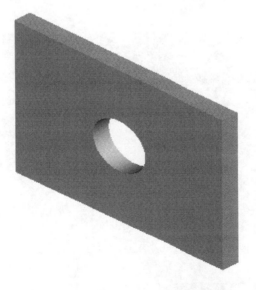

Using a Sketch File

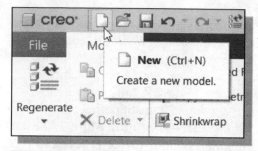

1. Click the **Create New** icon in the *Quick Access* toolbar. (We can also use the key combination Ctrl-N to start a new object.)

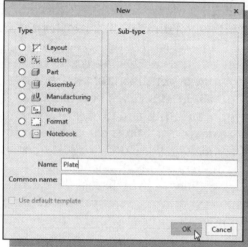

2. In the *New* form, pick **Sketch** in the **Type** list.

3. Enter **Plate** as the filename and click **OK**. (The associated file is a *Creo Parametric* sketch file, **Plate.sec**.)

❖ The *Creo Parametric Sketcher* appears. We have entered *Creo Parametric's 2D Sketcher* mode directly. The concept of this option is to allow users to utilize *Creo Parametric* in all aspects of the design process. In this case, as a 2D sketching pad.

4. On your own, use the **Rectangle** and **Circle** commands and create a sketch roughly as shown. (Do not be concerned with the displayed dimensional values; we will adjust them in the following sections.)

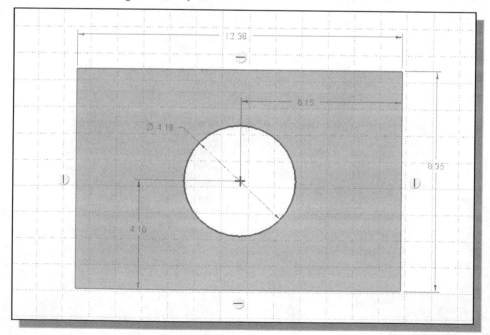

Dimensional Values and Dimensional Variables

Initially in *Creo Parametric*, dimensional values are used to reflect the size or distance between geometric entities created in the *Sketcher*. Each dimension is also assigned a name that allows the dimension to be used as a control variable. The default format is "**sdxx**," which stands for *Sketcher dimension*, and the "**xx**" is a number that *Creo Parametric* will increment automatically.

Let us look at our current design, which represents a plate with a hole at the center. The dimensional values describe the size and/or location of the plate and the hole. If a modification is required to change the width of the plate, the location of the hole will remain the same as described by the two dimensional values. This is okay if that is the *design intent*. On the other hand, the *design intent* may require (1) keeping the hole at the center of the plate and (2) maintaining the size of the hole to be one-third of the height of the plate. We will establish a set of parametric relations using the dimensional variables to capture this design intent.

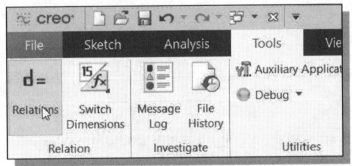

1. Pick **Relations** in the **Tools** pull-down menu.

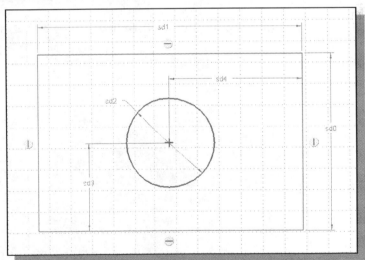

❖ Notice in the display area the *dimensional variables* are displayed instead of the *dimensional values*. (The dimensional variables are in the **sdxx** format, which represents the dimensions are *2D Sketcher* dimensions.)

2. In the *Relations* window, enter **sd3=sd0/2** as the first equation. This equation sets the vertical location dimension (**sd3**) of the hole to be one-half of the height of the rectangle (**sd0**). The variable names on your sketch might be different than what is displayed here; use the corresponding variable names to establish the parametric relation.

3. Inside the *Relations* window, enter **sd4=sd1/2** as the second equation. This equation makes sure the horizontal location dimension (**sd4**) of the hole is always one-half of the width-dimension of the rectangle (**sd1**). Use the corresponding variable names in your sketch to establish the parametric relation.

4. Enter **sd2=sd0/3** as the third equation. This equation sets the size of the hole (**sd2**) to be always one-third of the height of the rectangle (**sd0**). The variable names on your sketch might be different than what is displayed here. Use the corresponding variable names to establish the parametric relation.

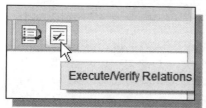

5. Click the **Execute/Verify** button in the *Relations* toolbar as shown.

6. Click **OK** to close the *Verify Relations* window.

7. Click **OK** to close the *Relations* window.

8. On your own, modify the width of the rectangle to **8**, and the height of the rectangle to **6**.

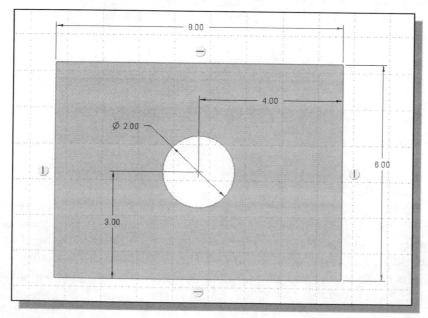

9. On your own, modify the width of the rectangle to **10** and the height of the rectangle to **7.5**.

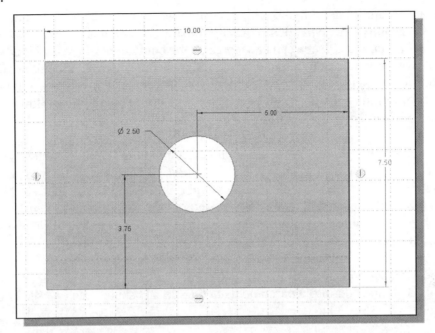

❖ *Creo Parametric* automatically adjusts the dimensions of the design, and the parametric relations we entered are also applied and maintained. The dimensional constraints are used to control the size and location of the hole. The design intent, previously expressed by statements (1) and (2) at the beginning of this section, is now embedded into the model.

Save the Sketch File and Exit the Sketcher Mode

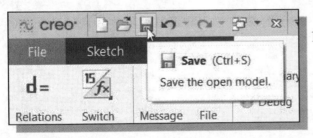

1. From the icon panel, select the **Save** icon. You can also use **Ctrl-S** to save the model.

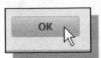

2. Click on the **OK** button or hit the **ENTER** key to accept the **Plate.sec filename** and proceed to save the file.

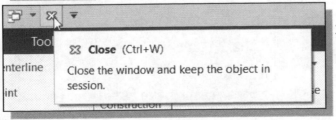

3. Click the **Close** button in the **quick access** toolbar to end the *Creo Parametric Sketch file*.

Sketch Files vs. Part Files

The main advantage of creating sketch files is that we do not need to go through all the steps that are required for part files in order to create 2D sketches in *Creo Parametric*. The *Creo Parametric Sketcher* can be viewed as the equivalent of the designer's sketchpad. Sketch files can be used to create rough sketches during the initial design stage. Modifications and changes can be done quickly and effectively without worrying about the interactions between features or parts. The use of sketch files allows us to concentrate on one feature at a time.

Once the sketches are finalized, we can then incorporate the finalized sketches into any part file. This approach allows us to reuse any of the 2D sketches that have been created. Note that *Creo Parametric* also allows us to export any of the 2D sketches that we have created in any part file. In fact, each time we entered the *2D Sketcher*, *Creo Parametric* created a sketch file automatically. The automatically generated sketch file is discarded when we exit *Creo Parametric* or when we issue the **Erase** command.

In the following section, we will illustrate the procedure for incorporating an existing sketch from a sketch file into a part model, so that the 2D sketch can be used to create a solid feature.

Starting a new *Creo Parametric* Part

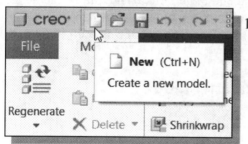

1. Click the **New** icon in the *Quick Access* toolbar. (We can also use the key combination Ctrl-N to start a new object).

2. In the *New* form, enter **Plate** as the solid part file **Name**. (The associated filename is *Plate.prt*, which is not the same as the *Plate.sec* sketch file.)

❖ Note the **Use default template** option is switched *on*. This will use whatever default unit that was selected during installation.

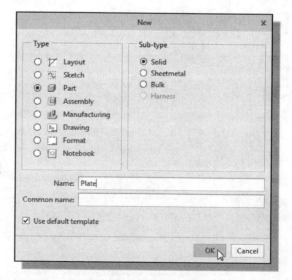

3. Click **OK** to continue. A set of predefined features, three datum planes and a *coordinate system*, appeared on the screen. The template can help reduce a considerable amount of repetitions that are performed each time we start a new drawing.

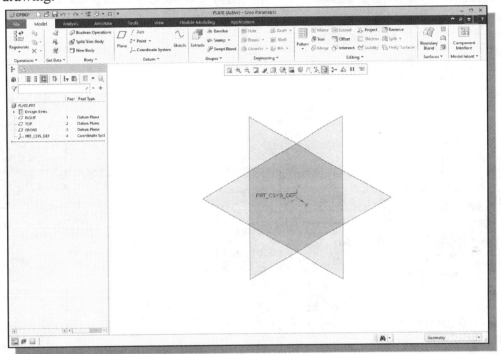

Create the Base Feature

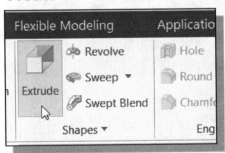

1. In the *Shapes* toolbar, select the **Extrude** tool option as shown.

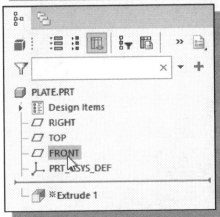

2. In the *Model Tree*, pick the **FRONT** datum plane in the sketching plane as shown.

➤ Note that we are using the auto-selection feature of *Creo Parametric*. By selecting the sketching plane, both **RIGHT** and **TOP** datum planes are selected as sketching references.

Placing the 2D Sketch

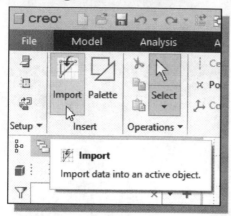

1. Pick **Import** in the *Get Data* toolbar as shown.

2. In the *Open* form, select
   Plate.sec in the file list.

3. Pick **Open** to proceed with
   importing the 2D sketch into the
   current model file.

4. Click near the intersection of the
   two references to place a copy of
   the 2D sketch into the *Sketcher*.

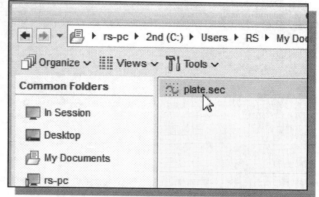

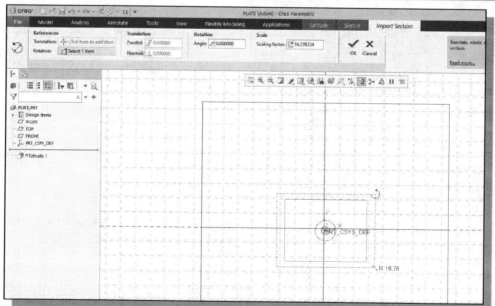

❖ The copy of the selected **sketch file** is placed into the *Sketcher* with three control
   handles: **Locate**, **Scale** and **Rotate**. The **Locate**, **Scale** and **Rotate** controls can
   also be adjusted through the dashboard panel.

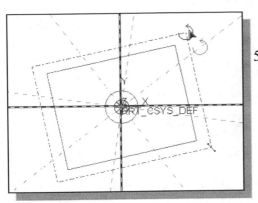

5. Drag with the left-mouse-button on the **Rotate** handle to rotate the 2D section.

6. On your own, use the **Scale** handle to adjust the scale of the 2D section. Notice both the rotated angle and new scale factor are also displayed in the *Scale Rotate* window to the right of the main window as we use the *adjusting handles* with the mouse.

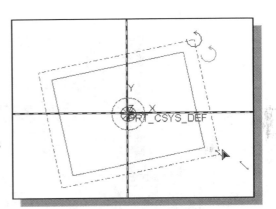

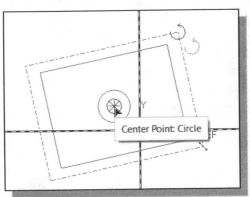

7. The **Locate** handle, the circle mark at the center of the 2D section, can be used to move the 2D section. Note that the cursor will automatically snap to the references.

8. On your own, experiment with the three handles, but place the 2D sketch back at the **origin** before continuing to the next step.

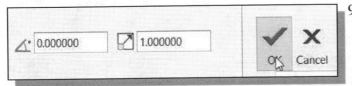

9. In the *Scale Rotate* window, enter **1** as the new scale factor and set the rotating angle to **0.00** as shown.

10. Click on the **OK** button to proceed with the insertion of the 2D section.

11. Use the **Refit** command to have *Creo Parametric* adjust the zoom scale so that all objects are displayed on screen.

➤ We have successfully transferred in the 2D sketch done previously. We remain in the *Creo Parametric Sketcher* as if we had just created the 2D sketch. (On your own, apply **Coincident** constraint to align the *center of the plate* to the two references.)

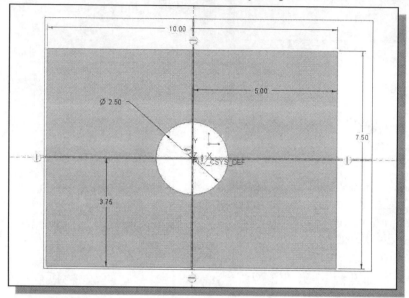

➤ On your own, try to adjust the hole diameter dimension (2.5). And notice the message area displays "*This dimension is governed by a relation and can not be modified.*" The set of parametric relations we have established in the sketch file is also transferred into the current part file.

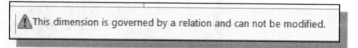

Display the Parametric Relations

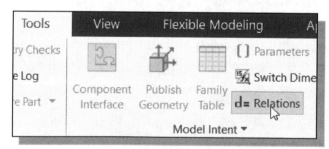

1. Pick **Relations** in the **Tools** tab through the *Ribbon* toolbars.

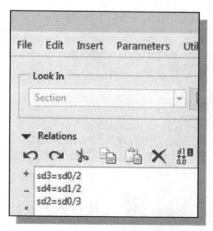

➤ The *Relations* window appears and the parametric equations defined are displayed. Note that the parametric relations are imported from the saved sketch file.

2. Pick **OK** to exit the *Relations* window and return to the *Sketcher*.

❖ On your own, modify the overall width of the rectangle to **8** and confirm the parametric relations are maintained. Reset the value back to **10** before continuing to the next section.

Complete the Extrusion Feature

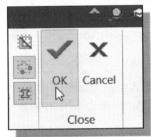

1. Now that the 2D sketch is completed, we will proceed to the next element. In the *Ribbon* toolbar, click on the **OK** button to end the *Creo Parametric Sketcher* mode.

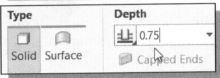

2. In the *feature option area*, enter **0.75** as the depth of extrusion and click **OK** to create the solid.

3. On your own, use the *Dynamic Rotate* function [**middle-mouse-button**] to view the 3D model.

Parametric Relations at Different Levels

In *Creo Parametric*, **parametric relations** can also be defined within a part or among parts of an assembly. So far, we have demonstrated the use of parametric **section relations**, which are applied to the 2D section within a solid feature. Similar procedures can also be applied to the feature, part, and assembly levels to establish specific parametric relations.

In *Creo Parametric*, five types of **parametric relations** are available:

- **Assembly Relations** relate different part parameters to each other in an assembly model.

- **Part Relations** relate different feature parameters to each other in a single model. Part relations can include feature relations which are within one feature in the model.

- **Feature Relations** relate specific parameters within one feature in the model. A feature relation can be for the 2D section only or for the entire feature.

- **Section Relations** relate specific parameters within one 2D sketch in the model.

- **Pattern Relations** relate specific parameters within a pattern in the model.

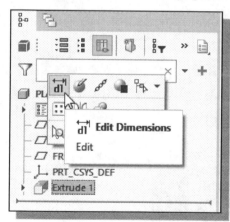

1. In the *Model Tree* window, move the cursor on top of Extrude 1 and click once with the left-mouse-button to bring up the option menu.

2. In the option menu, select **Edit Dimensions**.

➤ Notice that all dimensions related to the feature are displayed, including the extrusion depth (0.75).

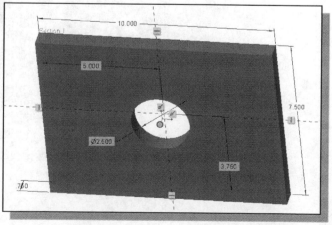

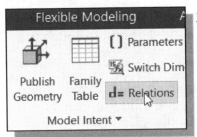

3. Pick **Relations** in the **Tools** pull-down menu.

❖ The *Relations* window appears and in the graphics area the *dimensional names* are displayed instead of the *dimensional values*. (The dimensional variables are in the **dxx** format, which represents the dimensions are in *Part level* or *Assembly level*.)

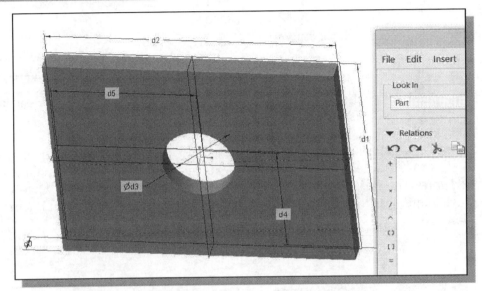

4. Choose **Section** in the Object Type option of the *Relations* window as shown.

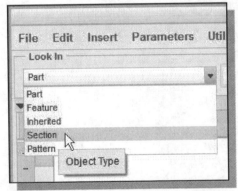

5. Choose any surface of the *Plate* model. (There is only one *section* in the model.) Note the parametric relations we set up in the sketch file are displayed.

6. On your own, use the **Switch Dim** option to examine the existing *section relations*.

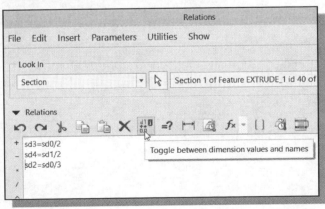

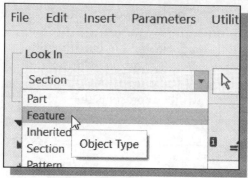

7. Choose **Feature** in the Look In option of the *Relations* window as shown.

8. Choose any part of the *Plate* model. Note that there is only one solid feature in the model. Also note the selected feature information is displayed in the *Relations* window.

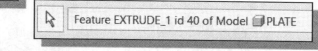

9. Use the **Switch Dim** option to toggle the display of dimensional values and names.

10. In the message area, *Creo Parametric* expects us to input mathematical equations. Enter **d0=d1/10**, where **d0** is the depth of extrusion and **d1** is the height dimension of the *Plate*. The variable names on your screen might be different than what is displayed here. Use the corresponding variable names to establish the parametric relation.

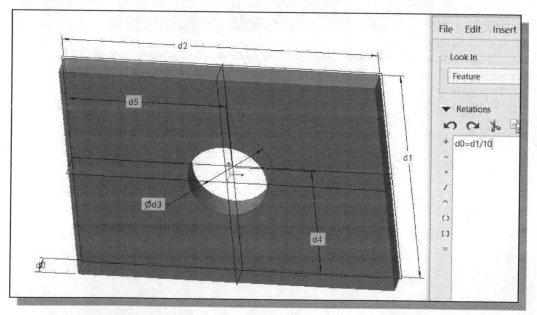

11. Click the **OK** icon to accept the entered equation and close the *Relations* window.

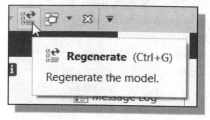

12. On your own, observe the changes of the model by adjusting the height of the *Plate*. (Use **Regenerate** to update the model.)

❖ The thickness of the part is updated based on the height dimension of the model. The parametric relations are maintained by the system so that an intelligent part is created.

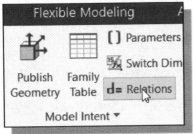

13. On your own, open the *Relations* window and delete the equation (**d0=d1/10**) we entered at the *Feature level*.

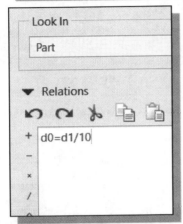

14. Place the same equation (**d0=d1/10**) at the ***Part level*** as shown.

15. On your own, **Edit** the **height dimension** of the model and observe the relations established within the part.

16. On your own, create additional solid features and experiment with establishing different relations at different levels.

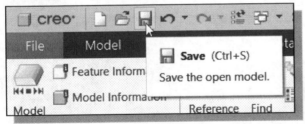

17. Select **Save** in the *Quick Access* toolbar.

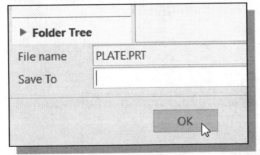

18. Click **OK** to proceed with saving the model.

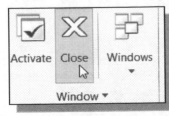

19. Click the **Close** button in the **View** tab to close the *Plate* model file. Note the windows option next to the icon, which allows us to switch between any opened files.

Review Questions:

1. What is the difference between *dimensional values* and *dimensional variables*?

2. How do we **disable** *geometric constraints* while sketching?

3. How are *geometric constraints* applied in *Creo Parametric Sketcher*?

4. Describe the different types of **parametric relations** available in *Creo Parametric*.

5. How do we create parametric relations in *Creo Parametric*?

6. How do we assure two points are *lined up horizontally*?

7. Create binary tree sketches showing the steps you plan to use to create the two models shown on the next page:

Ex.1)

Ex.2)

Exercises: (Establish three parametric relations for each of the designs and experiment with the different modification techniques illustrated in this lesson.)

1. **Wedge Block** (Dimensions are in inches.)

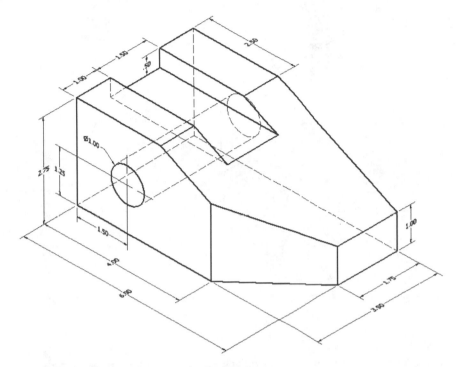

2. **Hinge Guide** (Dimensions are in inches.)

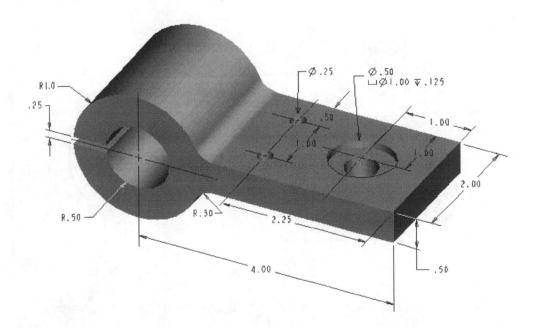

3. **Slide Mount** (Dimensions are in inches.)

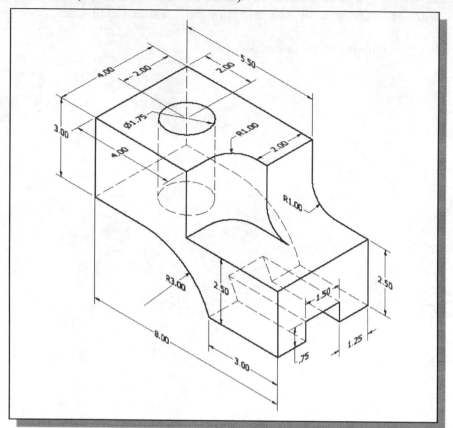

4. **Swivel Base** (Dimensions are in millimeters. Base thickness: 10 mm, Boss height: 5 mm.)

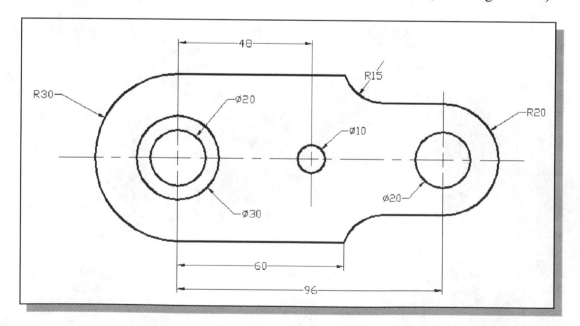

5.   **Block Base** (Dimensions are in inches. Plate Thickness: 0.25)

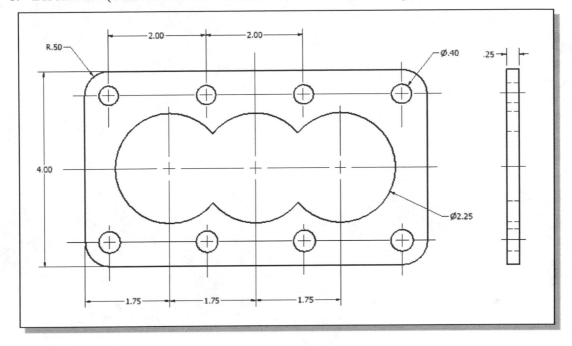

Notes:

Chapter 5
Parent/Child Relationships

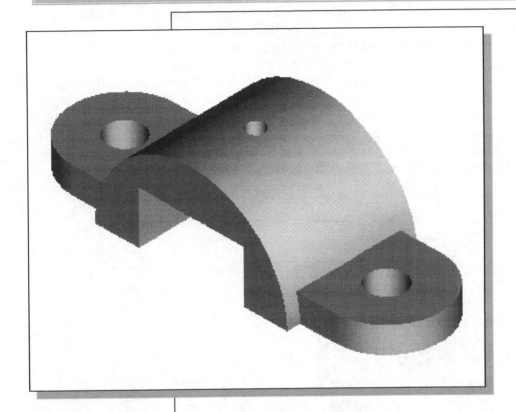

Learning Objectives

- ◆ **Understand the Parent/Child relations between Features**
- ◆ **Basic Modeling Considerations**
- ◆ **Drag & Drop Editing**
- ◆ **Apply Constraints Explicitly**
- ◆ **Rename Features**
- ◆ **Use Suppress/Resume Feature tools**
- ◆ **Embed Alternative Designs in the same Model**
- ◆ **Use the Model Player**

Introduction

The parent/child relationship is one of the most powerful aspects of *parametric modeling*. In *Creo Parametric*, each time a new modeling event is created, any of the previously defined features can be used to define information such as size, location, and orientation. The referenced features become **PARENT** features to the new feature, and the new feature is called the **CHILD** feature. The parent/child relationships determine how a model reacts when other features in the model change, thus capturing design intent. It is crucial to keep track of these parent/child relations. Any modification to a parent feature can change one or more of its children. As one might expect, parent/child relationships can become quite complicated when the features begin to accumulate. It is therefore important to think about modeling strategy before we start to create anything. The main consideration is to try to plan ahead for all possible design changes that might occur which would be affected by the existing parent/child relationships.

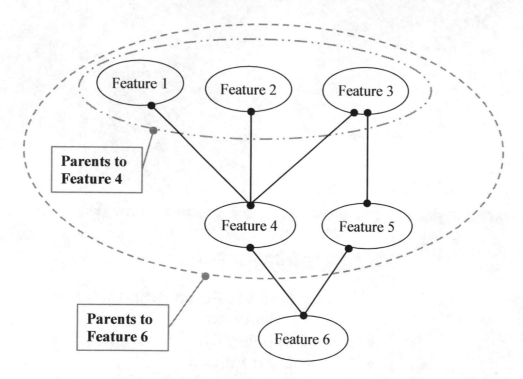

Parent/child relationships can be created *implicitly* or *explicitly*; implicit relationships are implied by the feature creation method and explicit relationships are entered manually by the user. We have been using implicit parent/child relationships in the previous lessons. For example, before creating a 2D section in the *Creo Parametric Sketcher* mode, we select a sketching plane and a reference plane. Both of these become parents of the new feature. If the sketching plane is moved, the child feature will move with it. The reference plane determines the orientation of the section on the sketching plane. If the orientation is changed, the orientation of the child feature is changed as well. The use of *explicit parent/child relationships* will be demonstrated in the following lessons.

The *U-Bracket* Design

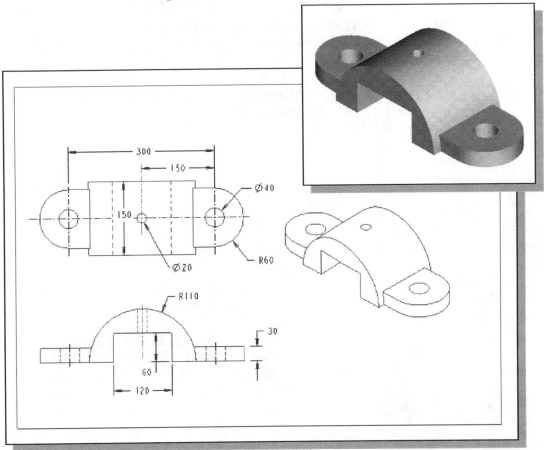

Preliminary Modeling Considerations

Prior to creating the model, a few minutes of planning and sketching on paper will save many hours of redefining the design on the computer. Two of the main modeling considerations are:

- **Design Intent** – Determine the functionality of the design; identify features that are central to the design.

- **Order of Features** – Consider the parent/child relationships necessary for all features.

> Based on your knowledge of *Creo Parametric* so far, how many features would you use to create the model? Which feature would you choose as the **base feature**? What is your choice for arranging the order of the features? Would you organize the features differently if the rectangular cut at the center is changed to a circular shape (Radius: 80mm)? Take a few minutes to consider these questions and do preliminary planning by sketching on a piece of paper.

Starting *Creo Parametric* and Using a Metric Template

1. Select the **Creo Parametric** option on the *Start* menu or select the **Creo Parametric** icon on the desktop to start *Creo Parametric*. The *Creo Parametric* main window will appear on the screen.

2. Pick the **New** icon in the toolbar. (You can also use the key combination **Ctrl-N** to start a new object.)

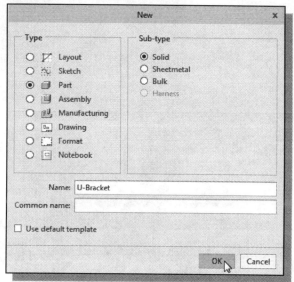

3. In the *New* form, enter **U-Bracket** as the solid part file **Name**.

4. Switch *off* the **Use default template** option as shown. (We will use one of the metric templates for the design.)

5. Pick **OK** to continue. The *Model Tree* window and the *Menu Manager* window appear on the screen.

6. In the *New File Options* window, select **mmns_part_solid_abs** as the template to use. This is the template for using with the metric system (length in **millimeters**).

❖ In *Creo Parametric*, a template file can contain a set of predefined settings, such as the datum planes, datum axes and units setup. The template files allow us to reduce startup repetitions.

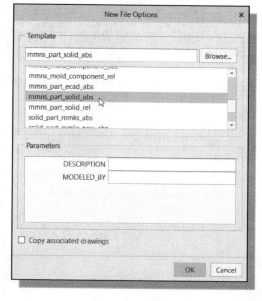

Create the Base Feature

❖ We will create the model using four features and the base feature will be the bottom section of the model. We will use *Creo Parametric* as a sketching tool for quickly changing the sketched geometric entities. Note that this is a common practice during the initial conceptual design stage.

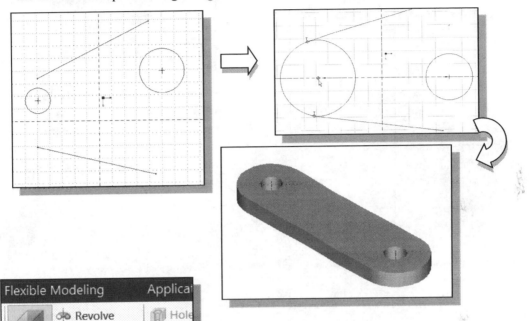

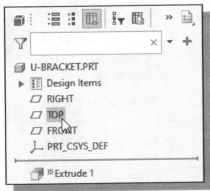

1. In the *Shapes* toolbar, select the **Extrude** tool option as shown.

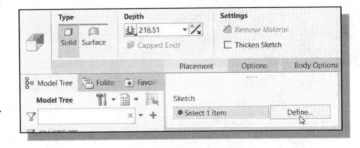

2. Click the **Placement** option and choose **Define** to begin creating a new *internal sketch*.

3. Select **TOP** by clicking on the Datum Plane name in the *Model Tree* window as shown.

❖ The feature names, listed in the *Model Tree* window, can also be selected just like selecting the features in the graphics window.

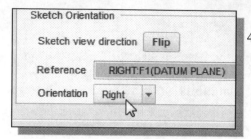

4. Confirm the **RIGHT** datum plane is facing the **right** edge of the computer screen and pick **Sketch** to enter the *Sketcher* mode.

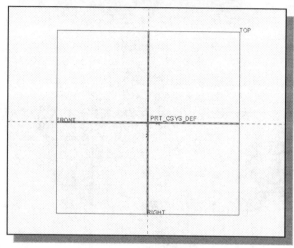

➤ Note that FRONT and RIGHT are pre-selected as the sketching references. In the graphics area, the two references are highlighted and displayed with two dashed lines.

Create and Transform a Rough Sketch

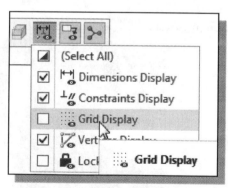

1. In the *Display Control* toolbar, click on the **Grid** icon to switch *on* the **grid display** as shown.

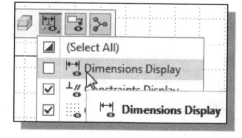

2. In the *Display Control* toolbar, also click on the **Disp Dims** icon to switch *off* the **dimension display** as shown.

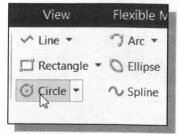

3. In the *Sketching* toolbar, select **Circle** as shown. The default option is to create a circle by specifying the center point and a point through which the circle will pass.

4. On your own, create two circles of different sizes, with the centers intentionally not aligned to the reference datum planes as shown.

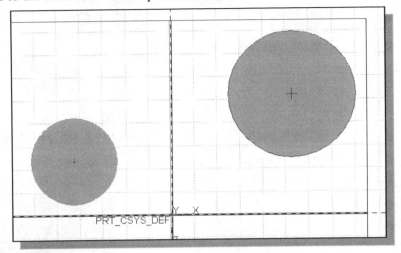

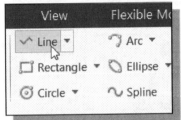

5. Choose the **Line** command in the *Sketching* toolbar.

6. Create two lines of arbitrary length as shown in the figure below. Again, we will intentionally not align the two lines to the references or the circles.

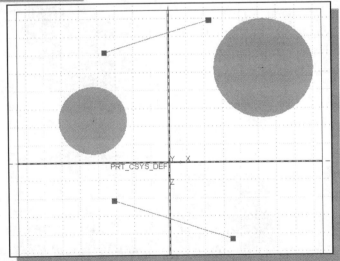

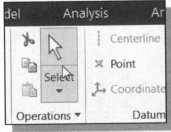

7. In the *Operations* toolbar, click on the **Select** icon.

8. On your own, move the created objects around by using the **drag and drop method** with the **left-mouse-button**.

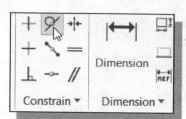

9. Pick **Tangent** in the *Constrain* toolbar.

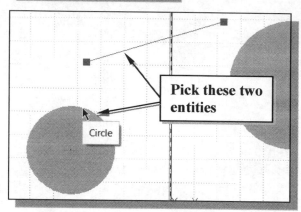

Pick these two entities

Circle

10. Pick the circle on the left and the line on top as shown.

11. Pick the circle on the left and the line at the bottom as shown.

❖ Note that *Creo Parametric* adjusts the two entities and displays the **Tangent** constraint symbol.

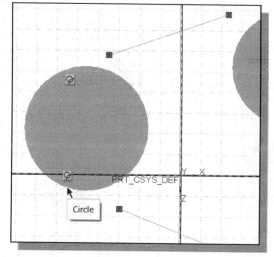

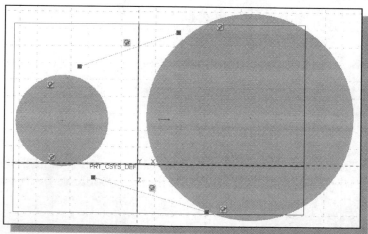

12. On your own, repeat the above process and apply two **Tangent** constraints to the two lines and the circle on the right.

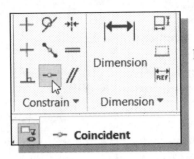

13. Click **Coincident** in the *Constrain* toolbar.

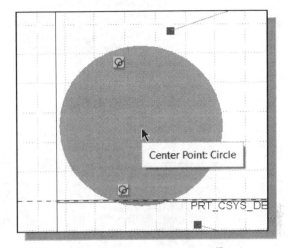

14. Pick the center point of the left circle as shown.

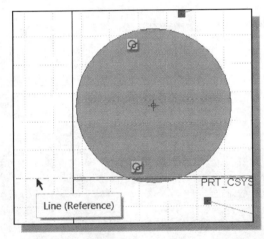

15. Pick the **horizontal reference** to align the center of the circle on the horizontal reference line.

❖ Note that *Creo Parametric* aligns the center of the circle to the horizontal reference and also maintains the applied Tangent constraints.

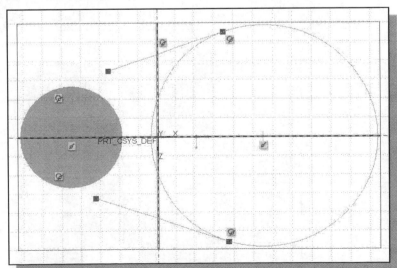

16. On your own, repeat the above process and align the circle on the right to the horizontal reference as shown.

17. Click once with the middle-mouse button to end the coincident constraint.

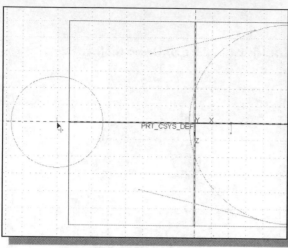

18. Drag the center point of the left circle toward the left side of the graphics window and observe all of the applied constraints are maintained.

19. On your own, drag the endpoints of the lines around and observe all of the applied constraints are maintained.

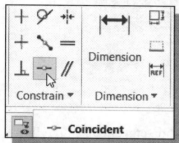

20. Click **Coincident** in the *Constrain* toolbar.

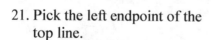

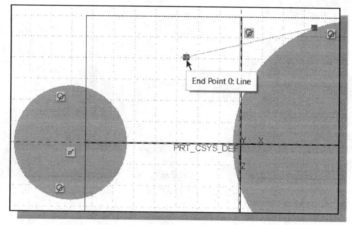

21. Pick the left endpoint of the top line.

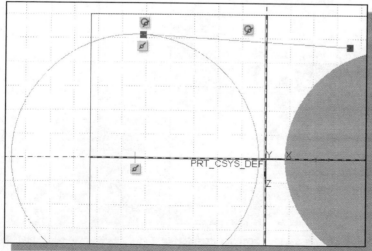

22. Pick circle on the left to apply the Coincident constraint.

❖ Note that *Creo Parametric* aligns the selected endpoint to the circle; the small coincident symbol signifies the applied constraint.

23. On your own, apply the proper **Coincident** constraint to align the endpoints to the circles as shown.

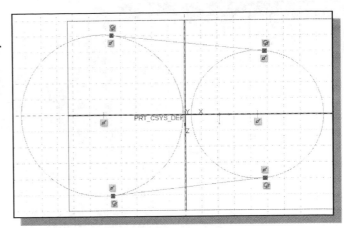

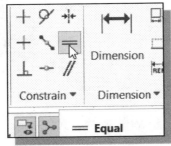

24. Click **Equal Length** in the *Constrain* toolbar.

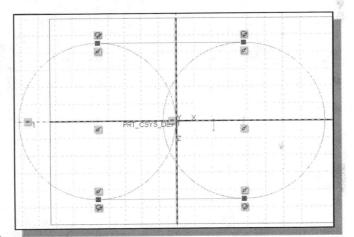

25. Pick the two circles to apply the **Equal Length** constraint.

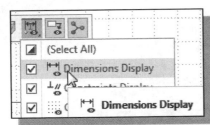

26. Toggle *on* the dimension display as shown.

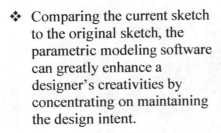

❖ Comparing the current sketch to the original sketch, the parametric modeling software can greatly enhance a designer's creativities by concentrating on maintaining the design intent.

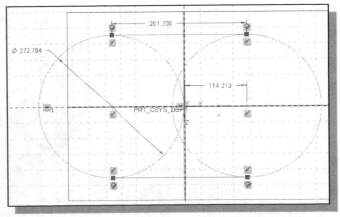

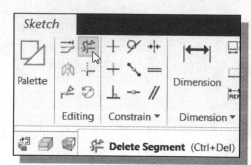

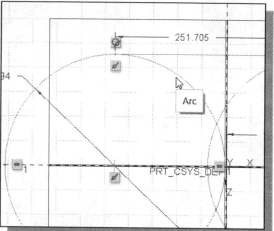

27. In the *Editing* toolbar, click **Delete Segment** as shown.

28. Select the portion of the circle that is on the inside of the sketch as shown.

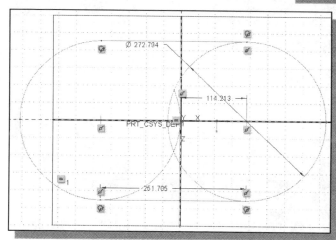

❖ Note that *Creo Parametric* trims the selected segment to the nearest intersections. Also note that some of the weak dimensions are also removed and/or applied automatically by the *Intent Manager*.

29. On your own, continue to remove the inside halves of the two circles as shown.

❖ Note that some of the applied constraints and dimensions have been removed and/or replaced, but the overall *design intent* is still maintained.

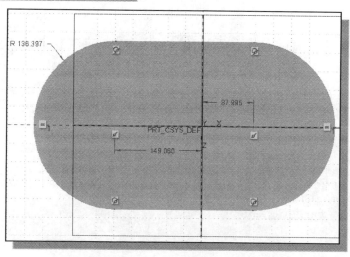

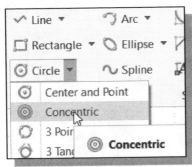

30. In the *Sketching* toolbar, select the **Concentric Circle** option.

31. Create two circles that are of the same size and concentric to the two existing arcs as shown.

32. On your own, **modify** the dimensions (**150, 60** and **40**) as shown in the figure below.

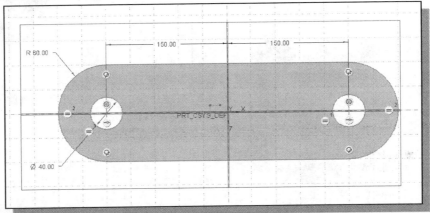

33. Now that the 2D sketch is completed, we will proceed to the next element. In the *Ribbon* toolbar, click on the **OK** button to end the *Creo Parametric Sketcher* mode.

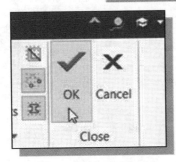

34. In the feature option area, enter **30** (mm) as the depth of extrusion and click **OK** to create the solid.

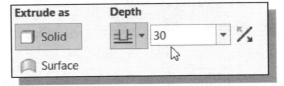

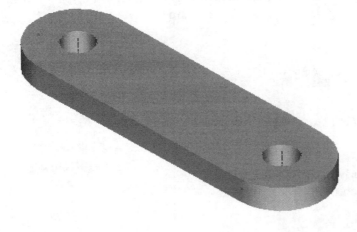

Create the Second Solid Feature

For the next solid feature, we will continue to use the **Extrude** tool.

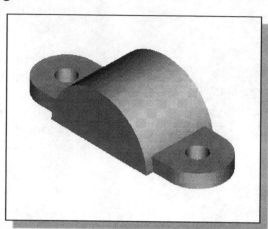

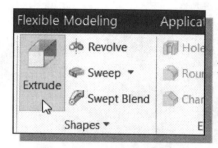

1. In the *Shapes* toolbar, select the **Extrude** tool option as shown.

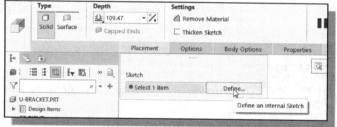

2. Click the **Placement** option and choose **Define** to begin creating a new *internal sketch*.

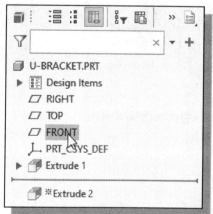

3. In the *Model Tree* window, select **FRONT** as the sketching plane.

4. Confirm the Sketch Orientation options are set to datum plane **RIGHT** and **Right** orientation as shown in the figure below.

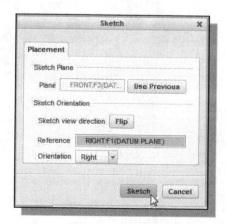

5. Click **Sketch** to enter the *Creo Parametric Sketcher* mode.

➢ Notice Creo automatically uses the RIGHT and TOP planes as the sketching references in the *Sketcher* mode.

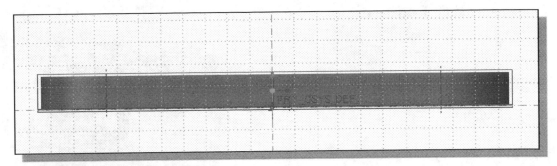

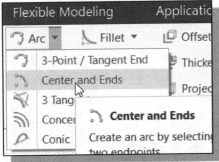

6. On your own, create the 2D section as shown. The sketch consists of an **arc** (arc-center aligned to the origin) and a **horizontal line**. The line is connected to the two ends of the arc to form a closed region.

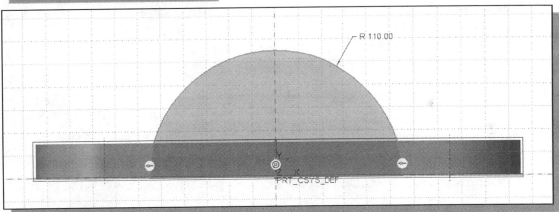

7. On your own, modify the radius of the arc to **110 mm**.

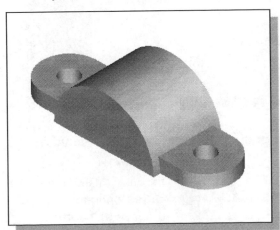

8. On your own, create the feature so that it is symmetrical to datum plane FRONT as shown. (Hint: Use the **Extrude Both Sides** option and set the depth of extrusion to **150** mm.)

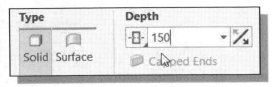

Create a CUT Feature

We will create a rectangular cut as the next solid feature.

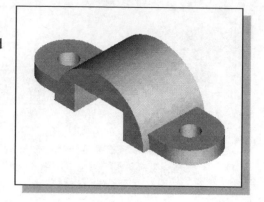

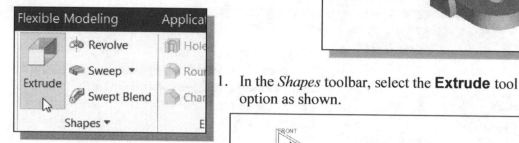

1. In the *Shapes* toolbar, select the **Extrude** tool option as shown.

2. In the graphics window, select the **Front Plane** of the model as the sketching plane.

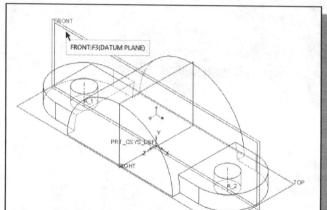

3. On your own, create a rectangle (**120 x 60 mm**) positioned as shown in the figure.

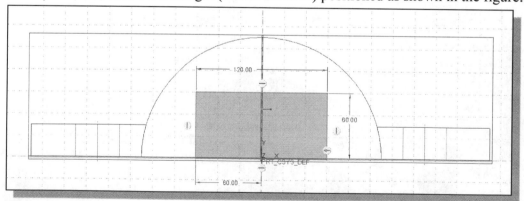

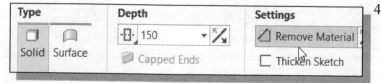

4. On your own, complete the cutout using the **Both Sides** option and a distance of **150mm**.

Create the CENTER_DRILL Feature

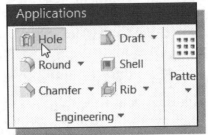

1. In the *Engineering* toolbar, select the **Hole** tool option as shown.

2. Select the horizontal face of the last cut feature as the **primary reference**, the *placement plane*.

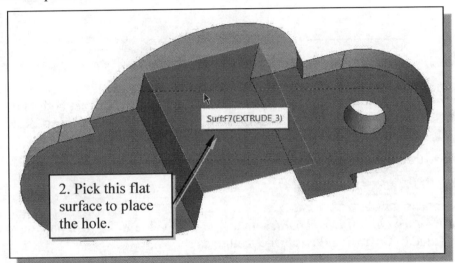

Surf:F7(EXTRUDE_3)

2. Pick this flat surface to place the hole.

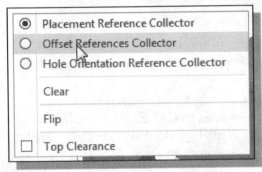

3. Inside the graphics area, press and hold down the **right-mouse-button** to bring up the option menu.

4. Switch to the **Offset References Collector** option as shown.

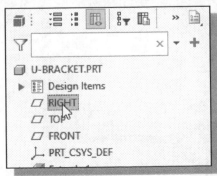

5. In the *Model Tree* window, pick **RIGHT** as the first *secondary reference*.

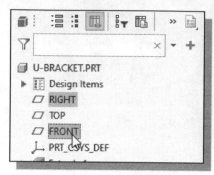

6. In the *Model Tree* window, hold down the **[CTRL]** key and pick **FRONT** as the second *secondary reference*. Note that the selected references appear in the Placement option list.

7. In the *Feature Option Dashboard*, select the **Placement** option if it is not activated.

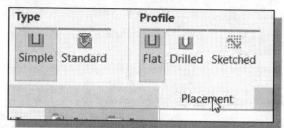

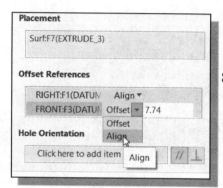

8. In the Placement option list, set both of the Offset References options to **Align** as shown in the figure.

9. In the *Feature Option Dashboard*, set the *diameter* value to **20 mm** and the extrusion *depth* value to **55mm** and create the feature as shown.

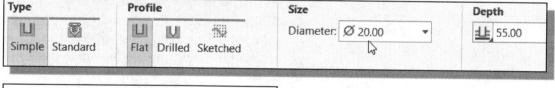

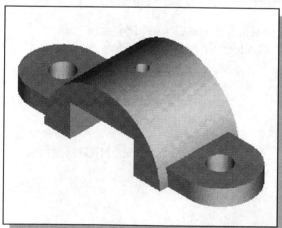

10. On your own, rename the feature names to BASE, MAIN_BODY, RECT_CUT and CENTER_DRILL as shown in the figure.

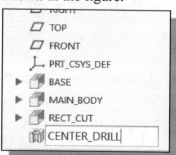

11. On your own, save the model by clicking the **Save** icon.

Examine the Parent/Child Relationships

The *Model Tree* window now contains seven items: three datum planes and four solid features. All of the parent/child relationships were established implicitly as we created the solid features. As more features are created, it becomes much more difficult to make a sketch showing all the parent/child relationships involved in the model. On the other hand, it is not really necessary to have a detailed picture showing all the relationships among the features. In using a feature-based modeler, the main emphasis is to consider the interactions that exist between the **immediate features**. Treat each feature as a unit by itself and be clear on the parent/child relationships for each feature. Thinking in terms of features is what distinguishes *feature-based modeling* from the previous generation solid modeling techniques. Let us take a look at the last feature we created, the **CENTER_DRILL** feature. What are the parent/child relationships associated with this feature? (1) Since this is the last feature we created, it is not a parent feature to any other features. (2) Since we used one of the surfaces of the rectangular cutout as the sketching plane, the **RECT_CUT** feature is a parent feature to the **CENTER_DRILL** feature. (3) We also used datum planes Right and Front as placement references; therefore, Right and Front are parents to the **CENTER_DRILL** feature.

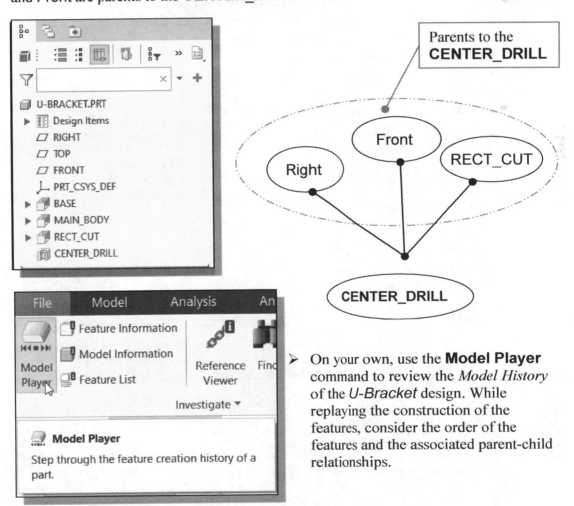

➤ On your own, use the **Model Player** command to review the *Model History* of the *U-Bracket* design. While replaying the construction of the features, consider the order of the features and the associated parent-child relationships.

Display Parent/Child Info

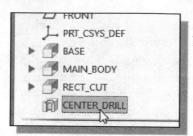

1. Pre-select the **CENTER_DRILL** in the *Model Tree* window.

2. In the **Tools** tab, select the **Reference Viewer** option in the *Investigate* toolbar. The *Reference Viewer* dialog box appears on the screen.

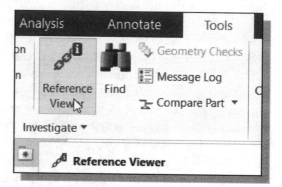

❖ In the *Reference* information window, **RIGHT**, **FRONT** and the **RECT_CUT** features are listed on the **left** area; these are parents to the **CENTER_DRILL**. Note that there is no child listed in the **Children** list area; this is what we have expected since this is the last feature of the model.

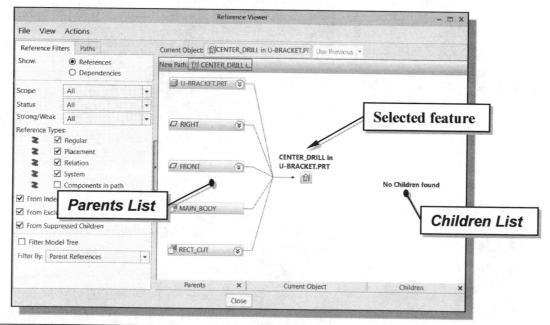

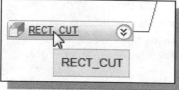

3. Double-click the **RECT_CUT** feature to examine its parent/child relations.

❖ On your own, examine the parent/child relations of the **MAIN_BODY** feature.

Modify a Parent Dimension

❖ Any changes to the parent features will affect the child feature. For example, if we modify the height of the **RECT_CUT** feature from 60 mm to 50 mm, the child feature (**CENTER_DRILL** feature) will be affected.

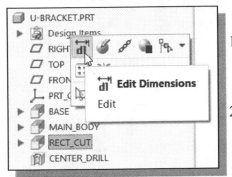

1. In the *Model Tree* window, move the cursor on top of RECT_CUT and click once with the left-mouse-button to bring up the option menu.

2. In the *option menu*, select **Edit Dimension.**

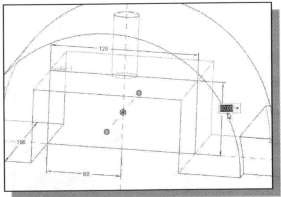

3. Double-click on the vertical height dimension text **60**.

4. Enter **50** as the new dimension value.

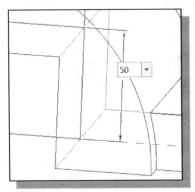

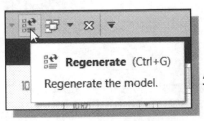

5. Click **Regenerate** in the *Quick Access* toolbar.

➤ The position of the **CENTER_DRILL** is moved downward by 10 mm since the placement plane is lowered by that amount; the drill does not go through the main body of the bracket anymore. If we are planning to make more dimensional changes to the rectangular cut, what would be your solution to make sure the drill will always go through the main body of the bracket?

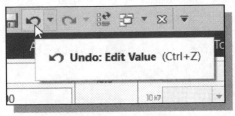

6. Click **Undo** once to set the height of the **RECT_CUT** feature back to 60 mm.

7. On your own, adjust the depth of the **CENTER_DRILL** feature to **Thru All**.

A Design Change

Engineering designs usually go through many revisions and changes. For example, a design change may call for a circular cutout instead of the current rectangular cutout feature in our model. *Creo Parametric* provides an assortment of tools to handle design changes quickly and effectively. In the following sections, we will demonstrate some of the more advanced tools available in *Creo Parametric*, which allow us to perform the modification of changing the rectangular cutout (120 x 60mm) to a circular cutout (radius: 80 mm).

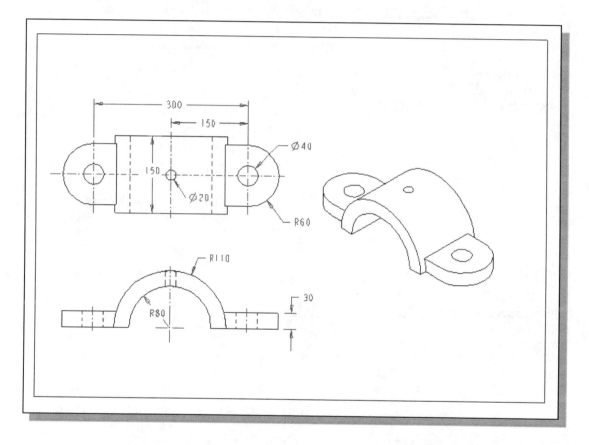

❖ Based on your knowledge of *Creo Parametric* so far, how would you accomplish this modification? What other approaches can you think of that are also feasible? Of the approaches you came up with, which one is the easiest to do and which is the most flexible? If this design change were anticipated right at the beginning of the design process, what would be your choice in arranging the orders of the features? Take a few minutes to consider these questions, and you are encouraged to perform the modifications prior to following through the rest of the tutorial.

➢ Before continuing to the next section, confirm the current **U-Bracket** model matches what is shown on page 5-3.

Feature Suppression

With *Creo Parametric*, we can take several different approaches to accomplish this modification. We could (1) create a new model, or (2) change the shape of the existing cut feature using the **Redefine** command, or (3) perform **feature suppression** on the rectangular cut feature and add a circular cut feature. The third approach offers the most flexibility and requires the least amount of editing to the existing geometry. **Feature suppression** is a method that enables us to disable a feature while retaining the complete feature information; the feature can be reactivated at any time. Prior to adding the new cut feature, we will first suppress the rectangular cut feature.

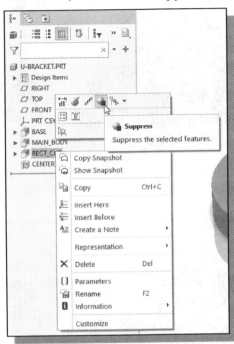

1. Move the cursor inside the *Model Tree* window. Click once with the **right-mouse-button** on top of RECT_CUT to bring up the option menu.

2. Pick **Suppress** in the pop-up menu. A pop-up window is displayed.

> In the display area and the *Model Tree* window, both the **RECT_CUT** feature and the **CENTER_DRILL** feature are highlighted. The child feature cannot exist without its parent(s), and any modification to the parent (RECT_CUT) influences the child (CENTER_DRILL).

3. Pick **OK** in the pop-up window to confirm the suppression of the RECT_CUT and CENTER_DRILL features.

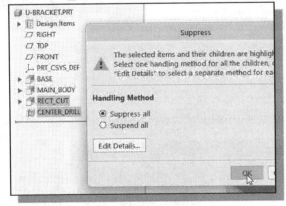

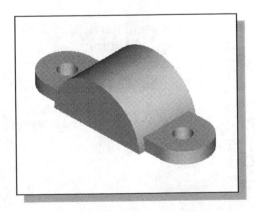

❖ We have literally *gone back in time*. The last two features of the design, RECT_CUT and CENTER_DRILL, have disappeared in the *Model Tree* window and the display area.

Create a Circular Cut Feature

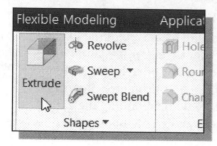

1. In the *Shapes* toolbar, select the **Extrude** tool option as shown.

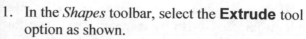

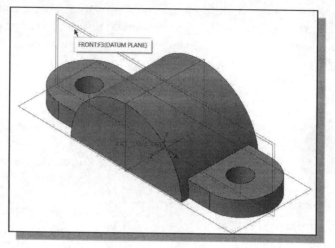

2. Pick the **Front Plane** of the model as the *sketching plane* and set the *extrusion direction* to point toward the center of the model.

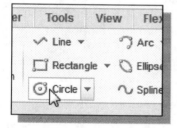

3. On your own, create a circle (diameter: **150 mm**) with the center aligned to the origin as shown.

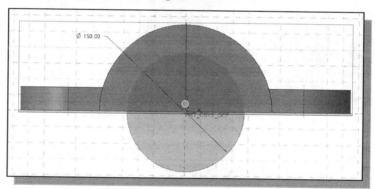

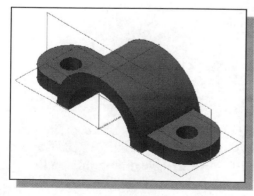

4. Complete the cutout using the **Both Sides** option and a distance of **150mm**.

5. On your own, set the name of the feature as **CIRCULAR_CUT**.

Listing Suppressed Features

1. In the *Model Tree* window, choose **Tree Filters** in the setting options as shown.

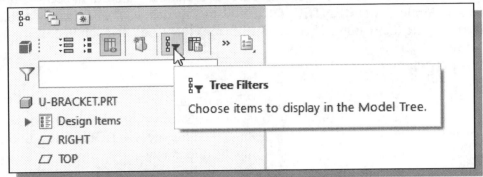

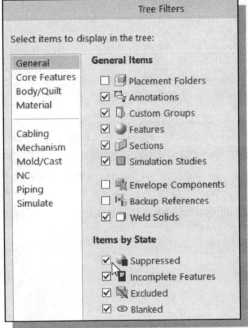

2. In the *Model Tree Items* window, click on the **Suppressed Objects** check-box to toggle *on* the display of suppressed features and components.

3. Click on the **OK** button to accept the settings.

❖ Notice in the *Model Tree* menu window a complete list of features, including suppressed features, is displayed. A small rectangular marker next to the feature signifies the feature is suppressed.

Suppressed features

Reactivate the CENTER_DRILL Feature

To complete the model, we still need the **CENTER_DRILL** feature at the center. It is not necessary to duplicate the **CENTER_DRILL** feature that already exists in the database. We will reactivate the feature. This is known as the **Resume** operation in *Creo Parametric*.

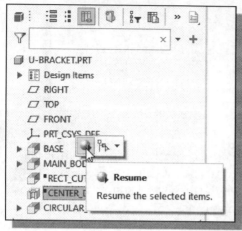

1. Move the cursor inside the *Model Tree* window. Click once with the left-mouse-button on top of **CENTER_DRILL** to bring up the option menu.

2. Pick **Resume** in the *option menu*.

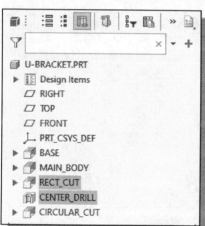

➤ Notice that both of the suppressed features (**RECT_CUT** and **CENTER_DRILL**) are reactivated, again. *The child feature cannot exist without its parents*. All of the model's features are displayed in the display area and in the *Model Tree* window.

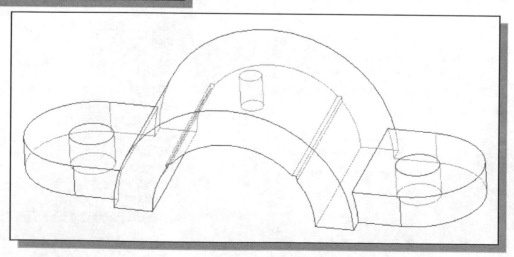

Reroute the CENTER_DRILL Feature

In order to suppress only the RECT_CUT feature, we will need to remove the parent/child relationship that exists between RECT_CUT and CENTER_DRILL. The parent/child relationship was established when we used one of the RECT_CUT surfaces as the sketching plane. We will need to select a different sketching plane for the CENTER_DRILL feature. In *Creo Parametric*, this is done by **Editing References** of the features.

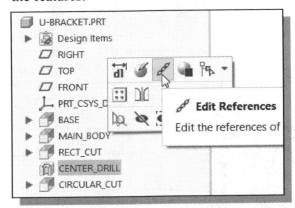

1. Move the cursor inside the *Model Tree* window. Click once with the left-mouse-button on top of **CENTER_DRILL** to bring up the option menu.

2. Pick **Edit References** in the option menu.

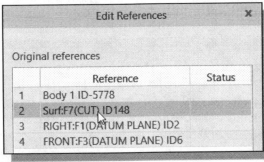

❖ The *Edit References* window allows us to examine and/or redefine the *sketching plane* and *reference planes* for the selected feature.

3. Click on the second item in the *Edit References* window as shown.

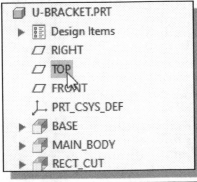

4. The selected item is the *placement plane* for the HOLE feature. Pick datum plane **TOP** in the *Model Tree* area.

5. Click **OK** to accept the change. *Creo Parametric* will regenerate the modified features and display the regenerated model. Everything looks the same, but RECT_CUT is no longer a parent feature to the CENTER_DRILL feature.

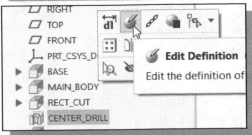

6. On your own, use the *Edit Definition* option and set the CENTER_DRILL depth option to **Through All**. If the hole appeared to be below the Top plane, use the **Flip** option to set the extrusion direction.

Suppress the RECT_CUT Feature

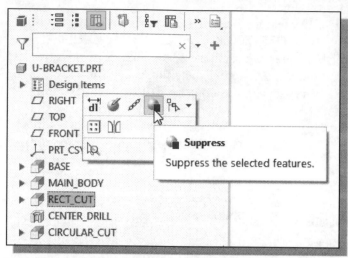

1. Move the cursor inside the *Model Tree* window. Click once with the right-mouse-button on top of RECT_CUT to bring up the option menu.

2. Pick **Suppress** in the option menu. A pop-up window is displayed.

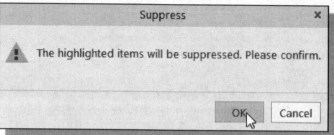

3. Pick **OK** in the pop-up window to confirm the suppression of the RECT_CUT feature.

❖ Now, there is only one suppressed feature shown in the *Model Tree* window. We have successfully updated the model by switching off the RECT_CUT feature but retained the complete feature information. On your own, do the design change that requires you to **Suppress** the CIRCULAR_CUT feature and **Resume** the RECT_CUT feature.

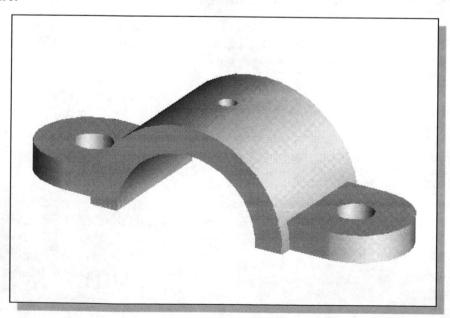

A Flexible Design Approach

In a typical design process, the initial design will undergo many analyses, testing, reviews and revisions. *Creo Parametric* allows the user to quickly make changes and explore different options of the initial design throughout the design process.

The model we constructed in this chapter contains two distinct design options. The *feature-based parametric modeling* approach enables us to quickly explore design alternatives and we can include different design ideas into the same model. With parametric modeling, designers can concentrate on improving the design and the design process much more quickly and effortlessly. The key to successfully using parametric modeling as a design tool lies in understanding and properly controlling the interactions of features, especially the parent/child relations.

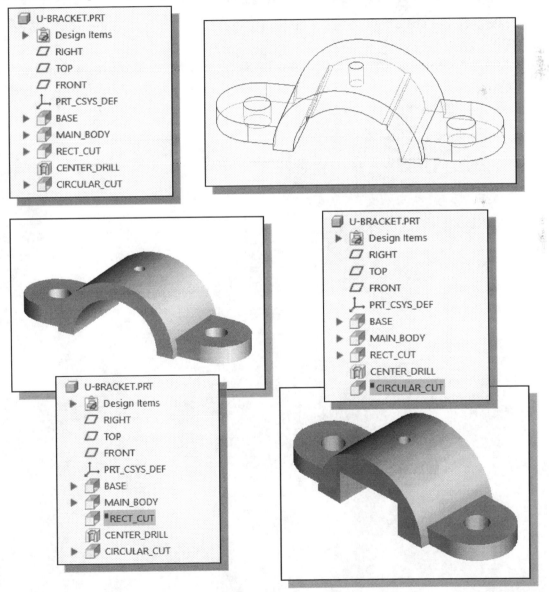

Review Questions:

1. How can we find out the order in which features are created in *Creo Parametric*?

2. In *Creo Parametric*, what does "rolling back the part" mean?

3. What is the difference between *Edit* and *Edit References*?

4. What determines how a feature reacts when other features in the model change?

5. Describe two methods in *Creo Parametric* to access the *Redefine* the existing feature references.

6. Create sketches showing the steps you plan to use to create the two models shown on the next page:

Ex.1)

Ex.2)

Exercises: (All dimensions are in inches.)

1. **V-slide Plate** (Dimensions are in inches. Plate Thickness: 0.25)

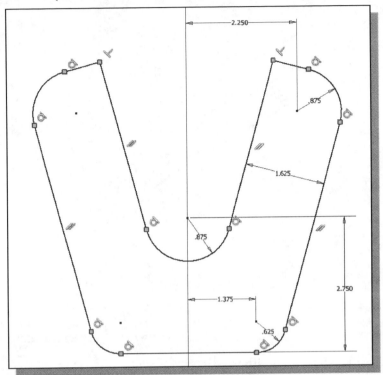

2. **Shaft Support** (Dimensions are in millimeters. Note the two R40 arcs at the base share the same center.)

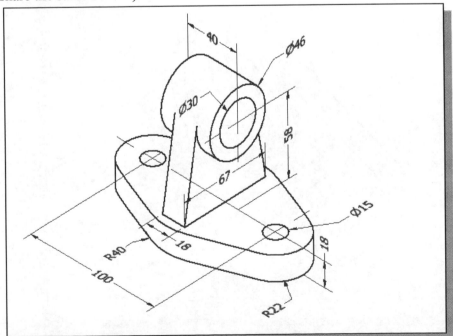

3. Connecting Rod (Material: **Carbon Steel**. Dimensions are in inches.)

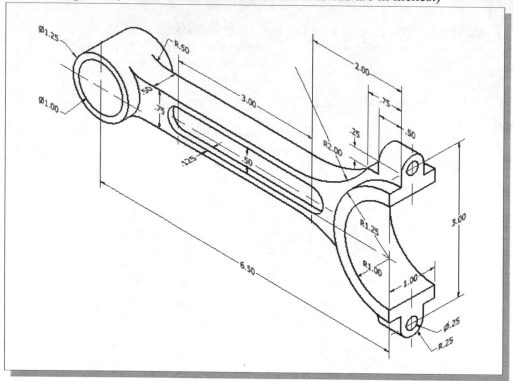

4. Tube Hanger (Dimensions are in inches.)

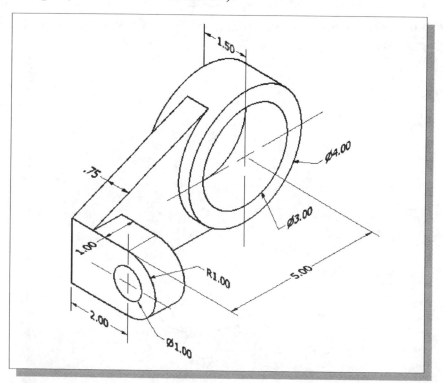

5. **Anchor Base** (Dimensions are in inches.)

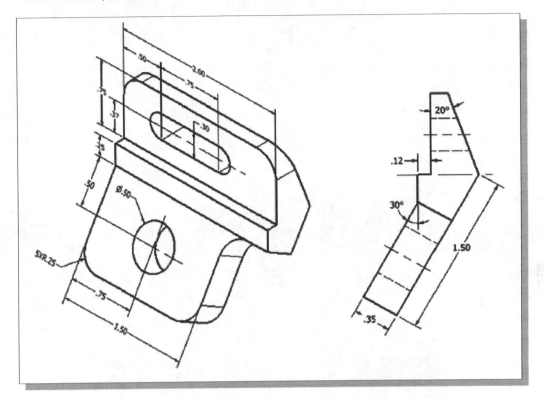

Notes:

Chapter 6
Datum Features, 3D Annotation and Part Drawings

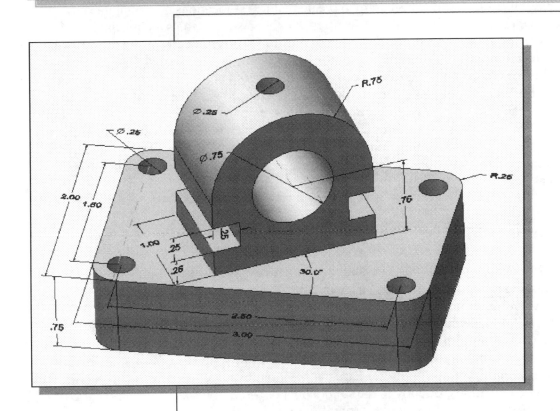

Learning Objectives

- ◆ **Understand the importance of Datum Features**
- ◆ **Use the Selection Filter**
- ◆ **Create Datum planes at an angle**
- ◆ **Create 2D drawings from 3D parts**
- ◆ **Use Creo Parametric Drawing Templates**
- ◆ **Create Standard views and Auxiliary Views in 2D drawings**
- ◆ **Use Feature/Driven Dimensions in 2D drawings**

Datum Features

Feature-based parametric modeling is a cumulative process. A new feature can use previously defined features to define information such as size, shape, location and orientation. At times, it is necessary to create features at locations that are not easily referenced by the existing surfaces/features. Although several tools are available in *Creo Parametric* to assist this process, the use of **datum features** is perhaps the most flexible and direct approach. For example, a datum plane can be created so that features can be sketched or placed on it. **Datum features**, such as datum points and datum planes, can be thought of as user-definable datum, which can be updated with the part geometry. Datum features can also be used to align features or to orient parts in an assembly. Note that in *Creo Parametric*, datum features can also be parametrically linked to the references, which means associative functionality exists between datum features and solid features. By creating parametric datum features, the established feature interactions in the CAD database also assure the capturing of the design intent.

3D Annotation

Creo Parametric provides users with the ability to access powerful digital product information for communication and in support of operations such as inspection, manufacturing, or marketing. With software/hardware improvements, it is now feasible to use the solid modeling software to document and communicate all production and manufacturing information in a three-dimensional (3D) environment. In *Creo Parametric*, exciting tools are now available for documenting and communicating product designs. We can apply 3D-based annotation, in conjunction with engineering drawing conventions, directly to the solid model or assembly. This new solids-based documentation approach provides the ability to create, manage, and deliver process-specific information without the need for paper-based documentation. With this new set of tools, the mental translation of 3D models to two-dimensional (2D) drawings for communicating information and then back to 3D models for manufacturing is replaced with a representative 3D prototype that provides all necessary information to communicate and manufacture the product. This approach means faster and more precise product communication, which can greatly improve the product development process. In *Creo Parametric*, the standard 2D drawing documentation symbols are also available in the 3D-based *Annotation* mode. In this lesson, the general procedure of creating 3D-based annotation is illustrated.

Drawings from Solid Models

With the software/hardware improvements in solid modeling, the importance of two-dimensional drawings is decreasing. Drafting is considered one of the downstream applications of using solid models. In many production facilities, solid models are used to generate machine tool paths for *computer numerical control* (CNC) machines. Solid models are also used in *rapid prototyping* to create 3D physical models out of plastic resins, powdered metal, etc. Ideally, the solid model database should be used directly to generate the final product. However, the majority of applications in most production facilities still require the use of two-dimensional drawings. Using the solid model as the starting point for a design, solid modeling tools can easily create all the necessary two-dimensional views. In this sense, solid modeling tools are making the process of creating two-dimensional drawings more efficient and effective. In this lesson, the general procedure of creating multi-view drawings from solid models is illustrated.

The *Rod-Guide* Design

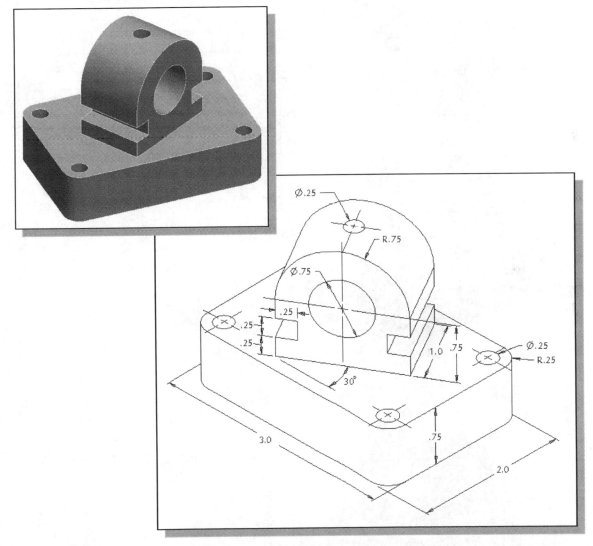

❖ Based on your knowledge of *Creo Parametric* so far, how would you create this design? What are the more difficult features involved in the design? Take a few minutes to consider a modeling strategy and do preliminary planning by sketching on a piece of paper. You are also encouraged to create the design on your own prior to following through the tutorial.

Modeling Strategy

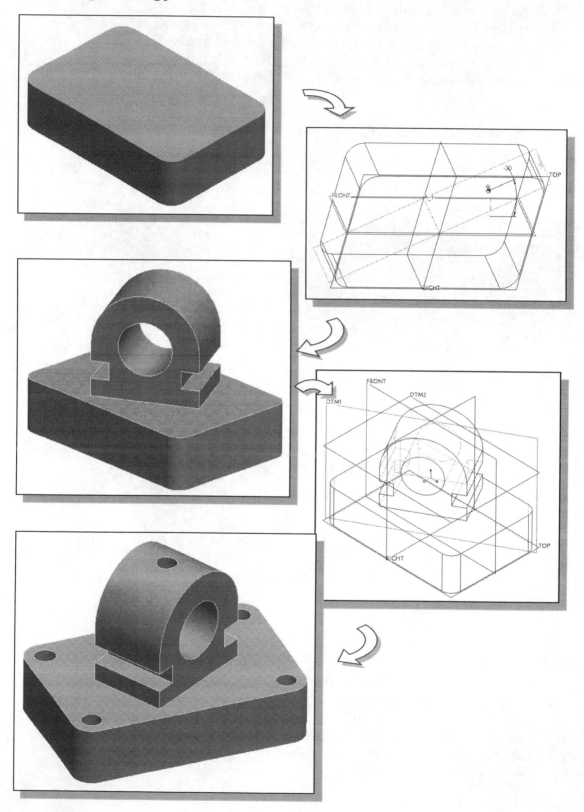

Starting Creo Parametric

1. Select the **Creo Parametric** option on the *Start* menu or select the **Creo Parametric** icon on the desktop to start *Creo Parametric*.

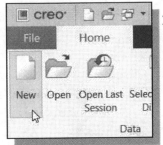

2. Click on the **New** icon, located in the *Ribbon toolbar* as shown.

3. We will start a new *solid part* file using *Rod-Guide* as the part Name.

4. Confirm the **Use default template** option is turned *on* so that the default **Inch-lbm-Second** template is used.

5. Click on the **OK** button to accept the settings.

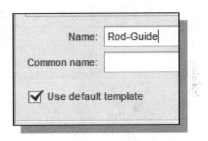

Create the Base Feature

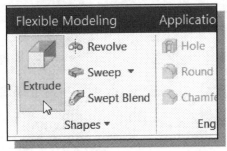

1. In the *Shapes* toolbar, select the **Extrude** tool option as shown.

2. Select **TOP** by clicking on the Datum Plane name in the *Model Tree* window as shown.

❖ Notice the default selection of the two datum planes, **RIGHT** and **FRONT**, as the references for the 2D sketch.

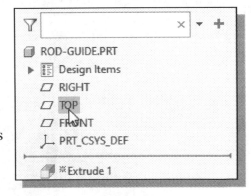

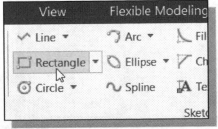

3. Choose the **Rectangle** command in the *Sketching* toolbar.

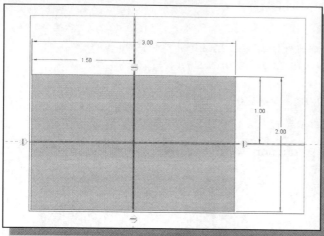

4. On your own, create a rectangle as shown in the figure.

5. Choose the **Fillet-Circular Trim** command in the *Sketching* toolbar.

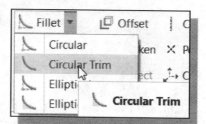

6. On your own, create the four rounded corners and modify the rectangle to be **3x2,** the radius dimension to **0.25** as shown in the figure below. Note the applied **Equal radii** constraint.

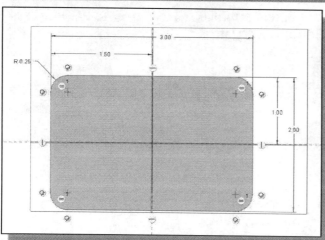

7. On your own, establish the *parametric relations* so that the two referenced datum planes pass through the center of the 2D sketch.

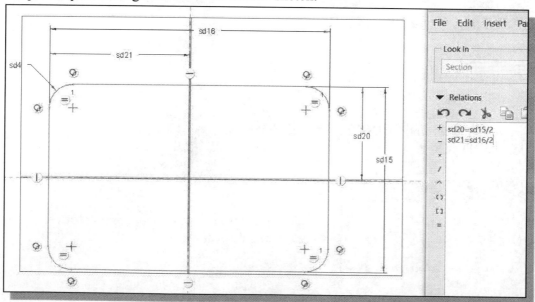

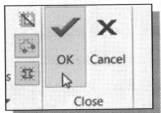

8. In the *Ribbon* toolbar, click on the **OK** icon to end the *Creo Parametric 2D Sketcher* and proceed to the next element of the feature definition.

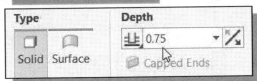

9. In the *depth* value box, enter **0.75** as the extrusion depth.

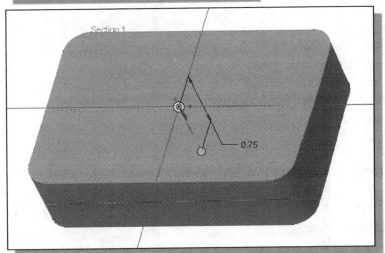

10. Click on the **Change depth direction** icon to reverse the extrusion direction as shown.

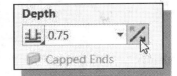

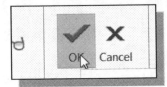

11. In the *Extrude Option Dashboard*, click the **OK** button to proceed with the solid feature creation.

12. Click on the **Refit** icon to adjust the display so that all objects fit inside the display area. Also, use the *Dynamic Viewing* function to adjust the viewing.

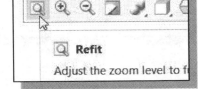

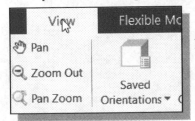

13. Click on the **View** tab in the *Ribbon* area.

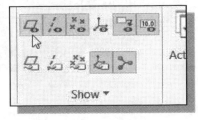

14. Click on the **Coordinate system display** icon once to toggle *off* the display of the coordinate system. Also turn *off* the display of the **Datum Planes**.

Create a Datum Axis

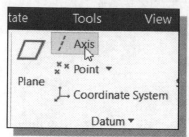

1. Switch to the **Model** tab in the *Ribbon* area.

2. In the *Datum* toolbar, select **Datum Axis** tool to start the creation of a new datum axis.

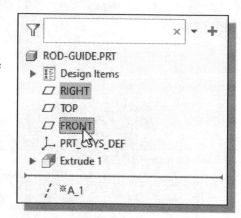

3. Inside the *Model Tree* area, hold down the [**CTRL**] key and select the **RIGHT** datum plane and the **FRONT** datum plane as the second Placement Reference.

❖ Note that the selected placement references are displayed in the *Datum Axis* window, and they are also highlighted in the *Model Tree* area.

4. Click **OK** to accept the creation of the datum axis A_1, which passes through the intersection of the two selected datum planes.

➢ Note the *display of the datum planes* was turned *off*, but the datum planes can still be selected through the *Model Tree* window.

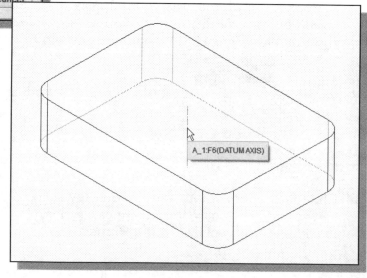

Create a Datum Plane at an Angle

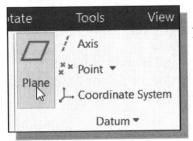

1. In the *Datum* toolbar, select the **Datum Plane** tool to start the creation of a new datum plane.

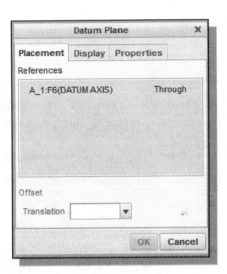

❖ Note that since datum axis A_1 was pre-selected, it is automatically placed in the *Datum Plane* window. *Creo Parametric* will attempt to create a datum plane passing through the selected datum axis. But more definitions are required to fully place the desired datum plane, thus the **OK** button is greyed out.

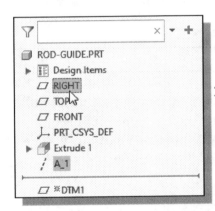

2. Inside the *Model Tree* area, hold down the [**CTRL**] key and pick the **RIGHT** datum plane as the second **Placement Reference**. Notice the selected reference plane is used, by default, as a rotation reference.

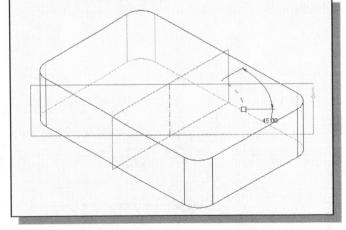

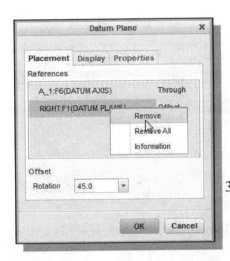

3. Inside the *Datum Plane* window, right-mouse-click once on the **RIGHT** datum plane and select the **Remove** option.

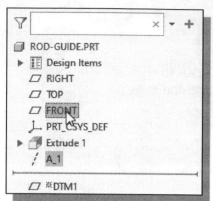

4. Inside the *Model Tree* area, hold down the [**CTRL**] key and pick the **FRONT** datum plane as the second Placement Reference.

5. Click once with the left-mouse-button on the **FRONT** datum plane inside the *Datum Plane* window as shown in the figure below.

6. Click on the *downward arrow* to display the available placement options as shown.

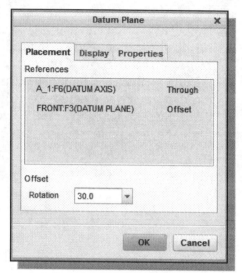

7. On your own, click on the **Normal** option and observe the placement of the datum plane with this option.

8. Change the **Placement** option to **Offset** and enter **-30** as the new Rotation angle as shown in the figure. (Note that the **negative value** will reverse the rotation direction.)

9. Drag, with the left-mouse-button, on the *adjust handle* and adjust the offset angle dynamically on the screen. (The adjust handle is the small box next to the dimension value.) Adjust the plane to the 30 degrees angle and direction as shown.

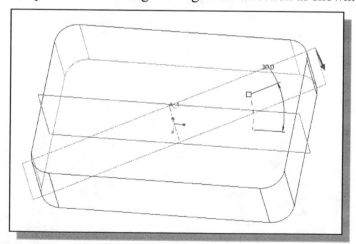

❖ Note that the *adjust handle* option in *Creo Parametric* allows the user to dynamically adjust dimension values without typing the values on the keyboard.

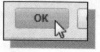

10. Click **OK** to accept the creation of the new datum plane **DTM1**.

Create the next Solid Feature

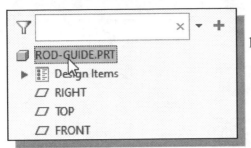

1. De-select **DTM1** by clicking the model file name in the *Model Tree* window.

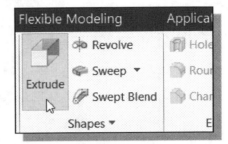

2. In the *Shapes* toolbar, select the **Extrude** tool option as shown.

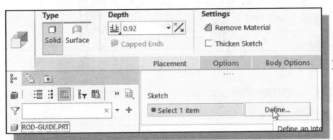

3. Click the **Placement** option and choose **Define** to begin creating a new *internal sketch*.

4. On your own, select the **DTM1** *datum plane* as the sketch plane and *datum plane* **TOP** as the sketch orientation reference as shown in the figure.

5. In the *sketch* dialog box, click on **Sketch** to enter the *sketcher* mode.

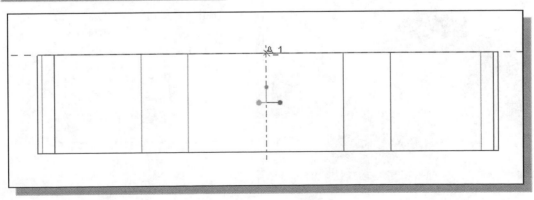

6. On your own, create the sketch and modify the dimensions as shown in the figure below. Note the different constraints used in the sketch.

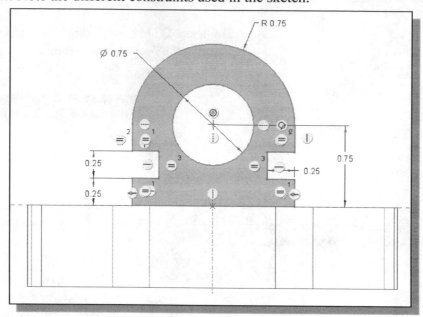

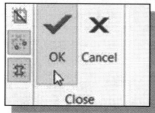

7. In the *Ribbon* toolbar, click on the **OK** icon to end the *Creo Parametric 2D Sketcher* and proceed to the next element of the feature definition.

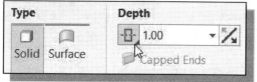

8. In the *Feature Options Dashboard*, choose the **Symmetric** option as shown.

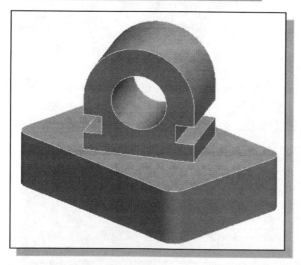

9. In the *depth* value box, enter 1.0 as the extrusion depth.

10. In the *Extrude Option Dashboard*, click on the **OK** button to proceed with the feature creation.

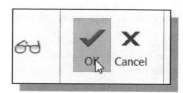

Create a Datum Plane Using the Filter Option

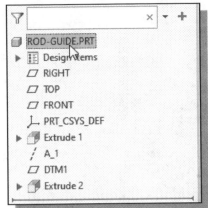

1. Inside the *Model Tree* area, select the **ROD-GUIDE** part name. By doing so, any pre-selected item (the solid feature we created in the previous section) is de-selected.

❖ Note that if we accept the default pre-selection to create the next datum plane, the sketching plane of the previous solid feature will be used as the first placement reference. To illustrate the use of the *Creo Parametric* **Selection Filter** option, we will first de-select any pre-selected item.

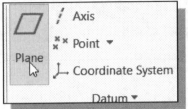

2. In the *Datum* toolbar, select the **Datum Plane** tool to start the creation of a new datum plane. The message *"Select up to 3 references, such as plane, surface, edge or point to place plane."* is displayed in the message area.

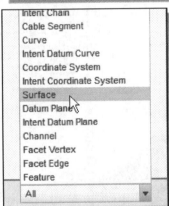

3. Click on the ***downward arrow*** of the *Selection Filter* option, which is located near the **lower-right corner** of the main screen, to display the different selection options as shown. (Note that the default option is set to **All**, which simply means that all objects are selectable.)

4. Select **Surface** in the option list as shown.

5. Move the cursor on top of the solid model and notice only surfaces of the model can be selected. (*Creo Parametric* automatically highlights selectable objects when the cursor is on top of the object.)

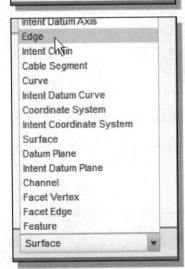

6. Adjust the *Selection Filter* option to the **Edge** option as shown.

❖ The *Selection Filter* option allows us to quickly select a specified type of entity, which can be difficult to do when entities are on top of each other.

7. Inside the *graphics area*, select the **front edge** of the top surface of the base feature as shown.

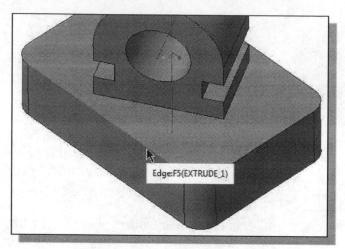

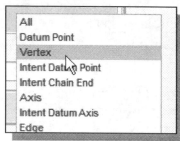

8. Change the *Selection Filter* option to the **Vertex** option as shown.

9. Hold down the [**CTRL**] key and select the top inside front right corner of the square notch on the right side of the second solid feature as shown in the figure below.

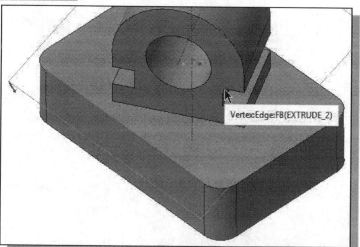

❖   The definition of a new datum plane can be accomplished by selecting different types of entities that belong to different features. Note that it is also possible to establish the same definitions with different combinations of selections.

10. On your own, use the *Dynamic Viewing* function and confirm the new datum plane passes through both the top front edge and the selected corner. (**Do not** end the *Datum Plane* window; more options are illustrated next.)

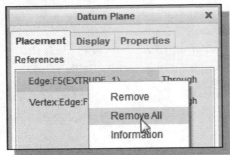

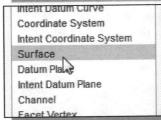

11. On your own, **Remove** both of the references in the *Datum Plane* window. (Hint: Use the right-mouse-button to bring up the option menu.)

12. Change the *Selection Filter* option to the **Surface** option as shown.

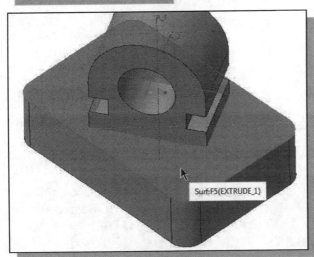

13. Select the **top surface** of the base feature when the surface is highlighted as shown.

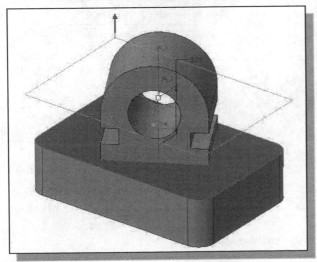

14. On your own, use the adjust handle and set the new datum plane to be **0.75** inches above the selected reference as shown.

❖ Note that negative values can also be used for the offset distance. A negative value simply means that the measurement is opposite to the positive side of the datum plane.

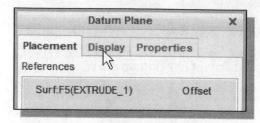

15. Inside the *Datum Plane* window, click the **Display** tab to switch to the display option page.

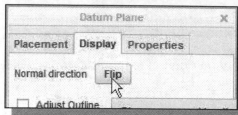

16. Click once, with the left-mouse-button, on the **Flip** button and notice the displayed arrowhead points in the opposite direction.

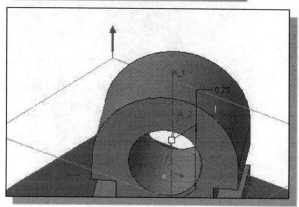

17. On your own, set the **arrow direction** as shown.

❖ In *Creo Parametric*, each plane/surface has a positive side and a negative side. The displayed arrowhead identifies the direction vector of the plane, which is defined as the positive side of the plane.

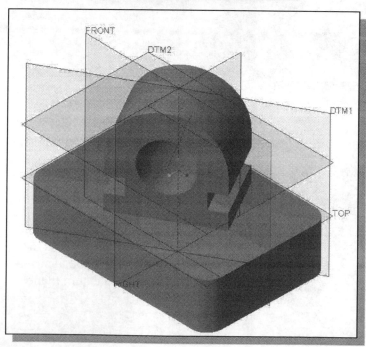

18. Click **OK** to accept the creation of the new datum plane **DTM2**.

19. Turn *on* the display of datum planes.

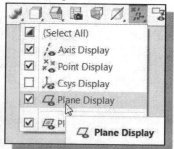

20. **Deselect** the **DTM2** plane by clicking in the blank area of the graphics window.

Create a Placed Feature

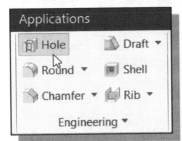

1. In the *Engineering* toolbar, select the **Hole** tool option as shown.

2. Select the datum axis **A_1** as the **primary reference**.

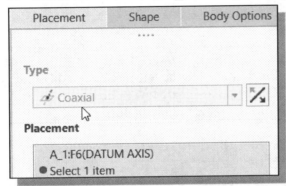

3. In the *Feature Options Dashboard*, select the **Placement** option to view the current setting. Note that since a datum axis has been selected as the primary reference, the **Coaxial** placement option is set. The center axis of the new hole feature is aligned to the selected datum axis.

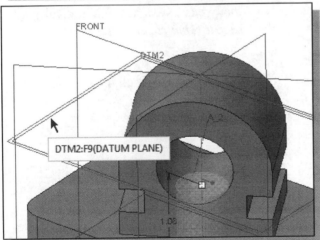

4. Hold down the [**Ctrl**] key and select datum plane **DTM2** as the secondary placement reference.

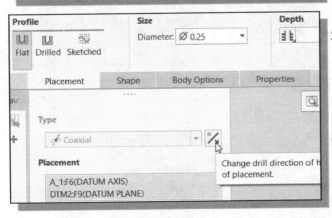

5. Set the *hole diameter* to **0.25** and the *depth option* to **Thru All.** If necessary, click on the **Flip** button once to switch the placement direction.

6. Click on the **OK** icon and proceed to create the hole feature.

Add Four Holes to the Base Feature

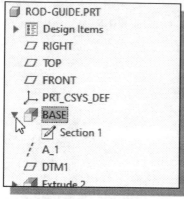

1. Inside the *Model Tree* area, click on the [**>**] marker in front of the first protrusion feature to expand the display of the feature list.

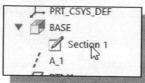

❖ The 2D section associated with the feature is displayed in the *Model Tree* area.

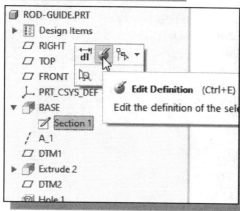

2. Click once with the left-mouse-button on the 2D section associated with the base feature to bring up the option menu.

3. Select **Edit Definition** in the option list as shown.

❖ Note that we are switched automatically into the *Creo Parametric Sketcher* mode.

4. On your own, create the four equal radii circles (diameter **0.25**) aligned to the centers of the rounded corners as shown in the figure below.

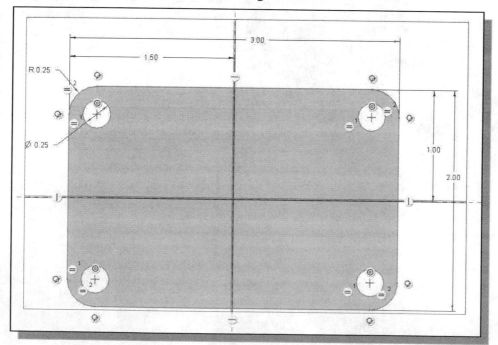

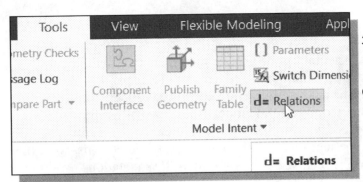

5. Click **Relations** in the **Tools** tab as shown.

6. Confirm the relations established earlier are still maintained.

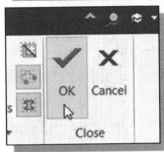

7. In the **Sketch** tab of the *Ribbon* toolbar, click on the **OK** button to accept the modification and exit *Creo Parametric Sketcher*.

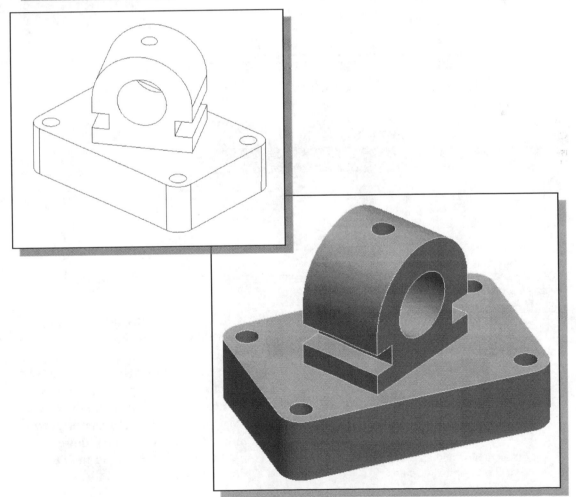

3D Annotation

In *Creo Parametric*, 3D-based annotation can be applied directly to the solid model or assembly. This new solids-based documentation approach provides the ability to create, manage, and deliver process-specific information without the need for paper-based documentation. With this new set of tools, a solid model can be used as a representative 3D prototype that provides all necessary information to communicate and manufacture the product. In *Creo Parametric*, besides creating additional driven dimensions, all of the dimensions in 2D parametric sketches can also be displayed as 3D annotations.

1. On your own, turn *off* the display of the *datum features* by clicking on the **Select All** icon in the *Display Control* toolbar as shown.

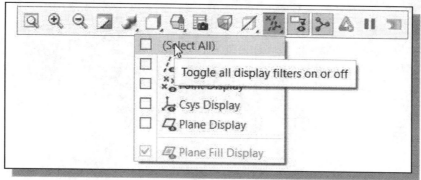

2. In the *Ribbon* area, select the **Annotate** tab to switch to the available annotation commands.

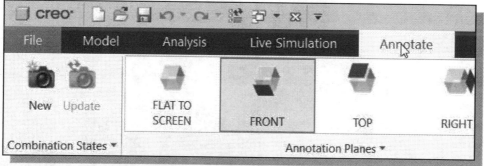

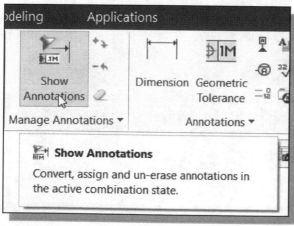

3. Click **Show Annotations** in the **Annotate** tab as shown.

➢ The **Show Annotations** command can be used to display existing dimensions used to create the model. Note that the annotations retrieved are not restricted to the required Annotation Plane setting in 3D annotation.

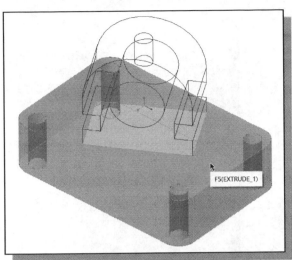

4. Select the **base feature** in the graphics area.

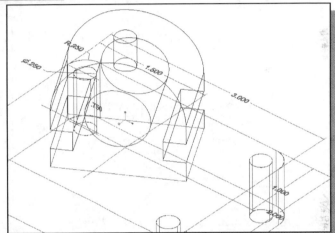

➢ Notice all of the dimensions, used to create the selected feature, appear on the solid model.

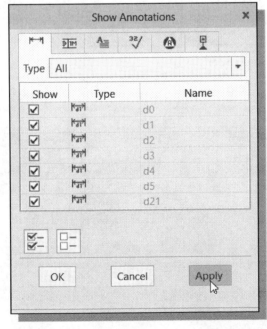

5. To keep the dimensions, use the check box in front of each dimension. In this case, keep all of the dimensions as shown.

6. Click on the **Apply** button to accept the selections.

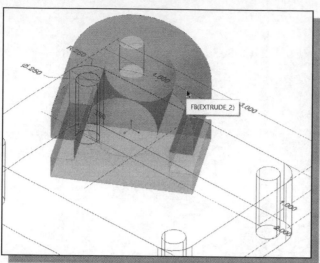

7. Select the **upper feature** in the graphics area.

8. Click on the **Select All** button to keep all of the dimensions.

9. Click **OK** to end the Show Annotations command.

10. On your own, use the *Dynamic Rotate* commands and examine the applied annotations.

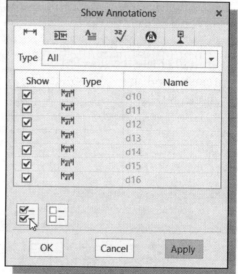

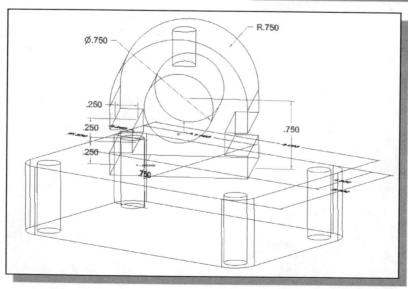

❖ Although the **Show Annotations** command provided the majority of dimensions necessary to document the design, some additional dimensions are still needed. This can be accomplished by creating additional **annotation features**. In *Creo Parametric*, a reference plane must be first selected before the annotation features can be created.

11. Select **TOP** in the *Annotation Planes* list as shown.

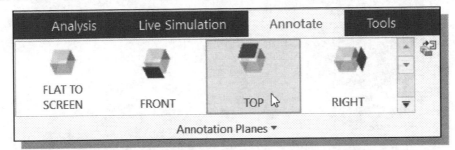

➢ Note that *Creo Parametric* provides seven basic annotation planes, corresponding to the standard 2D views typically used in a 2D drawing; and additional annotation planes can also be defined.

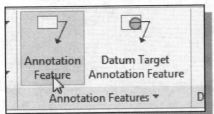

12. Click **Annotation Feature** to start the creation of an additional 3D annotation feature. The message "*Please add annotation elements.*" is displayed in the message area.

13. Activate the **Driven Dimension** option in the *Annotation Feature* dialog box as shown.

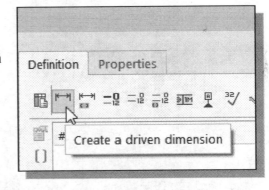

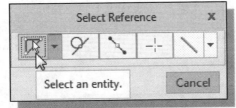

14. In the *Select Reference* window, confirm the **Select an Entity** option is activated as shown.

15. On your own, change the display to **Shading with Edges**.

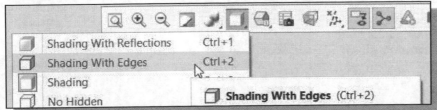

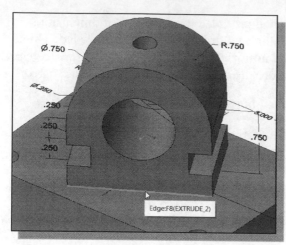

16. Select the **bottom edge** of the upper feature by clicking on the entity.

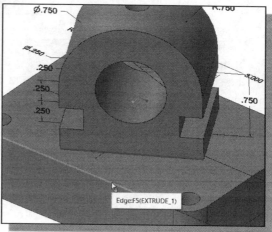

17. Hold down the **[Ctrl]** key and select the **top front edge** of the base feature by clicking on the entity.

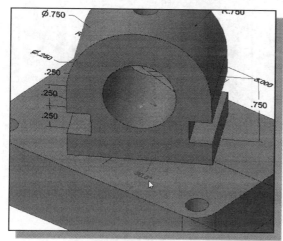

18. Move the cursor in between the two selected entities and click once with the **middle-mouse-button** to place the angle dimension.

19. Click **[Cancel]** to exit the selection option.

➢ Note the added driven dimension, along with the selected references, is listed in the *Annotation Feature* dialog box. These definitions can also be modified.

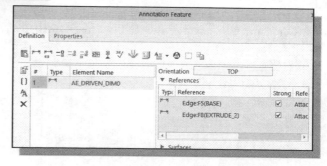

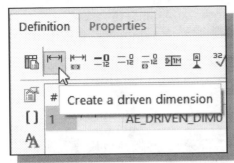

20. Activate the **Driven Dimension** option in the *Annotation Feature* dialog box to create another driven dimension.

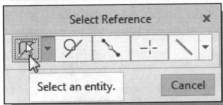

21. In the *Menu Manager*, confirm the ATTACH TYPE is set to **Entity** as shown.

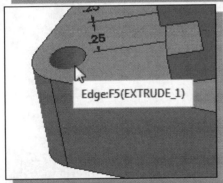

22. Click on the top edge of the front left hole as the first entity of the driven dimension.

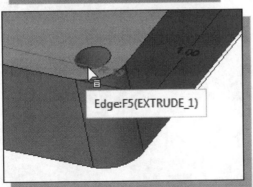

23. Hold down the **[Ctrl]** key and select the edge of the front right hole as the second entity of the driven dimension.

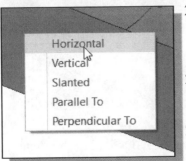

24. Press down the **right-mouse-button** to display the options for the new dimension and click **Horizontal** to set the orientation of the annotation.

25. On your own, click once with the **middle-mouse-button** to place the dimension.

26. In the *Selection Reference* toolbar, click [**Cancel**] to exit the selection option.

➢ All definitions of the created annotations can be modified through the *Annotation Feature* dialog box.

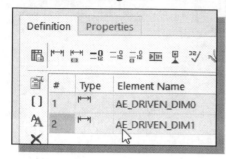

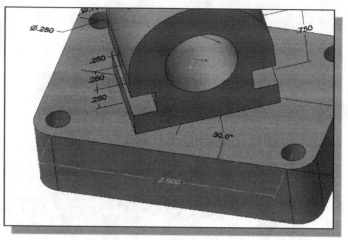

 27. Click **OK** to exit the Annotation Feature command.

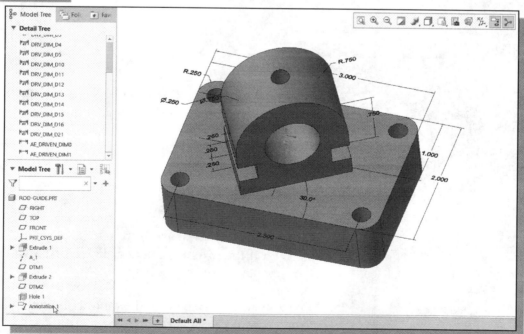

- The created annotation feature appears in the *Model Tree* window. Note that existing/additional annotated driven dimensions can be modified/added to the same annotation item.

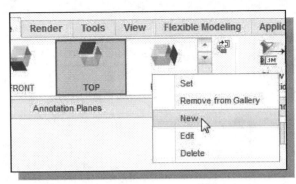

28. In the *Annotation Planes* list, press and hold the **right-mouse-button** to bring up the option menu and click **New** to define a new annotation plane.

➢ We will create a new plane that is parallel to the DTM2 datum plane.

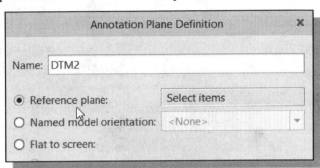

29. Change the plane option to **Reference Plane** as shown.

30. Enter **DTM2** as the name of the new annotation plane.

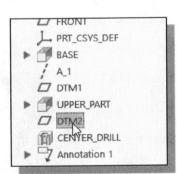

31. Select **DTM2** in the *Model Tree* window.

➢ The new annotation plane will be aligned in the direction parallel to DTM2.

32. Click **OK** to accept the setting.

33. Select DTM2 in the *Annotation Planes* list to set the annotation plane direction.

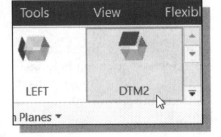

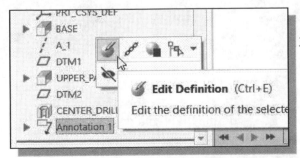

34. In the *Model Tree* window, press and hold the **left-mouse-button** on **Annotation 1** to bring up the option menu and click **Edit Definition** to bring up the *Annotation Feature* option form.

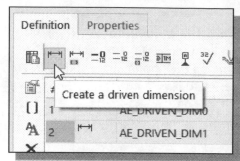

35. Activate the **Driven Dimension** option in the *Annotation Feature* dialog box to create another driven dimension.

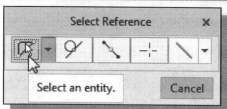

36. In the *Menu Manager*, set the ATTACH TYPE to **Select Entity** as shown.

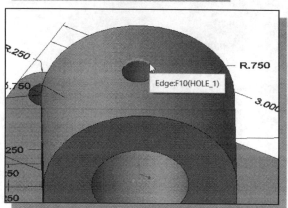

37. Click on the **top edge** of the left notch as the entity of the driven dimension.

38. Press and hold down the right-mouse-button to bring up the option list; select **Diameter**.

39. Move the cursor above the selected entities and click once with the **middle-mouse-button** to place the new driven dimension.

40. Click [**Cancel**] to exit the select reference option.

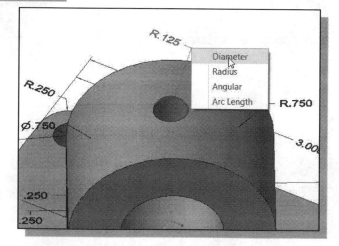

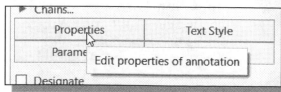

41. Click on the **Properties** button to edit the highlighted annotation.

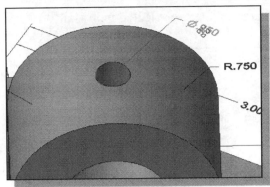

42. Note the **Move** icon is now available to adjust the current location of the created annotation.

43. On your own, drag and drop the dimension to a different location.

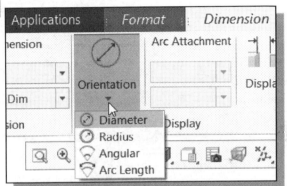

44. Click **Orientation** to view the different options available for the highlighted dimension.

45. Click Display to show dimension display options. Click **Flip** to flip the directions of the associated arrows.

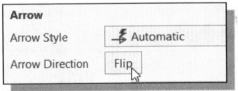

46. On your own, examine the other options available associated with the created annotation.

47. Exit the *Annotation Properties* options by clicking inside the graphics window.

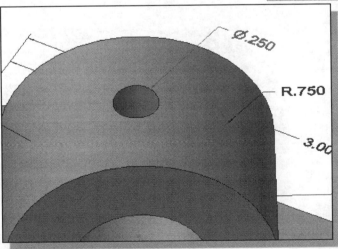

48. Click **OK** to exit the annotation feature.

49. Select the **height dimension** of the upper feature. Note the selected item is also highlighted in the *Detail Tree*.

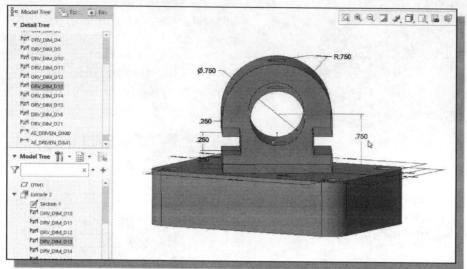

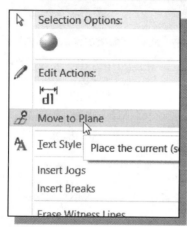

50. In the *graphics area*, press and hold the **right-mouse-button** to bring up the option menu and choose **Move to Plane** as shown.

➤ Note that the majority of editing options are available through the option menu.

51. Click on the front surface of the upper feature to align the selected dimension.

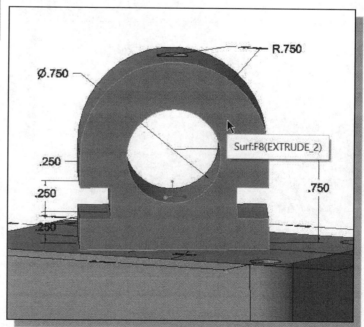

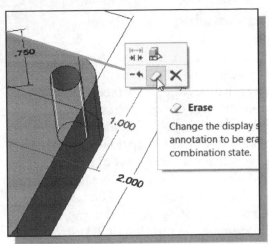

52. Select the linear dimension **1.00** of the base feature.

53. In the option list, choose **Erase** to remove the dimension from screen.

54. On your own, repeat the above steps and complete the 3D annotation of the model as shown.

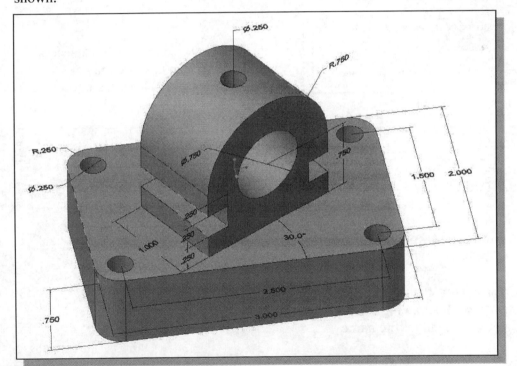

❖ Note that the values of dimensions created with the 3D Annotation command cannot be modified; they are used to document the completed solid model.

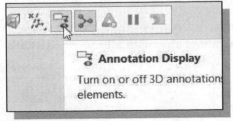

55. In the *Display Control* toolbar, click on the **Annotation Display** button to toggle the display of all the applied annotations.

Create a Multi-View Drawing using a Drawing Template

Creo Parametric allows us to generate 2D engineering drawings directly from the 3D solid models. An engineering drawing is a tool that can be used to communicate engineering ideas/designs to manufacturing, purchasing, service, and other departments. Until now we have been working in *Part Modeling* mode to create our design in ***full size***. We can arrange our design on a two-dimensional sheet of paper so that the plotted hardcopy is exactly what we want. This two-dimensional sheet of paper is known as the *Drawing* mode in *Creo Parametric*. In general, each company uses a set of standards for drawing content, based on the type of product and also on established internal processes. The appearance of an engineering drawing varies depending on when, where, and for what it is produced. However, the general procedure for creating an engineering drawing from a solid model is fairly well defined. In *Creo Parametric*, creation of 2D engineering drawings from solid models consists of four basic steps: drawing sheet formatting, creating/positioning views, annotations, and printing/plotting.

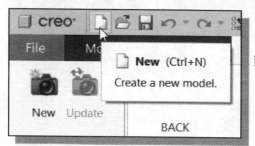

1. Pick the **New** icon in the quick access toolbar. (We can also use the key combination **Ctrl-N** to start a new object.)

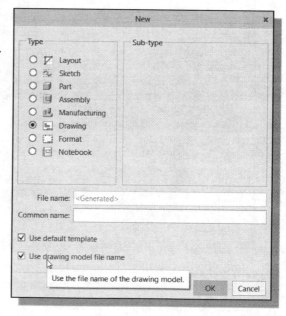

2. In the *New* form, select **Drawing** under the **Type** list.

3. Confirm the **Use default template** option is set.

4. Choose ***Use drawing model file name*** so the drawing file name is the same as the model file name.

5. Pick **OK** to start a new drawing file.

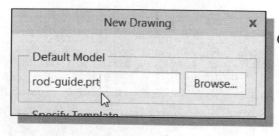

6. In the *New Drawing* form, the currently active part is automatically selected as the drawing model.

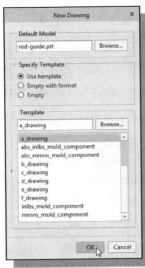

7. Confirm the Specify Template option is set to **Use template**.

8. Pick **a_drawing** size (8.5″ × 11″) in the Template list.

9. Pick **OK** to proceed with the drawing layout.

❖ Drawing templates can be used when creating a new drawing. The *Creo Parametric* drawing templates, such as the a_drawing template used in this example, automatically create standard views and set the view display options.

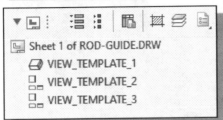

❖ A new drawing window will appear, with the *Drawing Tree* panel above the *Model Tree* navigator. Note that the part window is pushed into the background and can be made active at any time.

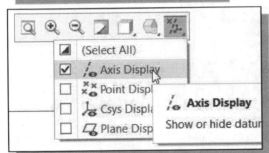

10. On your own, toggle *off* the display of the datum planes, datum points and coordinate systems as shown in the figure.

11. In the *Ribbon toolbar*, click **Repaint** to refresh the display in the graphics area.

❖ The **a_drawing** template is set to three standard views (**TOP, FRONT** and **RIGHT**) with the *View Display* option set to the **Hidden line** option as shown in the figure. Note the **annotations** are also displayed.

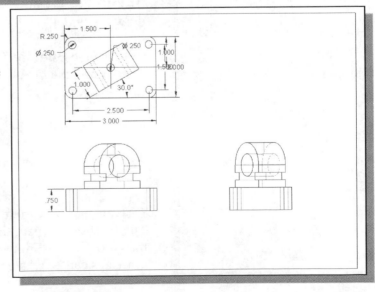

Create another Multi-View Drawing without Using a Template

For the **Rod-Guide** design, a drawing with an auxiliary view is needed for the angled feature of the design. We can either work on the current sheet or create a new sheet. In *Creo Parametric*, two options are available to create a new sheet; we can create a new drawing file or add a new sheet to the existing drawing file. In the following sections, both procedures are illustrated.

1. Pick the **New** icon in the *quick access toolbar*. (We can also use the key combination **Ctrl-N** to start a new object.)

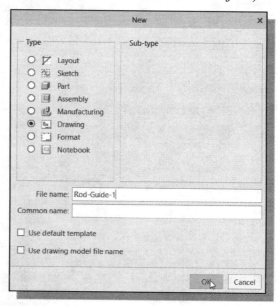

2. In the *New* form, select **Drawing** under the **Type** list.

3. Enter **Rod-Guide-1** as the drawing file **Name**.

4. Toggle *off* the **Use default template** option.

5. Pick **OK** to start a new drawing file.

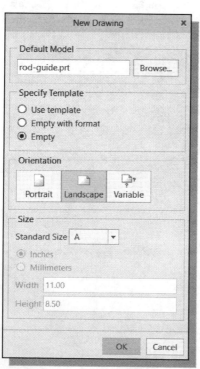

6. In the *New Drawing* form, the currently active part (Rod-Guide) is automatically selected as the drawing model.

7. Confirm the **Specify Template** option is set to **Empty** and the **Orientation** option is set to **Landscape**.

8. Pick **A** size (8.5″ × 11″) in the **Standard Size** list.

9. Pick **OK** to proceed with the drawing layout.

➢ A new blank sheet appears with the title "**Rod-Guide-1**." Note that we currently have three files opened: The *Rod-Guide* part, the *Rod-Guide* drawing and the *Rod-Guide-1* drawing. In Creo, multiple drawings can be created with the same solid model.

Adding a New Sheet in the First Drawing

In *Creo Parametric*, a drawing can contain multiple sheets; each sheet is independent of the others. This approach allows more flexibility in drawing management.

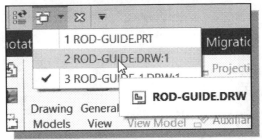

1. Switch to the **Rod-Guide.DRW** drawing by clicking the drawing name under the **Quick Access** list as shown.

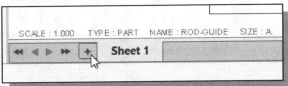

2. Click on the **New Sheet** icon, located next to the **Sheet 1** tab near the bottom of the graphics area, to add a new sheet.

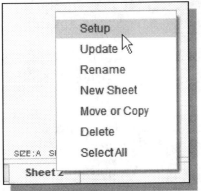

3. Click with the **right-mouse-button** on the **Sheet 2** tab to bring up the options list as shown.

4. Select the **Setup** option as shown.

❖ The Setup option allows us to adjust the basic settings of the sheet. By default, settings from the previous sheet are used.

5. Pick **OK** to exit the *Sheet Setup* window.

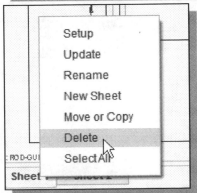

6. Click with the **right-mouse-button** on the **Sheet 1** tab to bring up the options list and select **Delete** as shown.

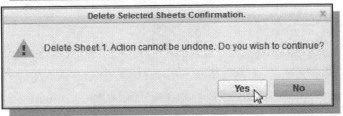

7. Click **Yes** to proceed with deleting Sheet 1.

Adding a Primary Main View

In *Creo Parametric Drawing* mode, the first drawing view we create is called a **main view**. A *main view* is the primary view in the drawing; other views can be derived from this view. When creating a main view, *Creo Parametric* allows us to specify the view to be shown. We can select any view to be used as the base view. Note that there can be more than one main view in a drawing.

1. In the **Layout** tab of the *Ribbon toolbar*, select the **General** view as shown.

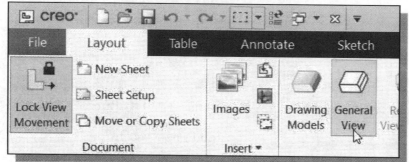

❖ The first view in a drawing is a main view which is created through the **General** view option.

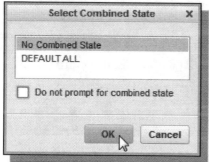

2. Click **OK** to accept not using the combined state option as shown.

- The combined state is to use the layer control features available in *Creo Parametric*.

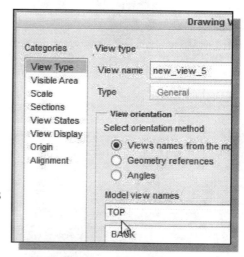

3. The message "*Select CENTER POINT for drawing view*" is displayed in the message area. Pick a location that is about **one-third** from the top left corner of the display window.

4. Choose **TOP** in the **Model view names** list as shown in the figure above.

5. Click on the **Apply** button to accept the settings and proceed with the creation of the main view.

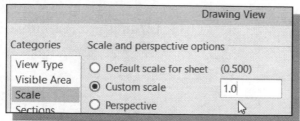

6. Set the **Scale** to **1.00** as shown.

7. Click on the **OK** button to exit the drawing view window.

Adding a Projected Auxiliary View

To create an auxiliary view, the definition of a view projection direction is required. In *Creo Parametric*, the **view projection direction** can be defined by selecting an edge, an axis or a datum plane of the primary view.

1. Click once with the left-mouse-button in an empty area to deselect the main view.

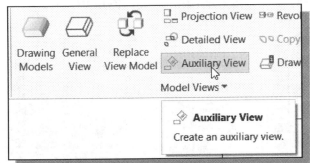

2. In the *Model Views* toolbar, select **Auxiliary** view as shown.

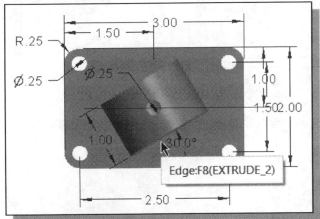

3. The message "*Select edge of or axis through, or datum plane as, front surface on main view*" is displayed in the message area. Pick the **front edge** of the angled feature as shown.

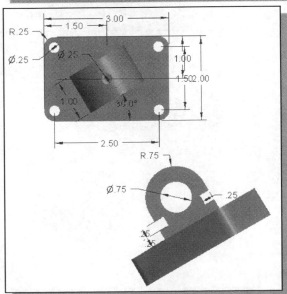

4. The message "*Select CENTER POINT for drawing view*" is displayed in the message area. Pick a location that is below and toward the right of the main view.

❖ An auxiliary view of the model, projected from the main view, is placed in the display window. The alignment of the auxiliary view to the main view is set automatically and maintained by the system.

Adjust the Overall Drawing Display

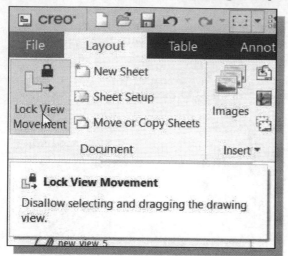

1. In the Ribbon toolbar, switch *off* the **Lock View Movement** option. (This option is set to *on* by default, which does not allow moving of the views.)

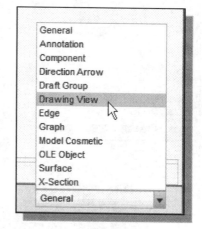

2. Set the *Selection Filter* option to **Drawing View** as shown.

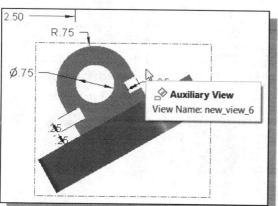

3. Pick the **auxiliary view** with the left-mouse-button; note the highlighted rectangle indicating the selection.

4. Move the cursor on the view and **drag** with the left-mouse-button to reposition the view. Note the position of the auxiliary view is restricted to alignment with the top view, the primary view.

5. On your own, reposition the *top view*. Note that the position of the auxiliary view is also adjusted to maintain its alignment to the top view.

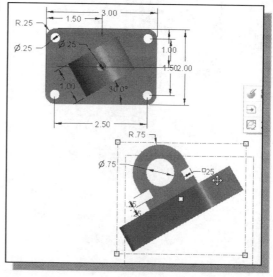

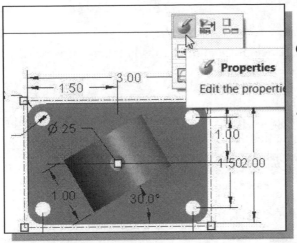

6. Pick the **top view** by clicking once inside the view with the left-mouse-button.

7. Select **Properties**, the first icon of the option list as shown.

8. In the *Drawing View* window, select **View Display** to change the view display settings.

9. In the **Display style** list, select **Hidden** to display hidden edges as dashed lines.

10. Click **OK** to accept the settings and exit the *Drawing View* window.

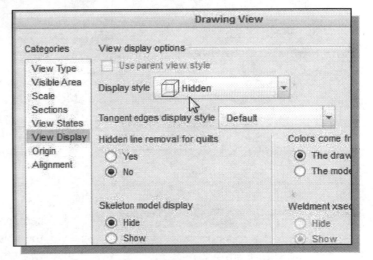

11. On your own, repeat the above procedure and adjust the display option of the *auxiliary view* to wireframe display with **Hidden Line**.

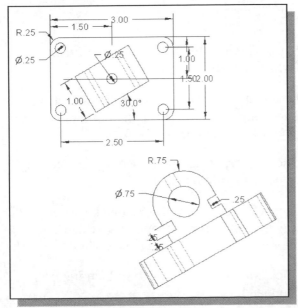

Dimension Appearances

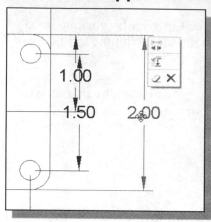

1. Select one of the dimensions in the **main view** of the design.

2. In the dashboard, a number of dimension related settings are available, such as setting the number of digits after the decimal point.

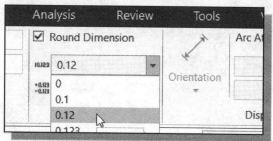

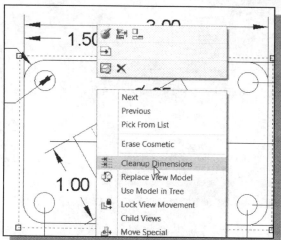

3. Select the **main view** of the design.

4. Inside the *display area*, press down the right-mouse-button and select the **Cleanup Dimensions** option.

❖ The displayed option menu lists only options that are applicable to the selected drawing dimensions. Note that the selected dimension can also be deleted through the option menu.

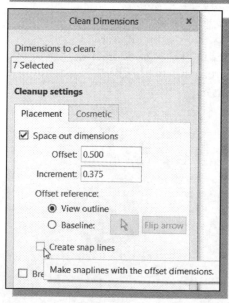

5. In the *Clean Dimensions* window, turn *off* the **Create Snap Lines** option.

❖ Note that only five dimensions are selected. The Clean Dimensions option is applicable only to the linear dimensions.

6. Click **Apply** to accept the settings and adjust the spacing of dimensions.

❖ Clean Dimensions offers a quick and easy way to Space out dimensions, which improves the legibility of the drawing.

7. Click **Close** to end the Clean Dimensions command.

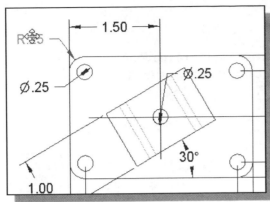

8. Select the radius dimension in the main view.

9. Move the cursor on top of the text to display the arrow icon.

10. Use the **left-mouse-button** and drag the text to reposition the dimension.

11. Select one of the linear dimensions in the main view. Move the cursor on top of one of the small markers to see the arrow icon.

12. Drag and drop one of the markers to adjust the alignment of the dimension.

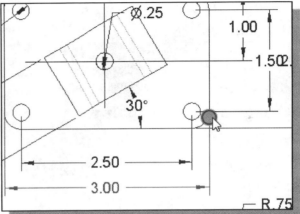

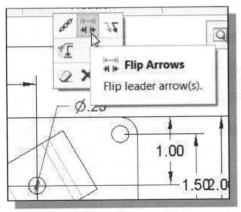

13. Select one of the **diameter** dimensions by clicking once with the left-mouse-button on the dimension text.

14. Select **Flip Arrows** in the option list as shown.

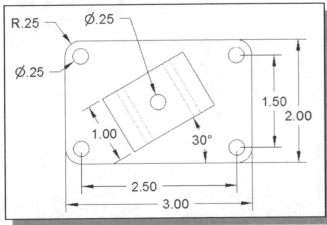

15. On your own, arrange the dimensions so that they appear as shown in the figure.

❖ Note that with the modern parametric modeling software, the feature dimensions used in the creation of the solid model are easily displayed in the *Drawing* mode.

Displaying Center Axes

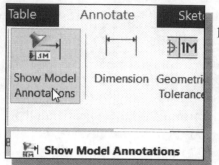

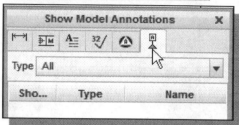

1. In the **Annotate** tab of the *Ribbon*, select the **Show Model Annotations** command.

2. Select the **Axes and Set Datum objects** tab in the *Show Model Annotations* window as shown.

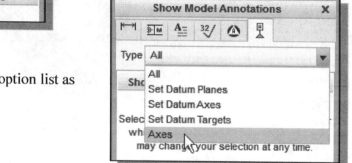

3. Choose **Axes** in the *Type* option list as shown.

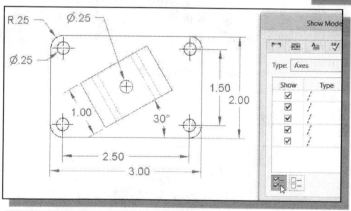

4. In the *message area*, the message "*Select model view or window.*" is displayed. Pick the features in the **top view** as shown.

5. In the *Show Annotation* form, click on the **Check All** button to keep all of the displayed centerlines.

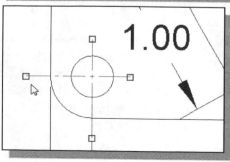

6. Pick the **OK** button in the Show Model Annotations command.

7. On your own, adjust the length of the displayed centerlines by selecting and dragging with the left-mouse-button as shown.

Annotation Appearances

1. On your own, adjust the locations of the dimensions so that they appear as shown in the figure below.

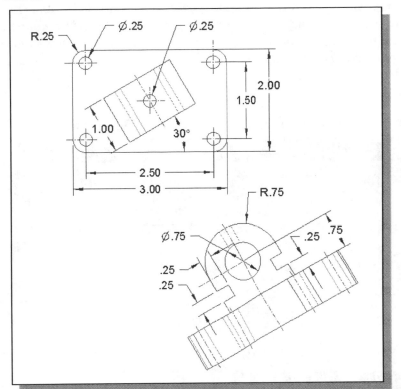

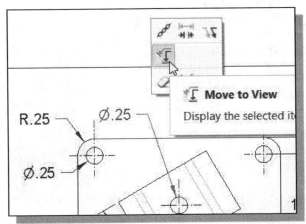

2. Pick the dimension text of the center hole (diameter **0.25**) in the main view as shown.

3. Select **Move Item to View** in the option list.

4. Pick the **auxiliary view** in the display area. Note the selected dimension now appears in the selected view of the drawing.

5. In the *Quick Access* toolbar, click **Undo** to reverse the last step.

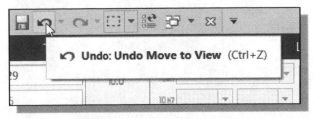

Adding/Deleting Dimensions – Feature/Driven Dimensions

Besides displaying the **feature dimensions**, dimensions used to create the features, we can also add additional **driven dimensions** to the drawing. *Feature dimensions* are used to control the geometry, where *driven dimensions* are controlled by the existing geometry. In the drawing layout, we can therefore **delete** the *driven dimensions*, but we can only **hide** the *feature dimensions*. One should try to use as many *feature dimensions* as possible and add *driven dimensions* only if necessary. It is also more effective to use *feature dimensions* in the drawing layout since they were created when the model was built.

1. In the **Annotate** tab of the *Ribbon* toolbar, select **Dimension - New References**.

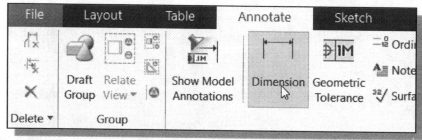

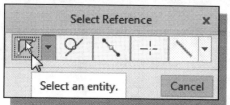

2. In the ATTACH TYPE submenu, confirm the selection of **On Entity**.

3. Pick the **right edge** of the base feature as shown.

4. Click once with the **middle-mouse-button** at a location next to the edge to create the dimension.

5. Click **Cancel** in the Select Reference command.

➤ Note that we add dimensions the same way we did in the *2D Sketcher*.

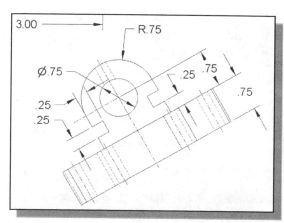

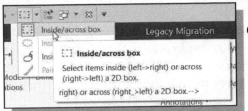

6. In the *Quick Access* toolbar area, confirm the box selection option is set to **Select items inside/across box** as shown.

7. Set the **Selection Filter** option to **Dimension**.

8. Inside the *display area*, select **all of the dimensions** by enclosing all dimensions inside a selection box with the left-mouse-button.

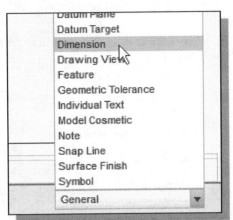

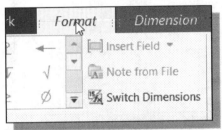

9. Inside the *ribbon toolbar area*, click the **Format** tab as shown.

❖ We will modify the dimension format for all of the selected dimensions.

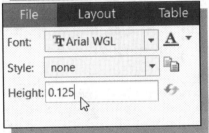

10. Enter **0.125** and hit the **[Enter]** key once to set the character height as shown.

11. Click inside the graphics area to accept the settings and exit the modify mode.

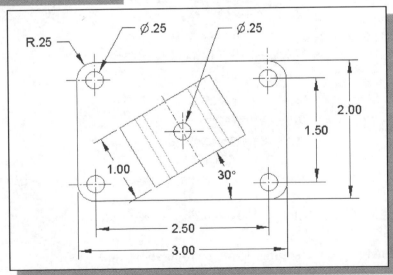

Add an Isometric View

In *Creo Parametric*, the default 3D orientation is set to display the model in the ***trimetric*** orientation. In order to add an ***isometric*** view in the *Drawing* mode, the easiest approach is to change the default orientation to *isometric* before creating the view.

1. Click on the **triangle** icon in the *Quick Access* toolbar. The *Customize Quick Access Toolbar* option list is displayed. Click on the **More commands** option; the *Creo Parametric* options dialog box appears, which allows us to adjust various settings.

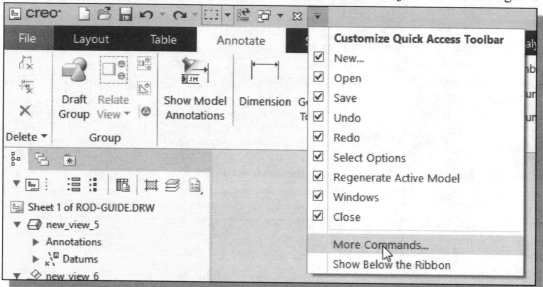

2. In the **Model orientation** section, choose **Isometric** as the Default model orientation.

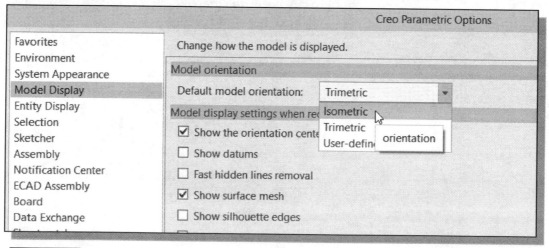

 3. Pick **OK** to exit the *Creo Parametric Options* form.

4. On your own, save the settings in the configuration file.

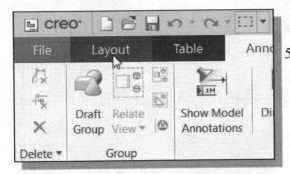

5. In the *Ribbon* toolbar, select the **Layout** tab as shown.

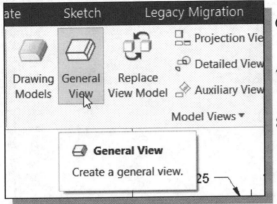

6. In the **Layout** tab area, click on the **General View** button as shown.

7. Click **OK** to accept the no combined state option.

8. The message "*Select CENTER POINT for drawing view*" is displayed in the message area. Pick a location that is to the upper right corner of the display window.

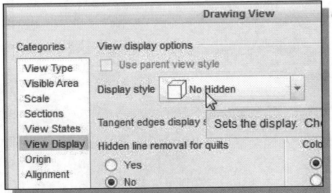

9. In the *Drawing View* window, adjust the **View Display** of the isometric view to **No Hidden** as shown.

❖ Note the default display option, **Follow Environment**, can be used to create shaded image.

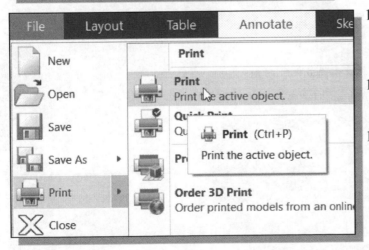

10. Click **OK** to accept the settings and end the Drawing View command.

11. In the **File** pull-down menu, select **Print**.

12. In the *Preview window*, use the **middle-mouse-button** to reposition the drawing and proceed with the print.

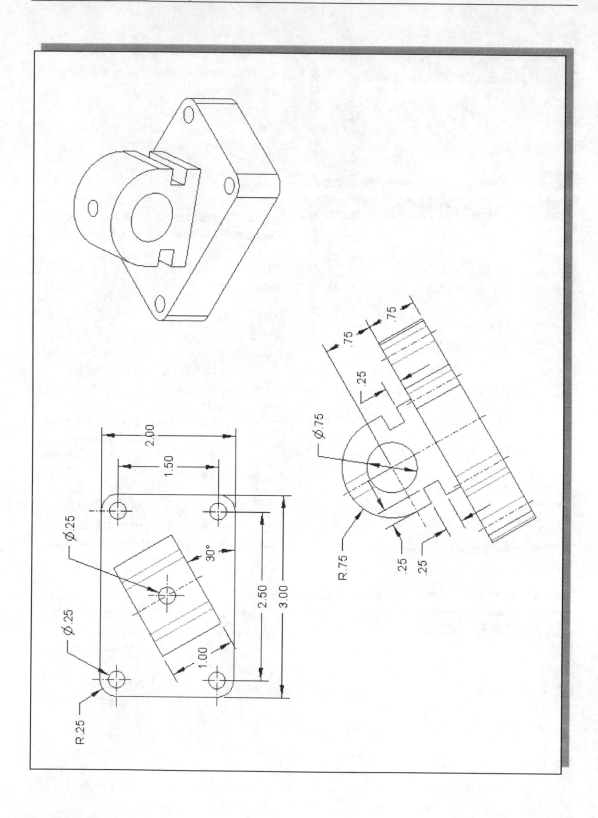

Review Questions:

1. How do we create geometry on a surface that does not already exist?

2. What are the advantages of using **Datum features**?

3. Describe the procedure to create an *angled datum plane*.

4. Describe the differences between **Feature** and **Driven Dimensions** in 2D drawings.

5. Describe the procedure to add an *Auxiliary view* in a drawing.

6. How do we access the **Show Annotation** command?

7. List the advantages of using **3D annotation**.

8. Create sketches showing the steps you plan to use to create the two models shown on the next page:

Ex.1)

Ex.2)

Exercises:

1. Angle Bracket (Material: **Carbon Steel**. Dimensions are in inches.)

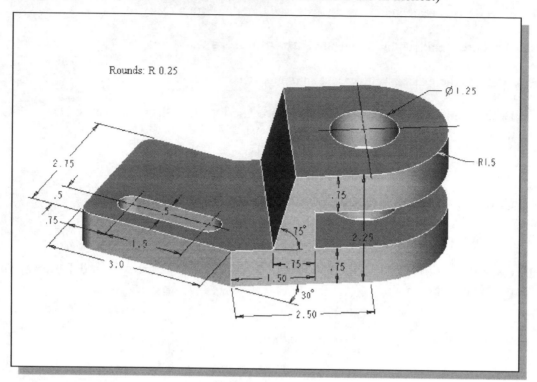

2. Shaft Guide (Dimensions are in inches.)

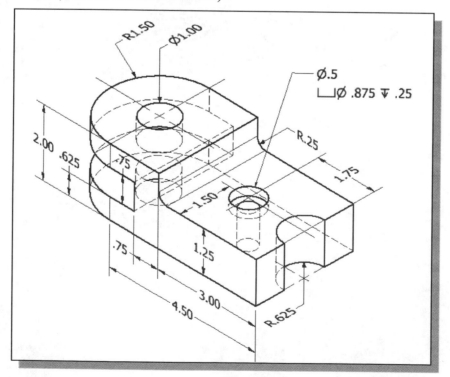

3. Angle Base (Dimensions are in Millimeters.)

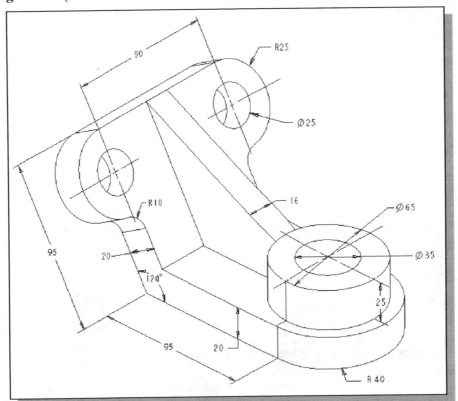

4. Bevel Washer (Dimensions are in inches.)

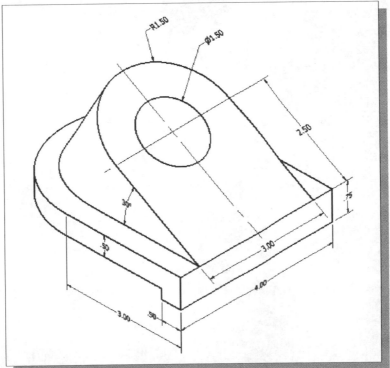

5. Angle V-Block (Dimensions are in inches.)

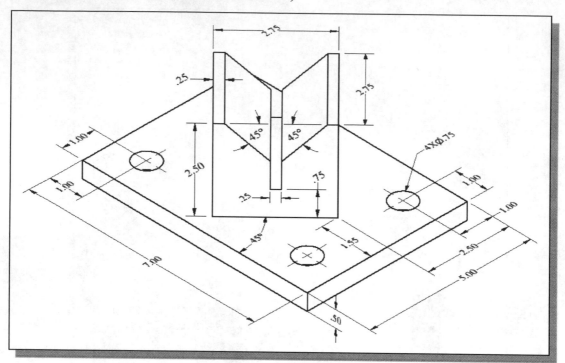

Chapter 7
Introduction to 3D Printing

Learning Objectives

- ♦ **Understand the History and Development of 3D Printing**
- ♦ **Be aware of the Primary types of 3D Printing Technologies**
- ♦ **Be able to identify the commonly used Filament types for Fused Filament Fabrication**
- ♦ **Understand the general procedure for 3D Printing**

What is 3D Printing?

3D Printing is a type of *Rapid Prototyping* (RP) method. Rapid prototyping refers to the techniques used to quickly fabricate a design to confirm/validate/improve conceptual design ideas. 3D printing is also known as "**Additive Manufacturing**" and construction of parts or assemblies is usually done by addition of material in thin layers.

Prior to the 1980s, nearly all metalworking was produced by machining, fabrication, forming, and mold casting; the majority of these processes require the removal of material rather than adding it. In contrast to the *Additive Manufacturing* technology, the traditional manufacturing processes can be described as **Subtractive Manufacturing**. The term *Additive Manufacturing* gained wider acceptance in the 2010s. As the various additive processes continue to advance and become more mature, it is quite clear that they will compete with material removal as the main manufacturing process for many applications in the very near future.

The basic principle behind *3D printing* is that it is an additive process. 3D printing is a radically different manufacturing method based on advanced technology that create parts directly, by adding material layer by layer at the sub millimeter scale. One way to think about 3D Printing is the additive process is really performing "**2D printing over and over again**."

A number of limitations exist to the traditional manufacturing processes, which can be labor intensive, requiring expensive tooling, designing of fixtures, and the assembly of parts. The *3D printing* technology provides a way to create parts with complex geometric shapes quite easily using thin layers. The traditional *subtractive manufacturing* can also be quite wasteful as excess materials are cut and removed from large stock blocks, while the *3D printing* process basically uses only the material needed for the parts. *3D printing* is an enabling technology that encourages and drives innovation with unprecedented design freedom while being a tooling-less process that reduces costs and lead times. The relatively fast turnaround time also makes *3D printing* ideal for prototyping. Components with intricate geometry and complex features can also be designed specifically for *3D printing* to avoid complicated assembly requirements. *3D printing* is also an energy efficient technology that can provide better environmental friendliness in terms of the manufacturing process itself and the type of materials used for the product. There are quite a few different techniques to 3D print an object. *3D Printing* brings together two fundamental innovations: the manipulation of objects in the digital format and the manufacturing of objects by addition of material in thin layers.

The term **3D-printing** originally referred only to the smaller 3D printers with moveable print heads similar to an inkjet printer. Today, the term **3D-printing** is used interchangeably with **Additive Manufacturing**, as both refer to the technology of creating parts through the process of adding/forming thin layers of materials.

Development of 3D Printing Technologies

The earliest 3D printing technology was first invented in the 1980s; at that time it was generally called **Rapid Prototyping (RP)** technology. This is because the process was originally conceived as a fast and time-effective method for creating prototypes for product development in industry. In 1981, Dr. Hideo Kodama of *Nagoya Municipal Industrial Research Institute* invented two methods of creating three-dimensional plastic models with photo-hardening polymer through the use of a UV Laser. In 1986, the first US patent for **stereolithography** apparatus **(SLA)** was issued to Charles Hull, who first invented his SLA machine in 1983. Chuck Hull went on to co-found *3D Systems Corporation*, which is one of the largest companies in the 3D printing sector today. Chuck Hull also designed the **STL (STereoLithography)** file format, which is widely used by 3D printing software performing the digital slicing and infill strategies common to the additive manufacturing processes. The first available commercial RP system, the **SLA-1** by *3D Systems* (as shown in the figure below), was made available in 1987.

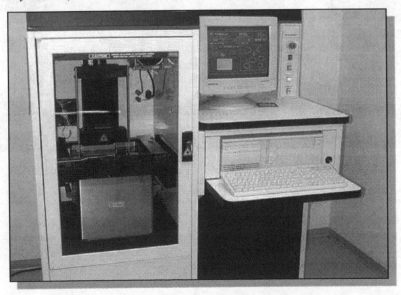

The 1980s also mark the birth of many RP technologies worldwide. In 1989, Carl Deckard of *University of Texas* developed the **Selective Laser Sintering (SLS)** process. In 1989, Scott Crump, one of the founders of *Stratasys Inc.*, also created the **Fused Deposition Modeling (FDM)**. In Europe, Hans Langer started *EOS GmbH* in Germany; the company focuses on the **Laser Sintering (LS)** process. The *EOS systems Corp.* also developed the **Direct Metal Laser Sintering (DMLS)** process. Today, *3D Systems*, *EOS* and *Stratasys* are still the main leaders in the *Additive Manufacturing* industry.

During the 1990s, the *3D printing* sector started to show signs of distinct diversification with two specific areas of emphasis which are much more clearly defined today. First, there was the high end of 3D printing, still very expensive systems, which were geared towards part production for relatively complex designs. For example, in 1995, *Sciaky, Inc.* developed an additive welding process based on its proprietary **Electron Beam Additive Manufacturing (EBAM)** technology. Many *RP* system companies, such as *Solidscape*, *ZCorporation*, *Arcam* and *Objet Geometries* were all launched in the 1990s. At the other end of the spectrum, some of the 3D printing system manufacturers started to develop smaller desktop systems in the 1990s.

The idea of creating low-cost desktop 3D printers also intrigued many technology professionals and hobby enthusiasts during the late 1990s. In 2004, a retired professor, Dr. Adrian Bowyer (the person on the left in the photo below), started the **RepRap** (*Replication Rapid-Prototyper*) project of an open source, self-replicating 3D printer (**RepRap 1.0 - Darwin**). This set the stage for what was to come in the following years. It was around 2007 that the open source *3D printing* movement started gaining visibility and momentum. In January of 2009, the first commercially available open source 3D printer, the **BFB RapMan** 3D printer, became available. *Makerbot Industries* also came out with their **Makerbot** 3D printer in April of 2009. Since then, a host of low-cost desktop 3D printers have emerged each year.

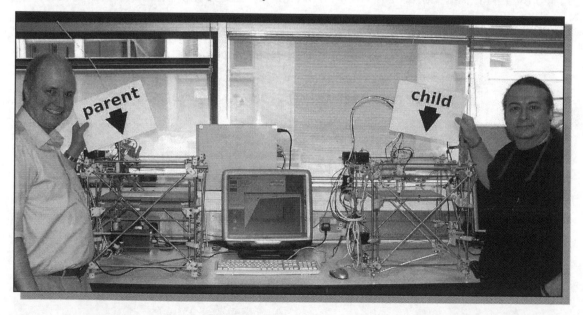

In the beginning of the 2010s, alternative 3D printing processes, such as using **Polymer Resins**, became available at the desktop level of the market. The **B9 Creator**, using **Digital Light Processing (DLP)** technology by *B9Creations*, came first in June of 2012, followed by the **Form 1** desktop printer by *Formlabs Inc.* Both 3D printers were launched via KickStarter's crowdfunding website and both enjoyed huge success. 2012 was also the year that many different mainstream media took note of the exciting 3D printing technology, which dramatically increased its visibility and awareness to the general public. 2013 was also a year of significant growth and consolidation; one of the most notable moves was the acquisition of Makerbot by Stratasys. In 2016, the new developments in 3D printing concentrated more on multi-color, multi-material using single or multiple extruder printers and new technologies to shorten the 3D printing time.

As a result of the market divergence, the price of desktop 3D printers continues to go down each year. Today, very capable fully assembled desktop 3D printers, such as Robo3D R1+, Prusa i3 MK2, can be acquired for under $1000. Fully assembled smaller desktop 3D printers, such as Xyz-printing's DA Vinci mini 3D Printer and M3D's Micro 3D can be acquired for less than $350. Unassembled desktop 3D printer kits can even be acquired for under $250.

Another trend that happened in the 2010s is the availability of **3D printing Services**. 3D printing services are growing quite rapidly in the US. For example, many public libraries, especially in California, are now providing 3D printing services to the general public and **UPS** started its worldwide 3D printing services in May of 2016. This trend is spreading throughout the US, with many more companies planning to provide 3D printing services in the very near future. It is now quite feasible, and perhaps more economical, to 3D print designs without owning or ever touching a 3D printer; but understanding of the technology is still needed to increase productivity.

As the exponential adoption rate continues on all fronts, more and more technologies, materials, applications, and online services will continue to emerge. It is predicted that the development of 3D printing will continue in the years to come and 3D printing will eventually become the mainstream manufacturing method in industries and homes.

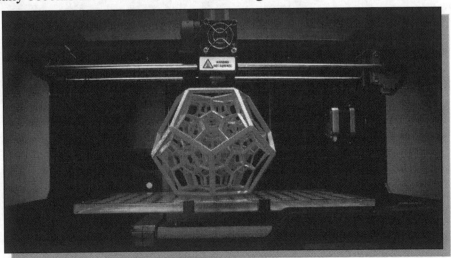

Primary Types of 3D Printing Processes

There are quite a few different techniques to 3D print an object. The different types of 3D printers each employ a different technology that processes different materials in different ways. For example, some 3D-printers can process powdered materials (nylon, plastic, ceramic and metal) which utilize a light/heat source to sinter/melt/fuse layers of the powder together in the defined shape. Others process polymer resin materials and again utilize a light/laser to solidify the resin in thin layers. **Stereolithography** (**SLA** or **SL**), **Fused Deposition Modeling** (**FDM** or **FFF**) and **Laser Sintering** (**LS** or **SLS**) represent the three primary types of 3D printing processes; the majority of the other 3D printing technologies are variations of the three main types.

Stereolithography

Stereolithography (**SLA** or **SL**) is widely recognized as the first 3D printing process; it was certainly the first to be commercialized. *SLA* is a laser-based process that works with photopolymer resins. The photopolymer resins react with the laser and cure to form a solid in a very precise way to produce very accurate parts. It is a complex process, but simply put, the photopolymer resin is held in a container with a movable platform inside. A laser beam is directed in the X-Y axes across the surface of the resin according to the 3D data supplied to the machine. The resin hardens precisely as the laser hits the designated area. Once the current layer is completed, the platform within the container drops down by a fraction (in the Z axis) and the subsequent layer is traced out by the laser. This 2D layer tracing continues until the entire object is completed.

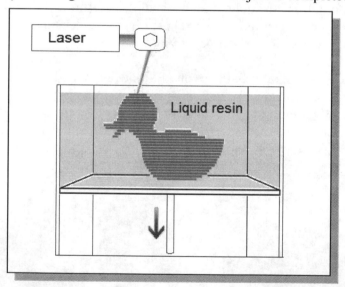

Because of the nature of the SLA process, support structures are needed for some parts, specifically those with overhangs or undercuts. These support structures need to be removed once the part is created. Many 3D printed objects using SLA need to be further cleaned and/or cured. Curing involves subjecting the part to intense light in an oven-like machine to fully harden the resin. SLA is generally accepted as being one of the most accurate 3D printing processes with excellent surface finish.

Fused Deposition Modeling (FDM) / Fused Filament Fabrication (FFF)

3D printing utilizing the extrusion of thermoplastic material is probably the most popular 3D printing process. The original name for the process is **Fused Deposition Modeling (FDM)**, which was developed in the early 1990s and is a trade name registered by *Stratasys*. However, a similar process, **Fused Filament Fabrication (FFF)**, has emerged since 2009. The majority of the desktop 3D printers, both open source and proprietary, utilize the FFF process, which is a more basic extrusion form of FDM.

The FDM and FFF processes work by melting plastic filament that is deposited, via a heated extruder, one layer at a time, onto a build platform according to the 3D data supplied to the 3D printer. Each layer hardens as it cools down and bonds to the previous layer.

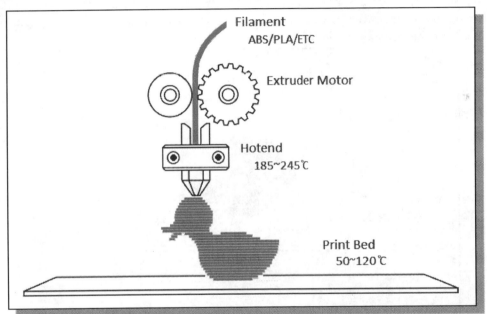

Stratasys has developed a range of proprietary industrial grade materials for its FDM process that are suitable for production applications. However, the most common materials for both FDM and FFF 3D-printers are **ABS (Acrylonitrile Butadiene Styrene)** and **PLA (Polylactic Acid)**. The FDM and FFF processes require support structures for any applications with overhanging geometries. This generally entails a second, typically water-soluble or breakaway material, which allows support structures to be easily removed once the print is complete.

The FDM and FFF printing processes can be slow for large parts or parts with complex geometries. The layer-to-layer adhesion can also be a problem, resulting in parts that warp or separate easily. The surface finish of FDM and FFF printed parts might appear a bit rough as the thin layers are generally visible. To improve the appearance, several options are feasible, such as using acetone, sanding and/or spray paint.

Laser Sintering / Laser Melting

Laser Sintering (LS) or **Selective Laser Sintering (SLS)** creates tough and geometrically intricate parts using a high-powered CO_2 laser to fuse/sinter/melt powdered thermoplastics. The main advantage of *SLS 3D printing* is that as a part is made, it remains encased in powder; this eliminates the need for support structures and allows for very complex 3D geometries to be 3D printed. SLS can be used to produce very strong parts as exceptional materials such as Nylon and metal powders are commonly used.

Laser sintering refers to a laser-based 3D printing process that works with powdered materials. The laser is traced across a powder bed of tightly compacted powdered material, according to the 3D data provided to the machine, in the X-Y axes. As the laser interacts with the powdered material it sinters and fuses the particles to each other forming a solid. As each layer is completed the powder bed drops incrementally and a roller is used to compact the powder over the top surface of the bed prior to the next pass of the laser for the subsequent layer.

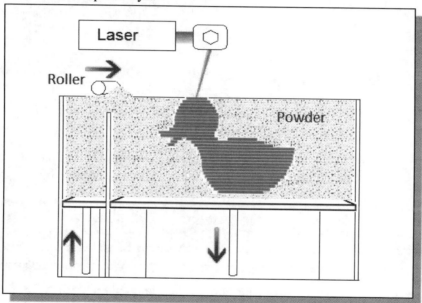

The build chamber is completely sealed as it is necessary to maintain a precise temperature during the process specific to the melting point of the powdered material of choice. One of the key advantages of this process is that the powder bed serves as an in-process support structure for overhangs and undercuts, and therefore complex shapes that could not be manufactured in any other way become possible with this process. Because of the high temperatures required for laser sintering, cooling can take a long time. Porosity is also a common issue with this process; additional metal infiltration processes may be required to improve mechanical characteristics.

Laser sintering can process plastic and metal materials, although metal sintering does require a much higher-powered laser and higher in-process temperatures. Parts produced with this process are much stronger than parts made with SLA or FDM, although generally the surface finish and accuracy is not as good.

Primary 3D Printing Materials for FDM and FFF

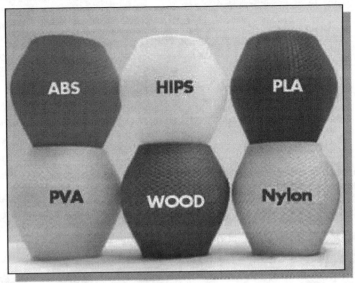

ABS (Acrylonitrile Butadiene Styrene)

ABS is a popular choice for 3D printing; it is a strong thermoplastic that is among the most widely used plastics. It is tough with mild flexibility, making it more durable to stress and has a higher heat resistance of up to 200 degrees Fahrenheit. However, this material has a tendency to shrink, which can affect the accuracy of designs. ABS has a pretty high melting point, and can experience warping if cooled while printing. Because of this, ABS objects are printed typically on a heated surface. ABS also requires ventilation when in use, as the fumes can be unpleasant. The aforementioned factors make ABS printing difficult for hobbyist printers, though it's the preferred material for professional applications.

PLA (Polylactic Acid)

PLA is a staple and it is becoming one of the most popular choices for 3D printing with good reason. In addition to the fact that it is a biodegradable thermoplastic derived from renewable resources such as corn starch, tapioca roots, chips or starch, or sugarcane, *PLA* is a very rigid material that is easy to use for 3D printing and it is able to withstand a good amount of impact and weight. It also has a glossier finish than ABS, and in most scenarios PLA is the preferred material for 3D printing large objects. The main disadvantage of PLA is it's not as heat resistance as ABS; it should not be placed in environments that exceed 140 degrees Fahrenheit.

Flexible (Thermoplastic Elastomer)

Flexible material is for applications that require incredible rubbery flex in their applications. Flexible filament goes beyond bending; it is more like rubber. When it comes to Flexible filament, it's all about finding a balance between flexibility (softness) and printability. This softness is sometimes indicated with a *Shore* value (like 85A or 60D). A higher Shore value means less flexibility. Harder filaments (less flexible) are easier to 3D print when compared to softer, more flexible filaments.

PETG (Polyethylene Terephthalate)

PETG is a material that is similar to *PLA*, with more attractive characteristics such as being generally a tougher and denser material with good heat resistance of up to 190 degrees Fahrenheit. It is reported to have the strength of *ABS*, while printing as easily as *PLA*.

HIPS (High Impact Polystyrene) and PVA (Polyvinyl Alcohol)

HIPS and *PVA* are relatively new materials that are growing in popularity for their dissolvable properties. They are used for creating support material. Their ability to dissolve in certain liquids means that they can be easily removed. These materials can be hard to print with because they don't stick well to the build plates. It is also important not to print *PVA* at too high a temperature, as it can turn into tar and jam the extruder.

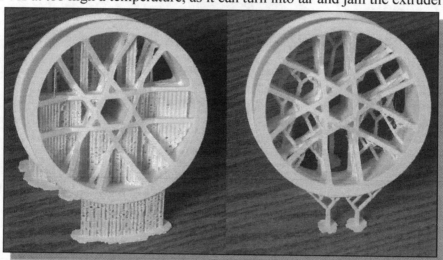

Wood Fiber

Wood Fiber filament contains a mixture of recycled wood with a binding polymer. Thus, a 3D printed object can look and smell like real wood. Due to its wooden nature, it's difficult to tell that the object is 3D printed. Using Wood filament is similar to using a thermoplastic filament like ABS or PLA. However, a 3D object having a wooden-like appearance can be created with this material.

From 3D Model to 3D Printed Part

To create a 3D printed part, it all starts with making a virtual design of the object. This virtual design may be created with a computer-aided design (CAD) package, via a 3D scanner, or by a digital camera and photogrammetry software. 3D scanning and photogrammetry software can process the collected digital data on the shape and appearance of a real object and create a digital 3D model. The 3D virtual design can generally be modified with 3D CAD packages, allowing verification of the virtual design before it is 3D printed.

Once the virtual design is verified, the 3D data will then be transferred to the 3D printing software. There is a multitude of file formats that 3D printing software supports. However, the most popular are the STL file format and the OBJ file format. The STL file format is the most commonly used file format for 3D printing. Most CAD software has the capability of exporting models in the STL format. The STL file contains only the surface geometry of the modeled object. The OBJ file format is considered to be more complex than the STL file format as it is capable of displaying texture, color and other attributes of the three-dimensional object. However, the STL file format holds the top spot for 3D printing, as this file format is simpler to use, and most CAD packages work better with STL files than OBJ files.

Once the 3D data of the virtual design is transferred into the 3D printing software, further examination and/or repair can be performed if necessary. The 3D printing software will also process the imported 3D data by the special software known as a **Slicer**, which converts the model into a series of thin layers and produces a G-code file containing instructions tailored to a specific type of 3D printer. G-code is the common name for the most widely used numerical control (NC) programming language. It is used mainly in computer-aided manufacturing to control automated machine tools. The generated G-code file can be sent to the 3D printer and create the 3D printed part.

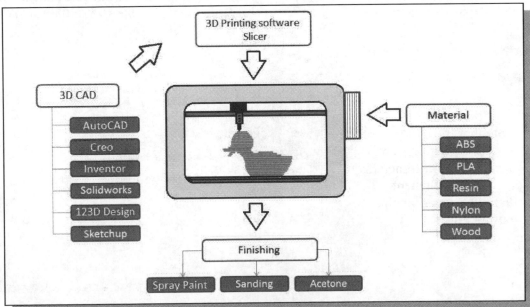

Starting Creo Parametric

1. Select the **Creo Parametric** option on the *Start* menu or select the **Creo Parametric** icon on the desktop to start Creo Parametric. The Creo Parametric main window will appear on the screen.

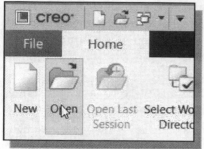

2. In the Creo Parametric *Startup* dialog box, select **Open** with a single click of the left-mouse-button.

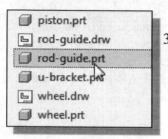

3. In the *Open* window, select the **Rod-Guide. PRT** file. Use the *browser* to locate the file if it is not displayed in the *File name* list box.

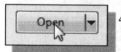

4. Click on the **Open** button in the *Startup* dialog box to accept the selected selection.

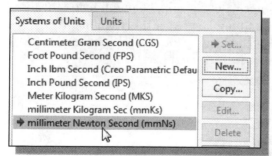

5. On your own, change the part *Units* to millimeter under the material properties as shown. (Hint: Use the *Convert dimension option* to maintain the size of the part.)

- Note that the majority of the 3D printer settings are measured in millimeters, such as the filament diameter, layer height, extruder size.

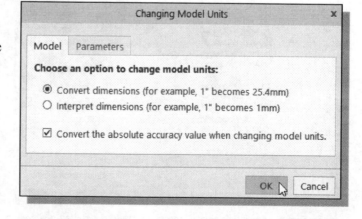

Prepare for 3D Printing

3D printers will generally accept a 3D model with the STL or OBJ file formats. In Creo Parametric, the **Prepare for 3D Printing** command is available to place multiple parts into an assembly, which can be 3D printed all at once. The *Prepare for 3D Print* command also provides patterning multiple copies of a part, printability validation and interference check prior to the parts being 3D printed locally or sending the assembly to online 3D printing services. Once the assembly is created, the **Save As** command can be used to very quickly export the 3D assembly as STL files.

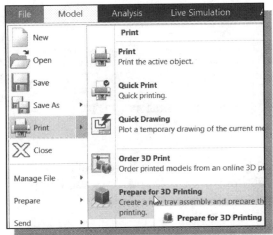

1. In the *File Toolbar,* select the **Print → Prepare for 3D Printing** command as shown.

   • Note the Order 3D Print command is also available.

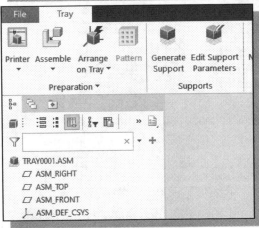

2. Note that Creo Parametric automatically creates a new assembly model, **TRAY0001.ASM**, which contains the Rod-Guide design. In the *ribbon toolbar area*, notice the different commands available to place multiple copies of parts, such as the *Assemble* command, *Pattern* command and *Nesting* command.

3. Click on the **Printer** icon and notice two items are listed: *PTC Generic* and the *i.materialise* online printing services. Additional local 3D printer information and online services can be added to the list.

4. Select **Printers List** in the pull-down menu to view the current settings.

- Note that information on the build space of the 3D printer and the URL of the online printing services can be entered here.

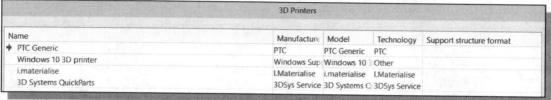

Name	Manufacture	Model	Technology	Support structure format
➔ PTC Generic	PTC	PTC Generic	PTC	
Windows 10 3D printer	Windows Sup	Windows 10 3	Other	
i.materialise	I.Materialise	i.materialise	I.Materialise	
3D Systems QuickParts	3DSys Service	3D Systems Q	3DSys Service	

5. Click **OK** to exit the *Printer list* dialog box.

6. By default, the parts placed inside the assembly are initially positioned outside the build space of the 3D printer. We can manually move the parts inside or use Creo Parametric's *Auto-Position* option to automatically arrange the parts.

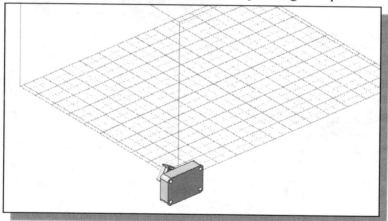

- To illustrate the usage of the Assemble and Auto-Position commands, we will add another copy of the rod-guide part into the current assembly.

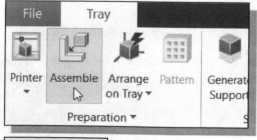

7. Click **Assemble** to add another model into the current assembly.

8. Select the *Rod-guide* part and click **Open** as shown.

9. Note the model is again placed outside of the build space. Click on the red ring to rotate the part about the x-axis. On your own, move and rotate the part by dragging & dropping the three arrows and rings.

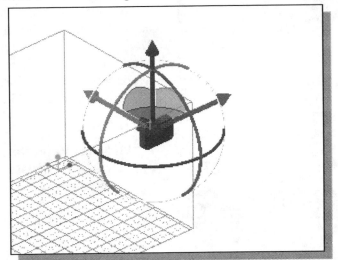

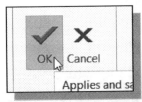

10. Click **OK** to end the manual placement of the part.

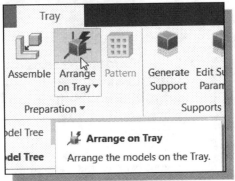

11. Click **Arrange on Tray** to have Creo automatically place the pieces inside the build space. Note that the parts are placed in a very compact fashion, which allows more efficient use of space and faster print time.

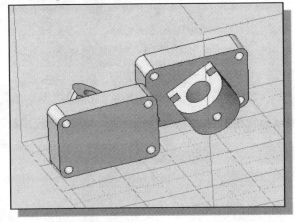

12. To manually reposition the parts at any time, use **Edit definition** of the part in the *Assembly model tree* as shown.

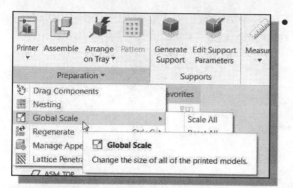

• Beside the Pattern and Nesting commands to place multiple copies of the same part within the same print, the Global Scale command can be used to adjust the size of the printed models. Note that most of these positioning, pattern and scale options are also available in the Slicer programs, which will be illustrated in this chapter.

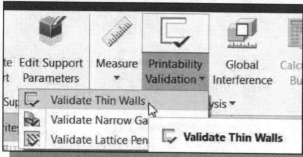

13. Select **Validate Thin Walls** to check for any issues with having too thin features within the parts.

14. Select one of the parts and set the minimum Wall thickness to 0.1 and click **Compute** to perform the check.

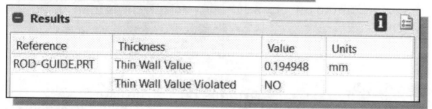

Reference	Thickness	Value	Units
ROD-GUIDE.PRT	Thin Wall Value	0.194948	mm
	Thin Wall Value Violated	NO	

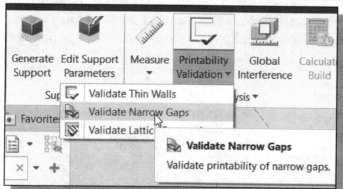

15. Select **Validate Narrow Gaps** to check for any issues with parts being placed too close to each other.

16. Select both of the parts and click **Compute** to perform the check.

Reference	Thickness	Value	Units
ROD-GUIDE.PRT	Small Gap Value	0.269049	mm
	Small Gap Value Violated	NO	
ROD-GUIDE.PRT	Small Gap Value	0.269049	mm
	Small Gap Value Violated	NO	
All References	Small Gap Value	0.269049	mm
	Small Gap Value Violated	NO	

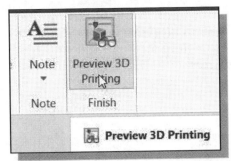

17. Select **Preview 3D Printing** as shown.

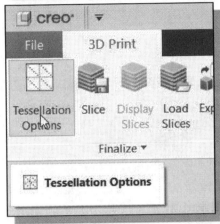

18. Select **Tesselate Options** to review the effect of adjusting the chord length for the STL models. Smaller numbers will result in a finer model.

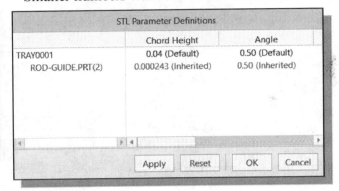

19. Select **Clipping** to view the effects of slicing in the Z axis.

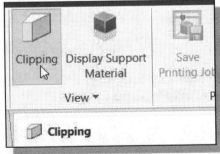

20. On your own, drag and drop the arrow to simulate the 3D printing process. Look for any potential issues of 3D printing the assembled models.

• Note that portions of the models are not touching the print-bed; they are not supported and will have problems during printing.

21. Click **Display Support Material** in the *Print3D* dialog box to switch on the addition of support material to the assembly.

22. On your own, drag and drop the arrow to simulate the 3D printing process. Notice the addition of the support material displayed in yellow.

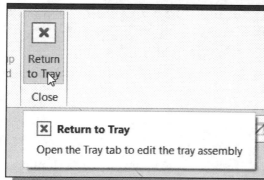

23. Click **Return to Tray** to end the **Preview 3D Printing** option.

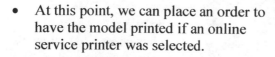

Return to Tray
Open the Tray tab to edit the tray assembly

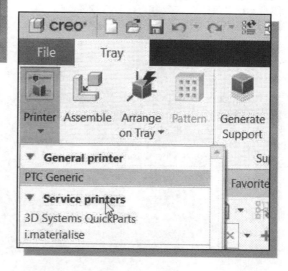

- At this point, we can place an order to have the model printed if an online service printer was selected.

Export the Design as an STL File

Note that it is feasible to export the design as an STL file and perform a majority of the validations in the Slicer program. We illustrate this by switching to the Rod-Guide model and prepare the model for local printing.

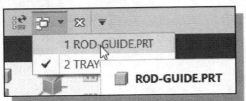

1. In the *Quick Access* area, switch to the **rod-guide** model as shown.

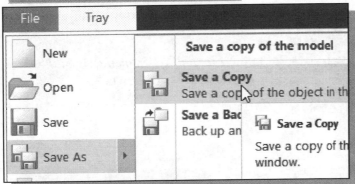

2. Select [**File→Save As**] as shown. Note that the *Save As dialog box* appears on the screen.

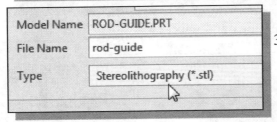

3. In the **Save As** dialog box, set the file format to STL as shown.

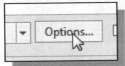

4. In the **Save As** dialog box, click on the **Options** button as shown.

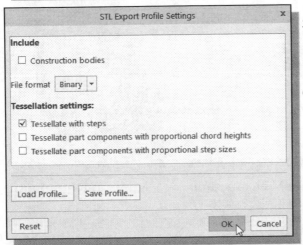

5. On your own, confirm the *output format* is set to **Binary** and examine the available *Tessellation* settings.

6. Click **OK** twice to proceed with saving the Rod-Guide.STL file.

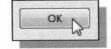

Using the 3D Printing Software to Create the 3D Print

To 3D print the model, we will open the STL file in the 3D printing software. We will use **Matter Control** to demonstrate the procedure. Note that *Matter Control* (Freeware) supports quite a few desktop 3D printers. The procedure illustrated here is also applicable to other similar software.

1. Start the **Matter Control** software.

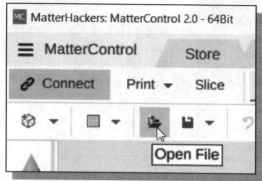

2. In the *File pull-down menu*, select **Add File to Queue** as shown.

3. On your own, switch to the saved STL file folder and select the Rod-Guide.STL file as shown.

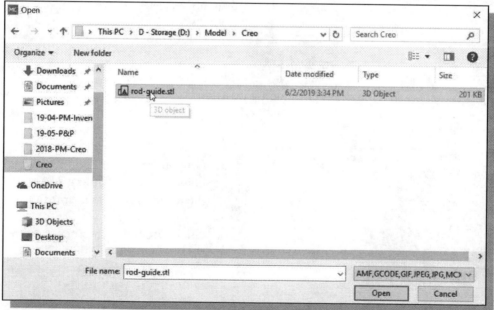

4. Click **Open** to import the STL file into Matter Control.

5. Once the STL file is imported into the program, the STL model is displayed in the *View* window. Most 3D Printers expect the model to be in millimeters. Note that most slicers allow conversion of units within if needed.

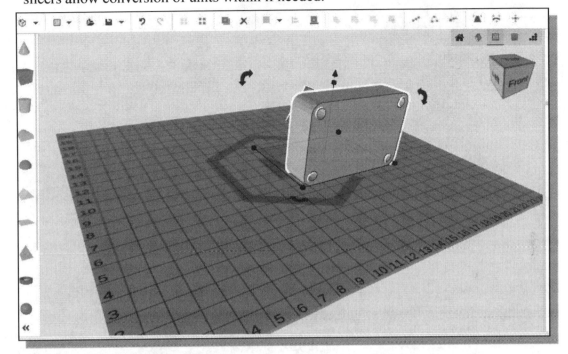

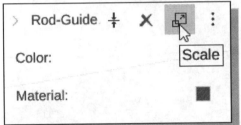

6. Click on **Scale** to expand the *Scale control panel*.

7. In the option list, different conversion options are available, such as **inches to mm**. Note that 1 inch equals 25.4 mm; MatterControl will adjust the model based on the selected scale.

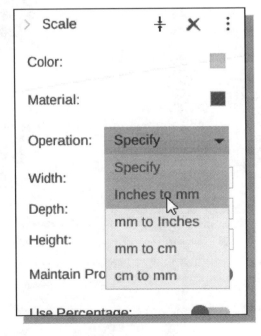

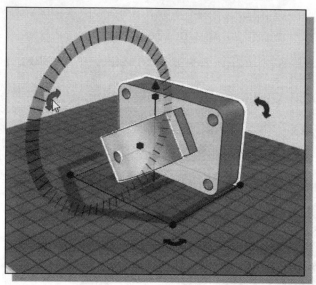

8. To **Rotate** the design, move the cursor on the curve-arrow icon and drag with the left mouse button to perform the rotate operation.

9. On your own, rotate the design 90 degrees, so that the flat face is aligned to the print-bed as shown.

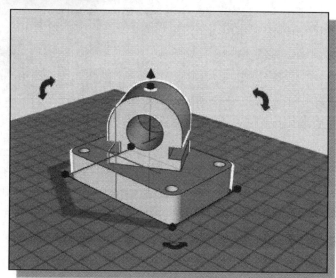

• Note the straight-arrows icons can be used to move/translate the design in X/Y/Z directions.

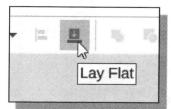

10. Click on the Lay Flat command to align the bottom face of the design to the print-bed.

11. On your own, examine the different display options, such as **Home** and **View Cube**, near the upper right corner of the main window.

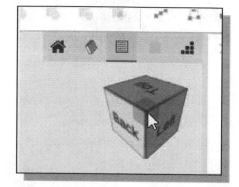

12. Use the right-mouse-button on the design to display the option list; note the different options available to modify the design.

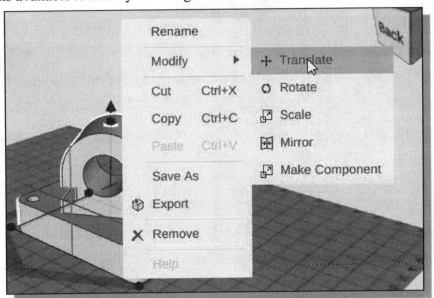

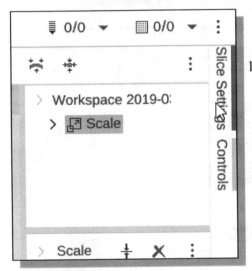

13. Near the upper right corner of the main window, click **Slice Settings** to review the Slicer program settings.

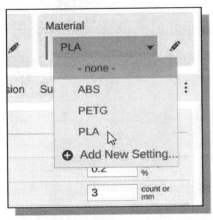

14. In the **Material list**, choose **PLA** as the type of filament properties. Note that the related properties, such as the diameter, extruder temperature, and bed temperature can be adjusted to match the actual filament being used.

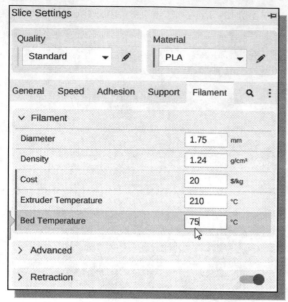

15. Note the different settings are organized as groups, which can be used to adjust the 3D printer's settings and the commands to directly control the movements of the extruder of the printer.

• Note that the temperatures are different based on the types of filaments, as well as the different brands and the type of printer bed. It may be necessary to do some testing and/or experimenting when a new roll of filament is used.

16. Switch to the **General** Tab, and examine the Layer control settings, such as the layer thickness, top layer and bottom layer thickness.

17. Also note the settings for **Infill Type**, the **Fill Density**.

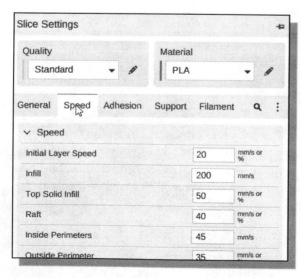

18. click on the **Speed** tab and examine the related settings to control the 3D printing speed.

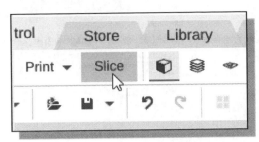

19. Click Slice to process the 3D model, which includes slicing and generating the G-code for the specific 3D printer.

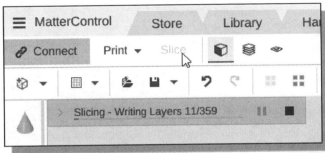

20. Matter Control will display a status bar showing the slicing progress.

- Note that Matter Control allows the use of other Slicers, such as Cura or Slic3r, which are the two extremely popular open source slicer programs.

21. On your own, drag the vertical slider to review the thin layers generated by the slicer.

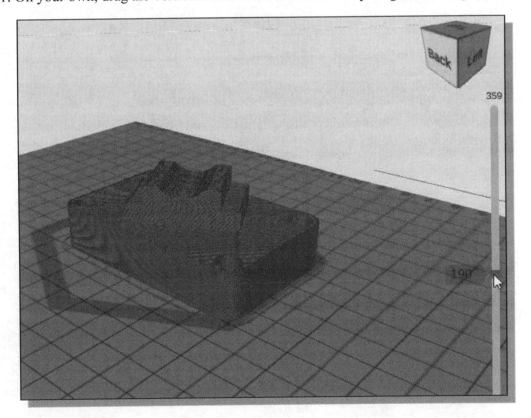

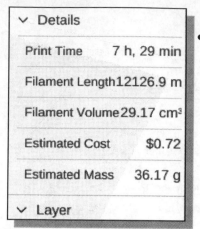

• Note that with the current settings, it will take 7 hours and 29 minutes to complete the print using 12126.9 mm of filament.

22. Click **Connect -→ Print** to start the printing of the 3D model.

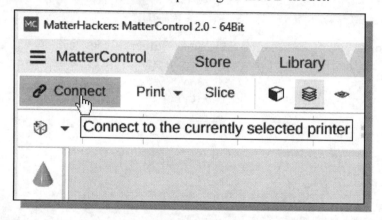

23. *Matter Control* will now begin printing; note the Print time for the model is displayed in the status window.

Questions:

1. What is the main difference between the **Additive Manufacturing** and the traditional **Subtractive Manufacturing** technologies?

2. Which 3D printing process is recognized as the first 3D printing process?

3. Describe the general procedure to create a 3D printed part.

4. What are the three primary types of 3D printing processes?

5. Which 3D printing process is the most popular 3D printing process?

6. What is the main advantage of using PLA over ABS for FFF process?

7. Which are the most popular file formats for 3D printing?

8. What is the main function of a **Slicer** program?

9. List and describe the print validation options available with the Creo Parametric **Print3D** command.

Notes:

Chapter 8
Symmetrical Features in Designs

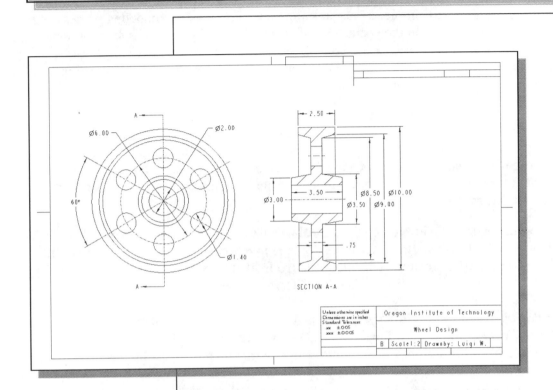

Learning Objectives

- ♦ **Create Revolved Features**
- ♦ **Use the Mirror Feature Command**
- ♦ **Create Circular Patterns**
- ♦ **Understand Associative Functionality**
- ♦ **Identify Symmetrical Features in Designs**
- ♦ **Create Sectioned Orthogonal Views**
- ♦ **Display and Hide Feature Dimensions**
- ♦ **Create Reference Dimensions**

Symmetrical Features in Designs

In parametric modeling, it is important to identify and determine the features that exist in the design. *Feature-based parametric modeling* enables us to build complex designs by working on smaller and simpler units. This approach simplifies the modeling process and allows us to concentrate on the characteristics of the design. Symmetry is an important characteristic that is often seen in designs. Symmetrical features can be easily accomplished by the assortment of tools that are available in feature-based modeling systems, such as *Creo Parametric*.

The modeling technique of extruding two-dimensional sketches along a straight line to form three-dimensional features, as illustrated in the previous chapters, is an effective way to construct solid models. For designs that involve cylindrical shapes, shapes that are symmetrical about an axis, revolving two-dimensional sketches about an axis can form the needed three-dimensional features. In solid modeling, this type of feature is called a **revolved feature**.

In *Creo Parametric*, besides using the **Revolve** command, several options are also available to handle symmetrical features. For example, we can create multiple identical copies of symmetrical features with the **Feature Pattern** command or create mirrored images of models or 2D sections using the **Mirror** command.

Creo Parametric provides full associative functionality in all *Creo Parametric* software modules. This functionality allows us to change the design at any level, and the system reflects it at all levels automatically. For example, the solid model can be modified in the standard *Part Modeler* and the system automatically reflects that change in the *Drawing* layout. And we can also modify a feature dimension in the *Drawing* layout, and the system automatically updates the solid model in all applications.

A Revolved Design: Wheel

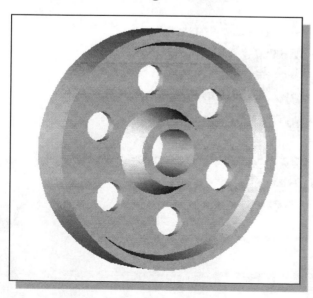

❖ Based on your knowledge of *Creo Parametric*, how many features would you use to create the design? Which feature would you choose as the **base feature** of the model? Identify the symmetrical features in the design and consider other possibilities in creating the design. You are encouraged to create the model on your own prior to following through the tutorial.

Modeling Strategy - A Revolved Design

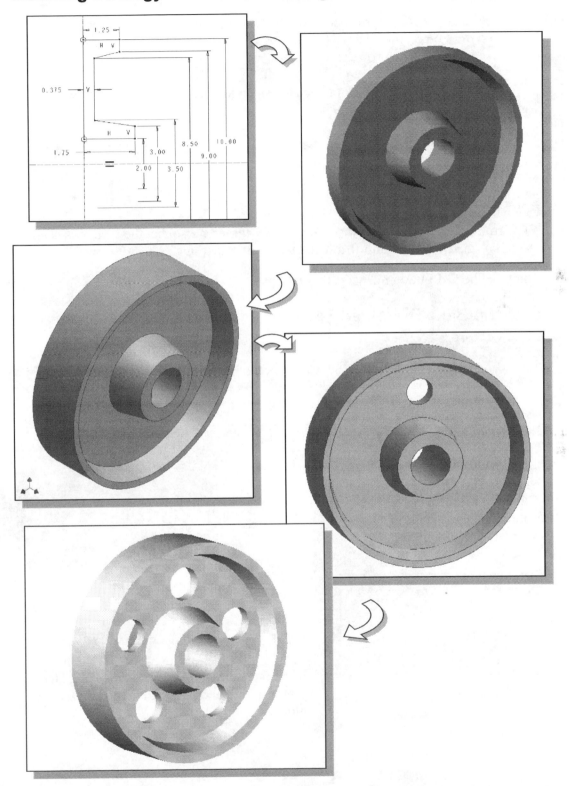

Starting Creo Parametric

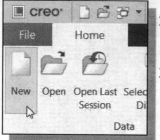

1. Select the **Creo Parametric** option on the *Start* menu or select the **Creo Parametric** icon on the desktop to start *Creo Parametric*. The *Creo Parametric* main window will appear on the screen.

2. Click on the **New** icon located in the *Ribbon toolbar* as shown.

3. On your own, start a new solid part file using **Wheel** as the PART Name.

4. Confirm the **Use default template** option is turned *on* so that the system is using the *Creo Parametric* default (**Inch-Ibm-Second**) settings.

5. Click on the **OK** button to accept the settings.

Create the Base Revolved Feature

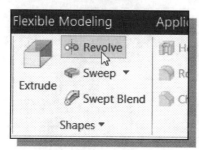

1. In the *Shapes* toolbar, select the **Revolve** tool option as shown.

2. Click the **Placement** option and choose **Define** to create a new *internal sketch*.

3. On your own, set up the **FRONT** datum plane as the sketch plane with the **RIGHT** datum plane facing the **right** edge of the computer screen and pick **Sketch** to enter the *Sketcher* mode.

❖ Note the default selection of the two datum planes, **RIGHT** and **TOP**, as the references for the 2D sketch.

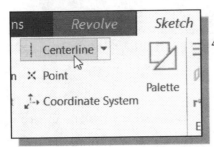

4. In the *Sketching* toolbar, select **Centerline**.

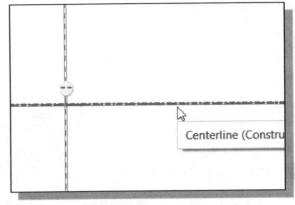

5. First, select a point on the *horizontal reference* (**TOP**).

6. Select a second point aligned to datum plane **TOP** to create a **horizontal centerline**.

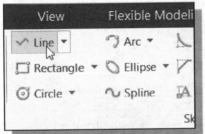

7. Next, switch to sketching regular line entities. In the *Sketching* toolbar, select **Line**.

8. On your own, create a closed region sketch approximately in shape as shown below. At this point, concentrate on the **shape** and **form** of the design. The dimension values may appear differently on your screen. Note there is no Equal Length and no Perpendicular constraints applied in the sketch.

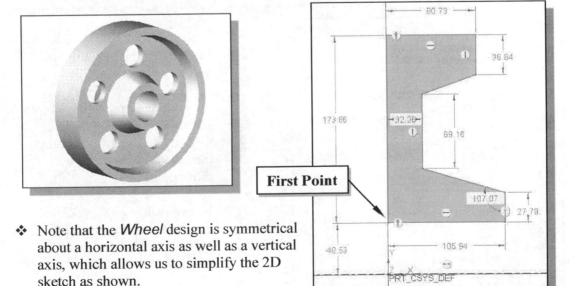

❖ Note that the *Wheel* design is symmetrical about a horizontal axis as well as a vertical axis, which allows us to simplify the 2D sketch as shown.

Create Diameter Dimensions

Creo Parametric provides a special option to create dimensions that will account for the symmetrical nature of the design; this is known as the ***diameter dimension***.
To create a diameter dimension: pick the entity, pick the centerline to use as the axis of revolution, and pick the entity again; and then place the dimension.

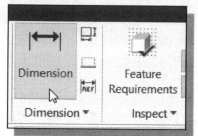

1. Select **Dimension** in the *Dimension* toolbar.

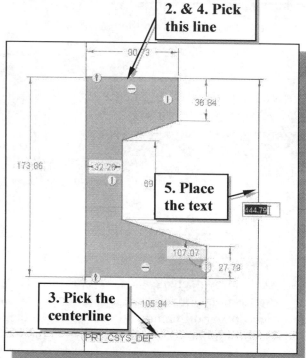

2. & 4. Pick this line

2. Pick the **top horizontal line** as the first entity to dimension.

3. Pick the *horizontal* **centerline**.

4. Pick the **top horizontal line** again.

5. Pick a location, with the *middle-mouse-button*, toward the right side of the sketch as shown.

5. Place the text

❖ The diameter dimension uses the centerline as a reference to calculate the dimension value.

3. Pick the centerline

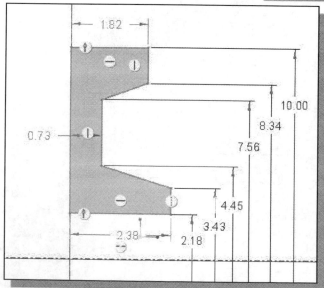

6. Enter 10 as the new dimension value.

7. On your own, repeat the above steps and create the desired dimensions as shown. Accept the default values of the dimensions.

❖ Notice many of the *weak* dimensions are removed by the system as the desired dimensions are added.

Modify the Values of Dimensions

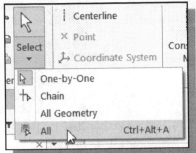

1. Choose **Select → All** in the *Operations* toolbar.

- Note that the quick-key combination, [CTRL]+[ALT]+[A], can also be used to quickly select all of the constructed objects.

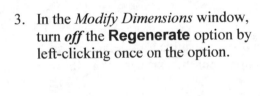

2. Choose **Modify** in the *Editing* toolbar.

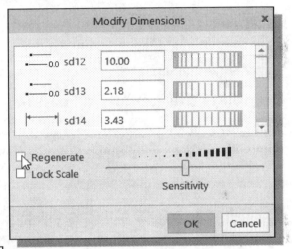

3. In the *Modify Dimensions* window, turn *off* the **Regenerate** option by left-clicking once on the option.

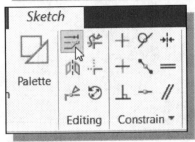

4. On your own, modify the dimension values so that they appear as shown in the figure.

5. Click the **OK** button to regenerate the sketch and exit the *Modify Dimensions* dialog box.

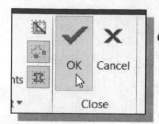

6. Click **OK** to accept the completed section.

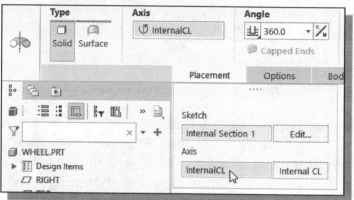

7. In the *Dashboard*, click **Placement** to show the sketch and axis options. The sketch contains a centerline, which is automatically selected to be used as the axis of rotation for the revolved feature.

Complete the Revolved Feature

For the **Revolve** operation, the 2D section needs to be a valid *section* (no self-intersecting edges), and the axis selected cannot cause the resultant model to intersect with itself. The axis of rotation can be one of the edges, but the axis cannot go through the *middle* of the section.

1. Confirm the *revolve angle* is set to **360** Degrees. Note the *flip direction* option is also available.

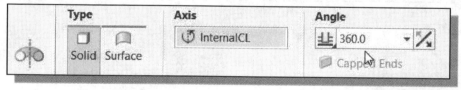

2. Pick **Accept** in the *Feature Option Dashboard* to create the solid feature.

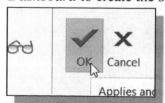

Using the Mirror Option

In *Creo Parametric*, features can be mirrored to create and maintain complex symmetrical features. We can mirror a feature about a datum plane or a specified surface. We can create a mirrored feature while maintaining the original parametric definitions, which can be quite useful in creating symmetrical features. For example, we can create one quadrant of a feature, then mirror it twice to create a solid with four identical quadrants.

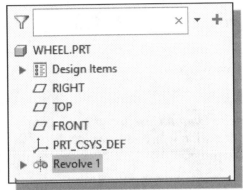

1. In the *Model Tree* window, confirm that the **Revolve 1** feature is pre-selected.

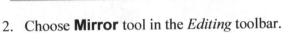

2. Choose **Mirror** tool in the *Editing* toolbar.

3. In the *message area*, the message "*Select a plane or create a datum to mirror about*" is displayed. Pick datum plane **RIGHT**, in the *Model Tree* window, as the plane about which to mirror.

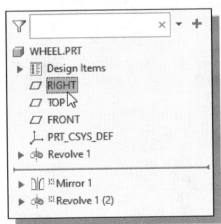

4. Pick **OK** in the *Feature Option Dashboard* to create the solid feature.

➤ Now is a good time to save the model (Quick key: **[Ctrl]** + **[S]**). It is a good habit to save your model periodically, just in case something might go wrong while you are working on the model. You should also save the model after you have completed any major constructions.

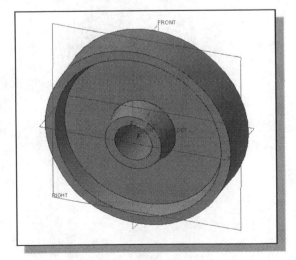

Add a Radial Hole Feature

We will create a circular hole as the next solid feature, which will be used as a ***pattern leader*** for the subsequent pattern feature.

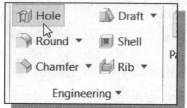

1. In the *Engineering toolbar*, select the **Hole** tool option as shown.

2. Select datum plane **RIGHT** (pick in the *Model Tree* window) as the ***primary*** placement reference.

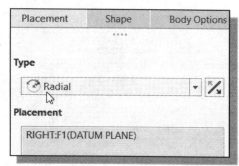

3. In the *Feature Option Dashboard*, click the **Placement** option icon and examine the placement settings.

4. Set the *hole placement* option to **Radial** as shown.

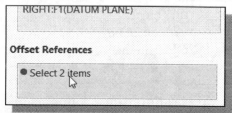

5. Click once inside the **Offset References** list area to activate the selection of secondary references.

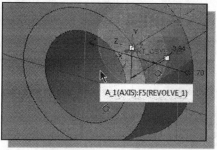

6. Select the **datum axis** located at the center of the model.

7. In the *Model Tree* area, hold down the [**CTRL**] key and select **FRONT** as a secondary reference.

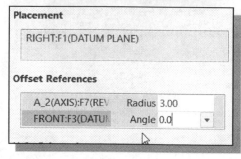

8. In the **Secondary References** list, adjust the **Radius** to **3.0** and **Angle** to **0.0** or **180**. (This hole is now aligned to the *Front plane*.)

❖ Note that the **Radial Hole** option uses a *polar coordinate system* to position the center of the hole feature.

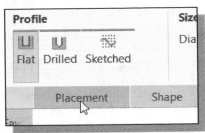

9. In the *Feature Option Dashboard*, click the **Placement** icon to close the **Placement** option list.

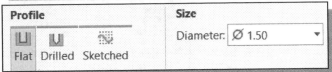

10. In the *Feature Option Dashboard*, set the feature *diameter* to **1.5**.

11. Set the *extrusion* option to **Symmetric**.

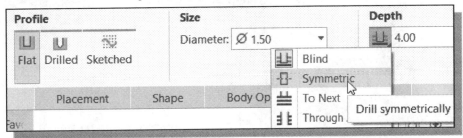

12. Set the *depth* value to **4.0**.

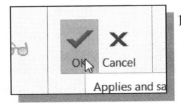

13. Click on the **OK** icon and proceed to create the hole feature.

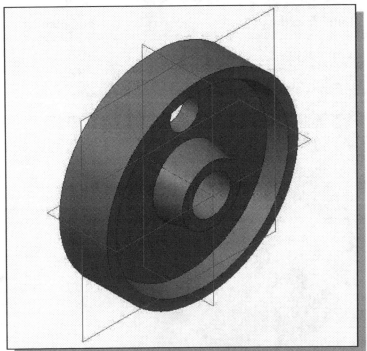

Create a Circular Pattern

1. In the *Editing* toolbar, select **Pattern**. (Note that the **Hole** feature is currently highlighted in the *Model Tree* area, which means it is pre-selected.)

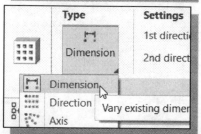

2. Select **Dimension** as the pattern option as shown. (Note that **Axis** is also a feasible option for our design; you are encouraged to repeat this section using the Axis option.)

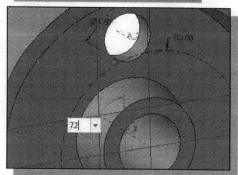

3. Pick the *angle dimension* (**0** or **180**).

4. Enter **72** in the value box as the *increment dimension*.

5. Enter **5** in the pattern number box for the total number of copies.

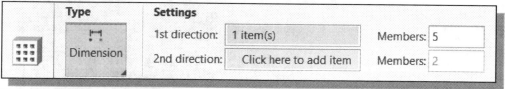

6. Click **OK** to create the pattern.

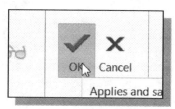

Parametric Relations

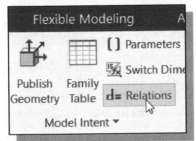

1. In the *Ribbon* area, select the **Tools** tab to switch to the available modeling commands.

2. Pick **Relations** in the *Model Intent* group.

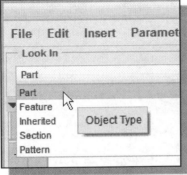

In part files, four main types of **parametric relations** can be established:

- **Part** relations relate different feature parameters to each other in a single model.
- **Feature** relations relate specific parameters within one feature in the model.
- **Section** relations relate specific parameters within one 2D sketch in the model.
- **Pattern** relations relate specific parameters within a pattern in the model.

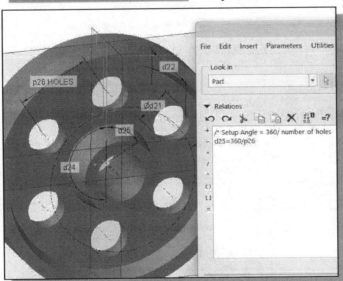

3. Click on one of the patterned holes.

4. Use **d25=360/p26** in the *Relations* window as the relation in between the patterned angle and the number of holes. Use the **smaller angle** appearing on your screen.

❖ The first line above the relation is a *note statement*. The characters *I** is used as a *note statement*.

5. Click on the **Execute/Verify** icon to confirm the entered relation is a valid relation.

6. The message "*Relations have been successfully verified.*" is displayed in the message window. Click **OK** to close the message window.

7. Press and hold down the [**Ctrl**] key and press the [**R**] key once. This is the quick-key combination to **Repaint** the display area.

8. Click **OK** to accept the relation settings and exit the *Relation* window.

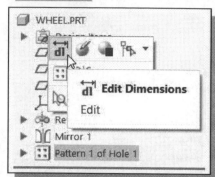

9. In the *Model Tree* area, left-mouse-click the **Pattern** feature and choose **Edit Dimensions** in the option list.

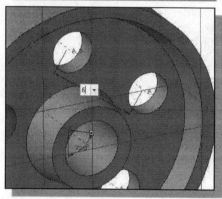

10. On your own, change the ***number of holes*** to **6**.

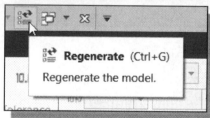

11. Pick **Regenerate** in the *Quick Access toolbar*.

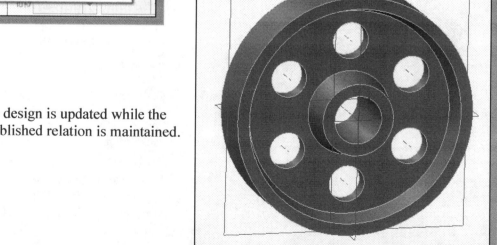

- The design is updated while the established relation is maintained.

Create a Multi-View Part Drawing

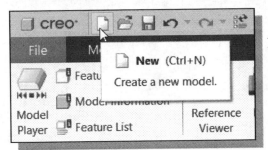

1. Pick the **New** icon in the toolbar. (We can also use the key combination **Ctrl-N** to start a new object.)

2. In the *New* form, select **Drawing** under the **Type** list.

3. Toggle *off* the **Use default template** option.

4. Activate *Use drawing model file name* so the drawing file name is the same as the model file name.

5. Pick **OK** to start a new drawing file.

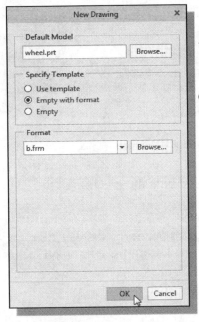

❖ In the *New Drawing* form, the currently active part is automatically selected as the drawing model.

6. In the **Specify Template** option, choose **Empty with format**.

7. Pick **b.frm** in the **Format** list. (Use the **Browse** button to locate the *Creo Parametric format files* if list contains no item.)

8. Pick **OK** to proceed with the drawing layout.

❖ A new window will appear with the title "*WHEEL.*" Note the *b.frm* format file sets the paper size to *B size*, orientation to *Landscape* and a *title block* is placed in the drawing area.

Add a Primary Main View

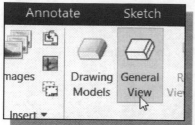

1. In the **Layout** tab area, click on the **General View** button as shown.

2. Click **OK** to accept the *No Combined State* option.

3. The message "*Select CENTER POINT for drawing view*" is displayed in the message area. Pick a location that is toward the left side of the display drawing sheet.

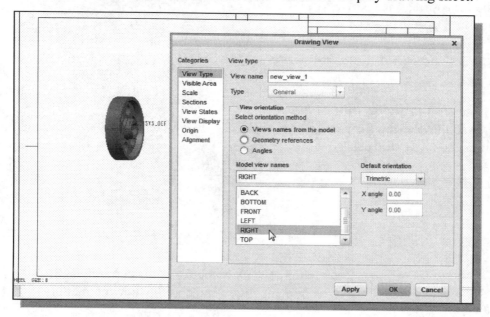

4. Choose **RIGHT** in the **Model view names** list as shown in the figure above.

5. Click on the **Apply** button to adjust the view orientation in the display area.

6. Click on the **OK** button to accept the settings and proceed with the creation of the main view.

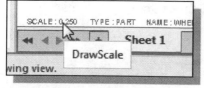

7. Move the cursor to the bottom of the display area and **double click** with the left-mouse-button on **DrawScale**.

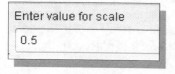

8. In the edit box, enter **0.5** as the new scale factor.

Add a Projected Sectional View

To create a section view, a view projection line is required. The **view projection line**, also known as the **cutting plane line**, defines a cutting line for the section view. Depending upon how a view projection line is drawn, the line can be used to define the type of section view or define a boundary for a partial view. If the view projection line is created outside the base view, it defines a plane from which to project an auxiliary view.

1. Click the **primary view** in the display area to **pre-select** the view.

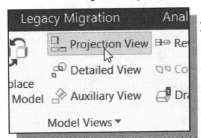

2. In the *Model Views* toolbar on the *Ribbon* toolbar, select the **Projection** view as shown.

3. The message *"Select CENTER POINT for drawing view"* is displayed in the message area. Pick a location that is toward the right side of the primary main view in the display area.

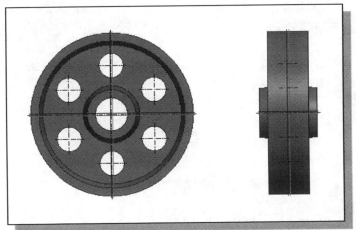

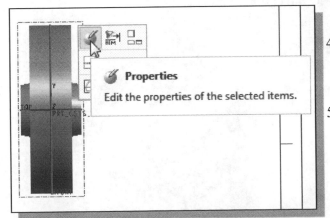

4. Select the projected **side view** by clicking once with the left-mouse-button on the view.

5. Inside the *option list*, select **Properties** list as shown.

6. In the *Drawing View* window, select **Sections** in the **Categories** list.

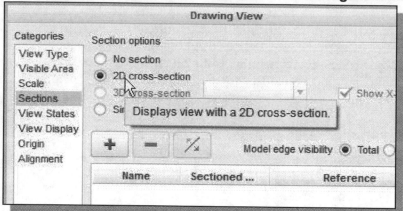

7. Under **Section options**, switch *on* the **2D cross-section** as shown in the figure above.

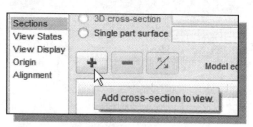

8. Click the **[+]** icon to **Add cross-section to view**.

9. In the *Menu manager*, confirm the setting is set to **Planar Single** as shown. Click **Done** to accept the settings.

10. Enter **A** as the name for cross section.

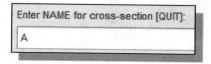

11. Pick the **FRONT** datum plane in the *primary main view* (front view) to use as the cutting plane location.

12. Click **Apply** to accept the current settings.

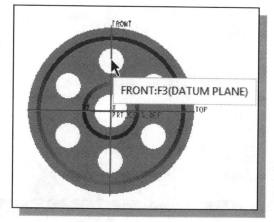

13. Select **View Display** in the Categories list.

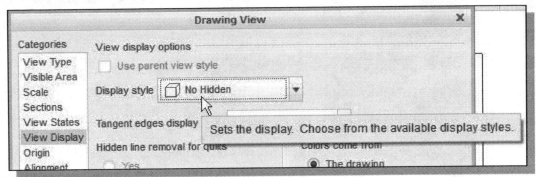

14. Set the Display style to **No Hidden** as shown in the above figure.

 15. Click **Apply** to accept the settings.

16. Click **Cancel** to end the Drawing View Properties command.

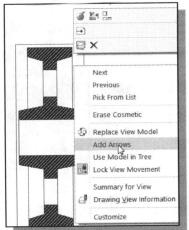

17. On your own, turn *off* all of the datum feature display.

18. Inside the display area, click once with the right-mouse-button on the projection view and select **Add Arrows** in the option list as shown.

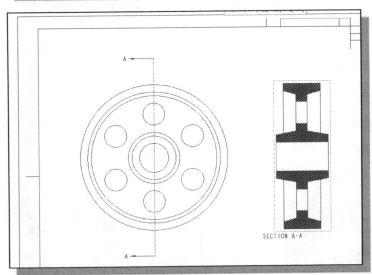

19. Pick the **primary view** by clicking on the view to add the cutting plane line and the two arrows.

20. On your own, set the display of the primary view to **Hidden** as shown.

Modify the Hatch Pattern in Section View

1. Pick the hatch pattern in the **section view** by clicking once on the view.

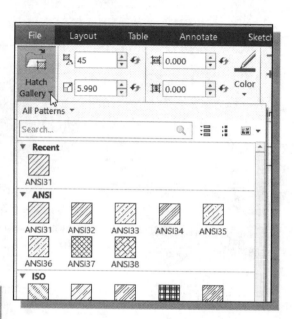

2. In the *dashboard* area, click the **Hatch Gallery** to display the available hatch patterns.

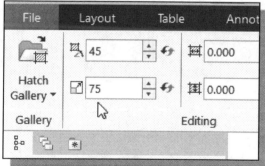

3. Enter **75** as the new Scale value as shown.

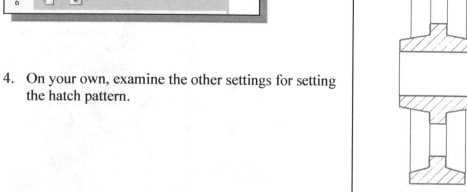

4. On your own, examine the other settings for setting the hatch pattern.

Modify the Overall Drawing Display

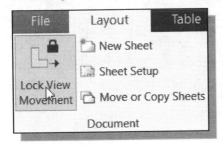

1. In the *Document* area, turn *off* the **Lock View Movement** option as shown. (This option is set to *on* by default, which does not allow the moving of views.)

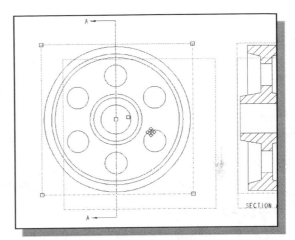

2. Pick the **primary view** by clicking once on the view. Note the cursor is changed to a four-arrow display indicating the moveable directions.

3. **Drag** with the left-mouse-button to reposition the view.

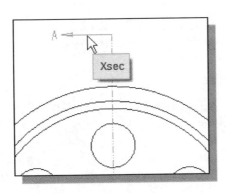

4. Click on the **upper arrow** to select the entity.

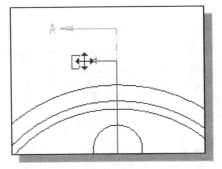

5. Reposition the location of the selected arrow by dragging the arrow with the left-mouse-button.

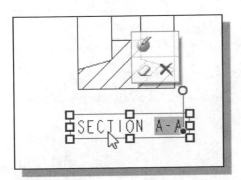

6. Pick the section view label **SECTION A-A**, next to the section view, with the left-mouse-button.

7. **Drag and drop** with the left-mouse-button to reposition the label.

8. On your own, repeat the above steps to adjust the views, and reposition the arrows and text as shown in the figure below.

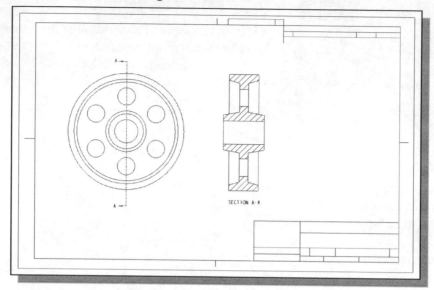

Display Feature Dimensions

By default, feature dimensions are not displayed in 2D views in *Creo Parametric*. We can change the default settings while creating the views or switch on the display of the parametric dimensions using the Show Model Annotations command.

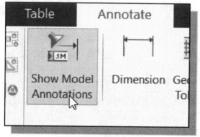

1. In the **Annotate** tab area, select the **Show Model Annotations** command.

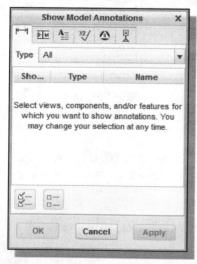

2. Confirm the **Regular Model Dimension** tab is activated in the *Show Model Annotations* window as shown.

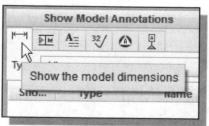

❖ Note that other options are also available under the different tabs.

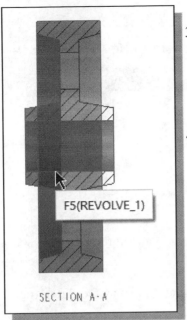

SECTION A-A

F5(REVOLVE_1)

3. In the *message area*, the message *"Select views, components, and/or features for which you want to show annotations"* is displayed. Pick the **main body** feature in the section view. (Note the feature label of the main body is REVOLVE_1.)

4. Click on one of the diameter dimensions and notice the check mark switches **on** as we click on the dimension.

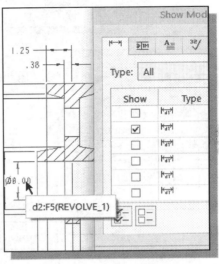

d2:F5(REVOLVE_1)

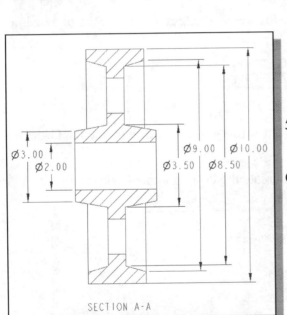

SECTION A-A

5. On your own, display all of the diameter dimensions as shown.

6. Click **OK** to accept the selections.

OK

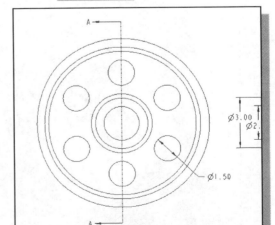

7. On your own, repeat the above steps and pick the **lower hole** on the right side of the hole pattern in the *front view*, and show the hole dimension.

8. Pick the **OK** button to exit the Show Model Annotations command.

Adding Additional Driven Dimensions

1. In the **Annotate** tab of the *Ribbon* toolbar, select **Dimension - New References**.

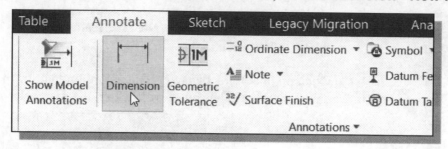

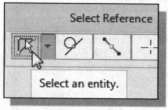

2. In the Select Reference menu, confirm the option is set to **Select an Entity**.

❖ Note that we add dimensions the same way we did in the *2D Sketcher*.

3. Pick the **two inside vertical edges** and create the dimension as shown.

Create this dimension.

4. Inside the display area, click once with the **middle-mouse-button** to end the Create Dimension command.

5. On your own, repeat the above steps and create the desired dimensions in the 2D views.

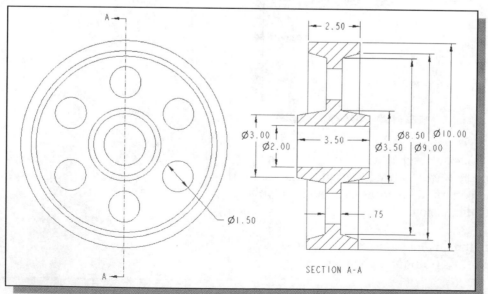

Adjust Dimension Appearances

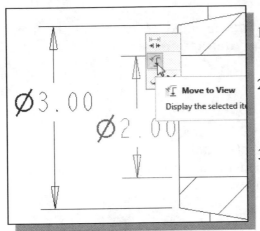

1. Pick the **dimension text** of the center hole (diameter 2.0) in the *section view* as shown.

2. Inside the *display area*, press down with the right-mouse-button to bring up the **option menu**.

3. Select **Move Item to View** in the option list.

4. The message *"Select model view or window"* is displayed in the message area. Pick the primary view, the *front* view, in the display area.

• The dimension is moved to the front view.

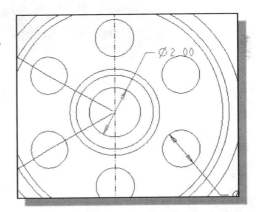

5. On your own, adjust the dimensions so that they appear as shown in the figure below.

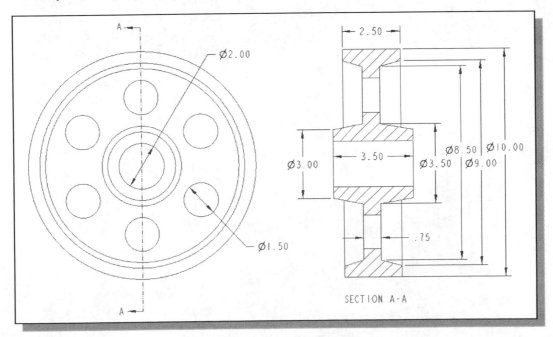

Display Center Axes

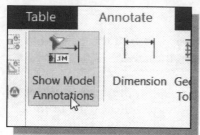

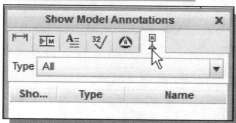

1. In the **Annotate** tab of the *Ribbon* toolbar, select the **Show Model Annotations** command.

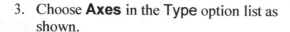

2. Select the **Axes and Set Datum objects** tab in the *Show Model Annotations* window as shown.

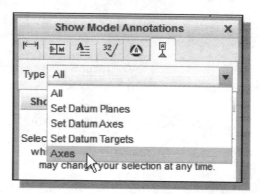

3. Choose **Axes** in the Type option list as shown.

4. In the message area, the message "*Select model view or window*" is displayed. Pick the large center **hole feature** in the *section view*.

5. Hold down the **[Ctrl]** key and pick the other two hole features to display the center axes.

6. Pick the **Check All** option to display all center axes.

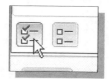

7. Pick the **OK** button to exit the Show Model Annotations command.

Adjusting the Length of Centerlines

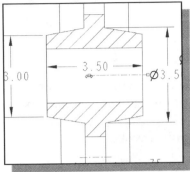

1. Pick the **axis** that is through the center of the *section view* as shown.

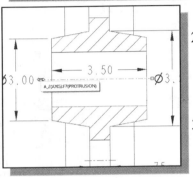

2. Two small **markers** appear on the centerline. Use the left-mouse-button to drag one of the markers to lengthen/ shorten the centerline.

3. On your own, repeat the above steps and adjust the **length** of the other displayed centerlines.

❖ Note that the same procedure is used to adjust the location of any dimensions as well as any of the dimension related entities.

Create Additional Centerlines

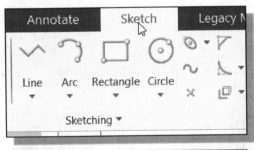

1. Switch to the **Sketch** tab in the *Ribbon* toolbar as shown.

- Note that the Sketch option allows us to create 2D entities that exist only in *Drawing* mode.

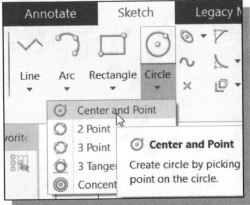

2. In the *Sketching* toolbar, select the **Circle: Center and Point** command.

- By default, Creo will use the XY Cartesian coordinate system to define the geometry location.

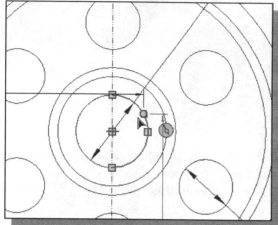

3. Move the cursor on top of one of the circular features of the main body; note that Creo will display several snap reference points as shown.

4. Select the **center snap reference point** to place the center of the circle as shown.

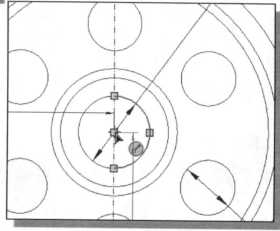

5. Move the cursor on top of one of the patterned **holes** to display the associated snap reference points.

6. Select the **center point** of the *Snapping Reference* we just identified as shown.

7. Click once with the middle mouse button, or press the [Esc] key once, to end the circle command.

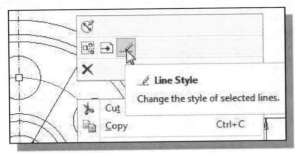

8. Inside the *display area*, click once with the **right-mouse-button on the circle we just created** to bring up the option menu.

9. Select **Line Style** in the option list as shown.

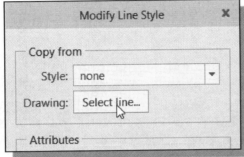

10. In the *Copy From* section of the Modify Line Style window, click **Select lines** as shown.

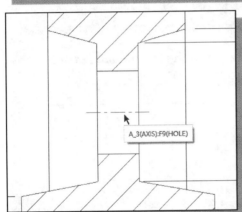

11. Select one of the center lines in the section view.

12. Click on the **Apply** button to accept the settings.

13. Click on the **Close** button to exit the **Modify Line Style** option.

- The *Copy From* option allows us to set the line type matching existing objects so that the line style is consistent for the drawing.

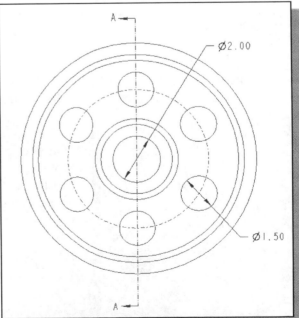

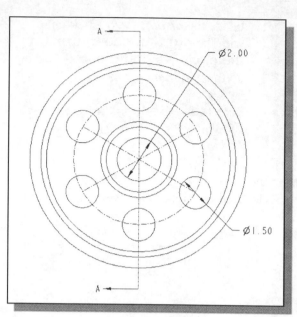

14. On your own, repeat the above steps and create the two centerlines connecting the hole pattern as shown.

15. On your own, single left mouse click & drag the endpoints to extend the lines.

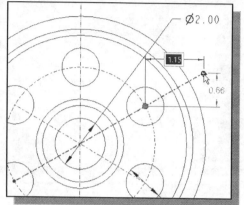

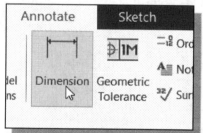

16. In the **Annotate** tab, activate the **Dimension** command as shown.

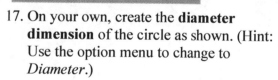

17. On your own, create the **diameter dimension** of the circle as shown. (Hint: Use the option menu to change to *Diameter*.)

• The drawing is set to half scale and the size of the created circle reflects that.

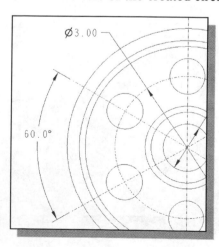

18. On your own, create the **Angle dimension** of the two patterned circles on the left as shown. (Hint: Hold the [Ctrl] key when selecting the two centerlines.)

Overriding a Dimension Value

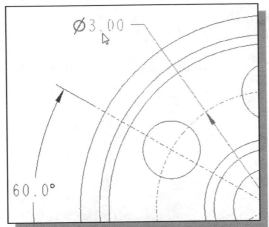

1. Select the **diameter dimension** as shown.

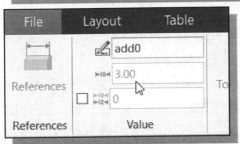

2. In the *Value area*, note the default value is 3.00 and the variable name is add0 as shown.

 • Note that we can adjust/change many settings of the selected dimension.

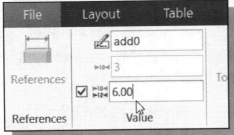

3. Switch *on* the **Override Value** option.

4. Enter **6.0** as the new dimension value.

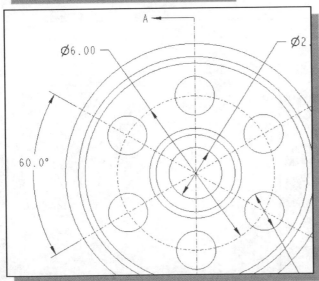

5. Click inside the graphics area to accept the settings and exit the modify mode.

 • The dimension for the centerline circle has been updated to reflect the drawing scale used.

Associative Functionality – Modifying Feature Dimensions

Creo Parametric's associative functionality allows us to change the design at any level, and the system reflects the change at all levels automatically.

1. **Double-click** with the left-mouse-button on the dimension text (**Ø 1.5**) of the hole pattern to enter the *Modify mode*.

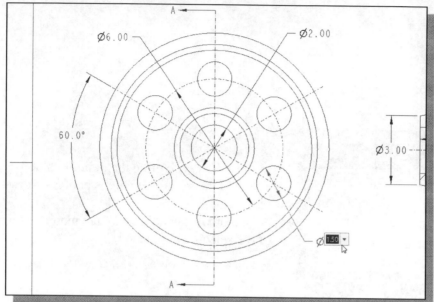

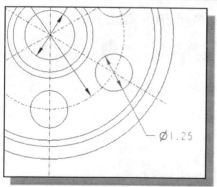

2. In the *Dimension Properties* window, enter **1.25** as the new **Nominal Value**.

3. Press the **Enter** key to accept the new value.

4. Pick **Regenerate** in the *Quick Access toolbar*. Note the hole size in the front view has been adjusted.

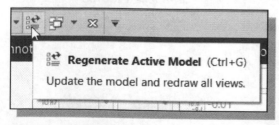

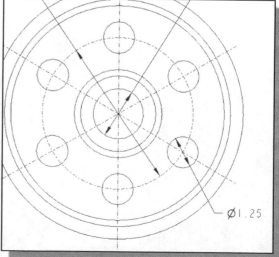

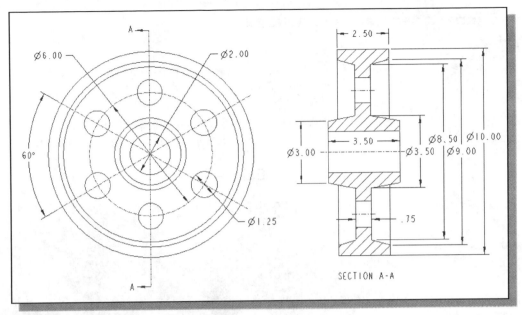

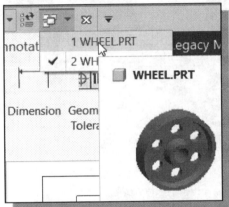

5. Select **WHEEL.PRT** in the *Quick access toolbar* as shown to switch to the 3D solid model.

6. On your own, display the hole dimension and confirm it has been updated.

➢ Note the modification was done in the *Drawing* mode; the feature in the solid model is also updated to the new dimension value.

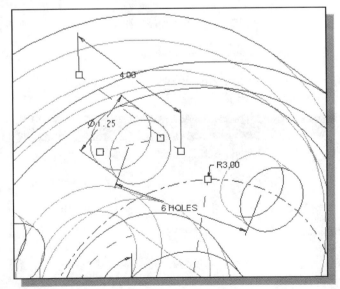

Bi-Directional Associative Functionality

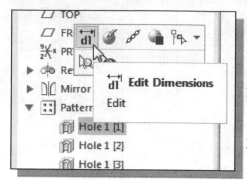

1. Expand the **Pattern** listing in the *Model Tree* window.

2. Click with the **left-mouse-button** on one of the hole features to bring up the option list.

3. Pick **Edit Dimensions** in the option list.

4. Use the drag point and **Modify** the hole diameter to **1.40** as shown.

5. Pick **Regenerate** in the *Quick Access* toolbar.

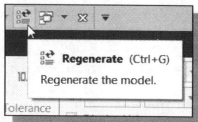

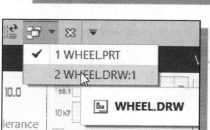

6. Select **WHEEL.DRW** in the *Quick access toolbar* to switch back to the 2D drawing as shown.

❖ Notice the drawing is updated automatically as the part was modified in the *Part Modeling* mode.

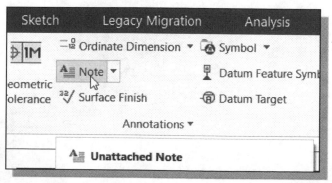

7. On your own, use the **Create Note** command to fill in the title block. Also, add the additional centerlines and a 3D isometric view so that the completed drawing appears as shown on the next page.

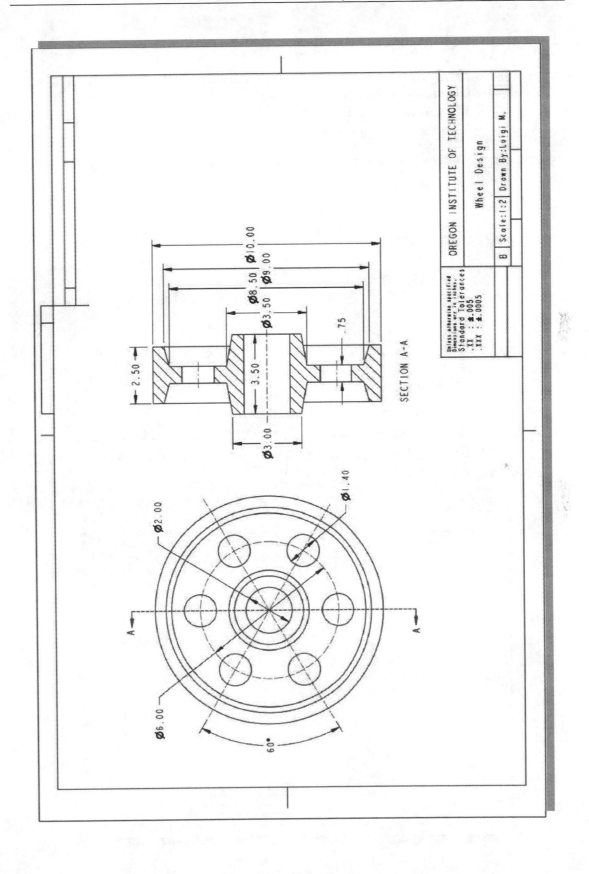

SECTION A-A

Ø10.00
Ø9.00
Ø8.50
Ø3.50
.75
2.50
3.50
Ø3.00

Ø2.00
Ø1.40
Ø6.00
60°
A
A

OREGON INSTITUTE OF TECHNOLOGY
Wheel Design
B   Scale:1:2   Drawn By:Luigi M.

Unless otherwise specified
Dimensions are in inches.
Standard Tolerances
.XX : ±.005
.XXX : ±.0005

Create a Custom Creo Title Block

A Creo *Format File* is a template file containing a drawing title block. Several options are available to create a custom title block in Creo; for example, we can import a 2D drawing made from AutoCAD or use a Creo 2D Sketch file.

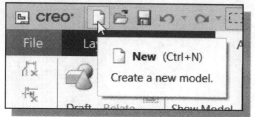

1. Pick the **New** icon in the toolbar. (We can also use the key combination **Ctrl-N** to start a new object.)

2. In the *New* form, select **Sketch** under the **Type** list.

3. Enter **A-H-Title** as the *Sketch file* Name.

4. Pick **OK** to start a new sketch file.

5. On your own, create the 2D drawing as shown. Note that all line segments are either horizontal or vertical.

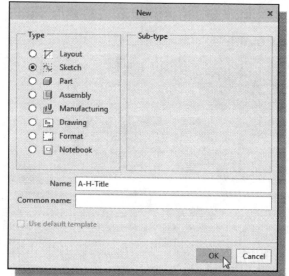

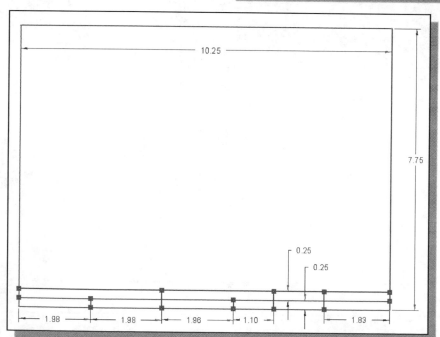

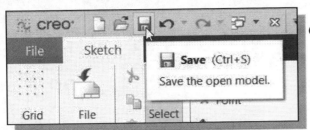

6. Click **Save** to save the *Sketch* file.

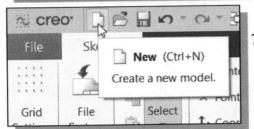

7. Click the **New** icon in the *Quick Access* toolbar.

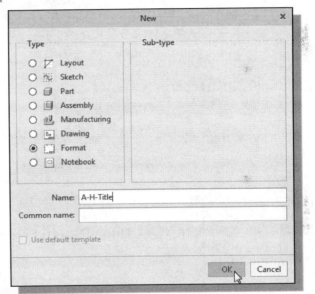

8. In the *New* form, select **Format** under the **Type** list.

9. Enter *A-H-Title* as the *format file* Name.

10. Pick **OK** to start a new format file.

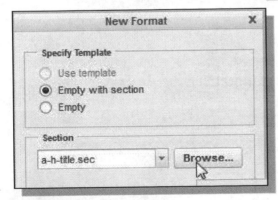

11. In the *New Format* form, set the **Specify Template** option to **Empty with Section**.

12. Use the *Browse* option and select the *A-H-Title.sec* file we just saved.

13. Click **OK** to proceed with the new file.

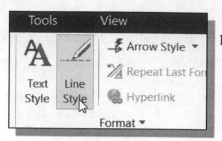

14. Select **Line Style** in the *Format toolbar* as shown.

15. Use a selection window and select all of the line segments in the graphics area.

16. Click once with the **middle-mouse-button** to accept the selection.

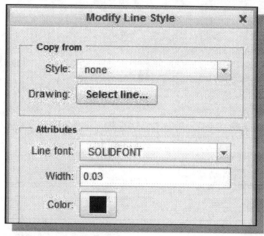

17. In the Attributes section, set the *Line Width* to **0.03** as shown.

18. Click **Apply** to adjust the selected line segments.

19. Click **Close** to exit the Modify Line Style form.

20. Click **Done/Return** to end the *Modify Line Style* command.

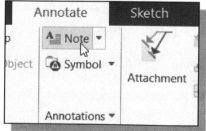

21. Click the **Annotate** tab and select the **Note** command.

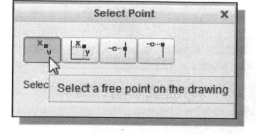

22. Confirm the select point option is set to **Select a free point** as shown.

23. On your own, place the text box near the lower left corner of the top box of the title block as shown.

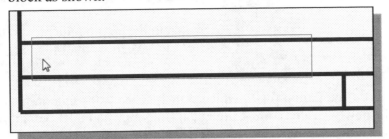

24. Set the font style to **Blueprint MT Bold** and text size to **0.125** as shown.

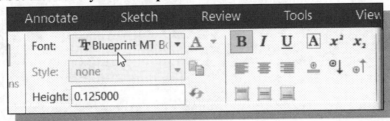

25. Enter your company or school name.

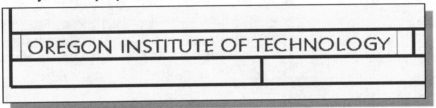

26. Hit the **Escape key** once to end the input.

27. Use the left-mouse-button, **drag and drop**, to reposition the text box.

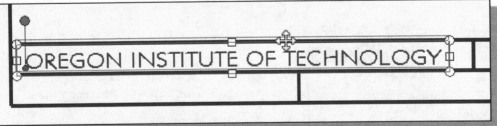

28. On your own, create the additional title block labels as shown.

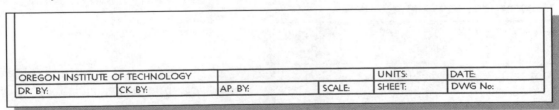

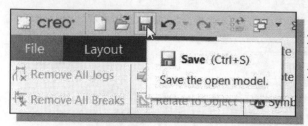

29. Click **Save** to save the *Format* file.

30. Switch to the proper folder and click **Save** to save the format file.

➢ Note that the default Creo format files are placed in the Creo 9 .0 common files folder. Placing the newly created format file in the Creo formats folder will allow this template to be available when using the **New Drawing** option.

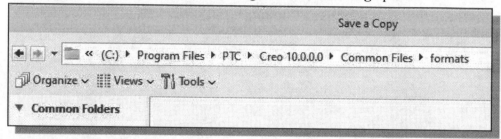

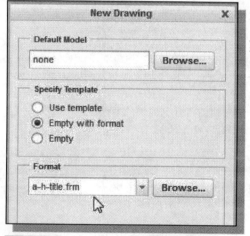

➢ The format template can now be used when a new drawing is created. If the format file does not show up in the list, you can also use the *Browse* option to select the ***A-H-Title.frm*** file.

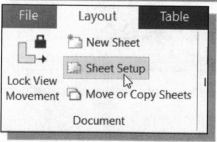

➢ To insert the template in an existing Creo drawing, use the **Sheet Setup** command.

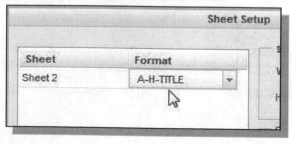

➢ Click on the **Format list** to expand and select from the available formats.

Create a Drawing Template with 2D Views and Title Block

A Creo template file can greatly reduce the amount of repetitive steps and make our work much more efficient. Using template files also helps us maintain consistent design and drafting standards. In this section, we will illustrate the procedure to set up a drawing template file that contains specific drawing settings, standard orthographic views and a title block.

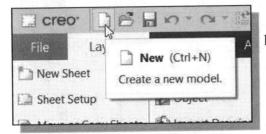

1. Pick the **New** icon in the toolbar. (We can also use the key combination **Ctrl-N** to start a new object.)

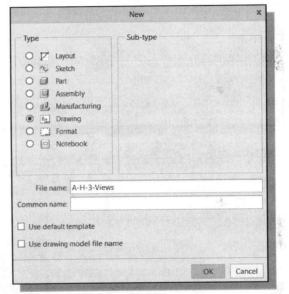

2. In the *New* form, select **Drawing** under the *Type* list.

3. Enter **A-H-3-Views** as the *Drawing Name*.

4. Uncheck both the **Use Default Template** and the **Use drawing model filename** options.

5. Pick **OK** to proceed to the next step.

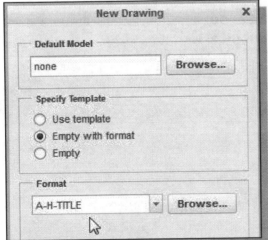

6. In the *New Drawing* form, set the *Default Model* to **none**.

7. In the *New Drawing* form, set the **Specify Template** option to **Empty with format**.

8. Select the **A-H-Title.frm** file as shown. If necessary, use the *Browse* option to find the file.

9. Click **OK** to proceed with the new file.

- Note the A-H-Title format is displayed on the screen.

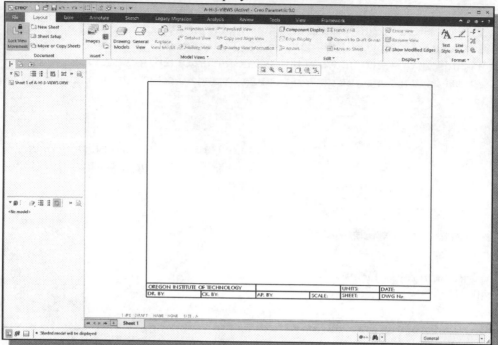

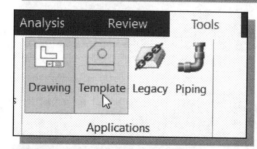

10. Click on the **Tools** tab and set switch to the **Template** mode as shown.

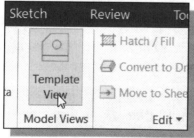

11. Click on the **Layout** tab and select the **Template View command** as shown.

12. Enter **Front** as the *Template View* name.

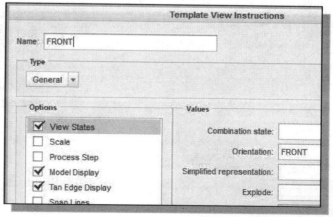

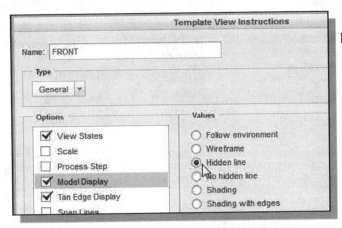

13. In the *options list*, select **Model Display** and set the display option to **Hidden Line** as shown.

14. In the *options list*, select **Tan Edge Display** and set the display option to **None** as shown.

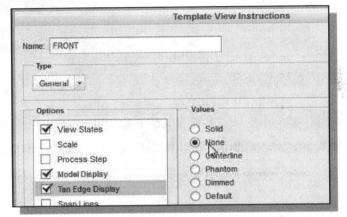

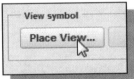

15. Click **Place view** near the bottom of the form.

16. Click a point near the lower left corner of the title block to place the view. Note that the view can be repositioned later.

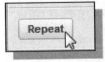

17. Click **Repeat** to start another *View Template*.

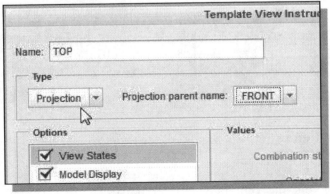

18. Enter **TOP** as the *Template View* name.

19. Choose *Projection* as the View type and set the Projection parent to **FRONT**.

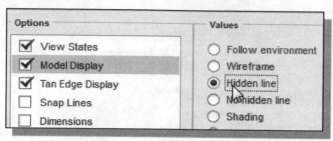

20. In the *options list*, select **Model Display** and set the display option to **Hidden Line** as shown.

21. In the *options list*, select **Tan Edge Display** and set the display option to **None** as shown.

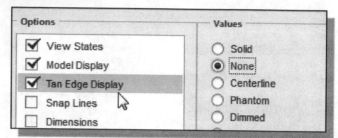

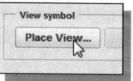

22. Click **Place view** near the bottom of the form.

23. Place the view by clicking a point above the front view.

24. Click **Repeat** to start another *View Template*.

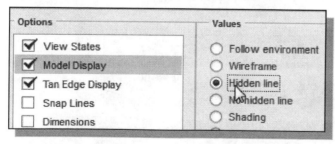

25. Enter **RIGHT** as the *Template View* name.

26. Choose **Projection** as the *View type* and set the *Projection parent* to **FRONT**.

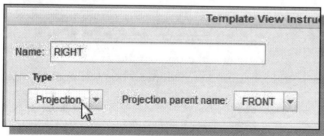

27. On your own, set the **Model Display** and set the **Tan Edge Display** options the same as the TOP view.

28. Click **Place view** near the bottom of the form.

29. Place the view by clicking a point to the right of the front view. Click **OK** to end the Template view command.

- We will also set up some of the dimensioning related properties to match better with the ANSI standard.

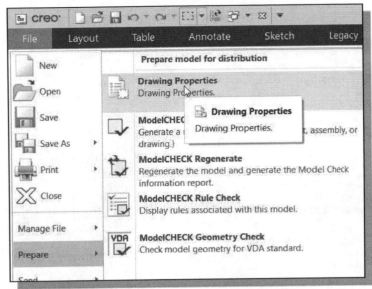

30. In the *File menu*, select **Prepare → Drawing properties** as shown.

31. In the *Drawing Properties* window, select **Change** to enter the *Detail Options* setting.

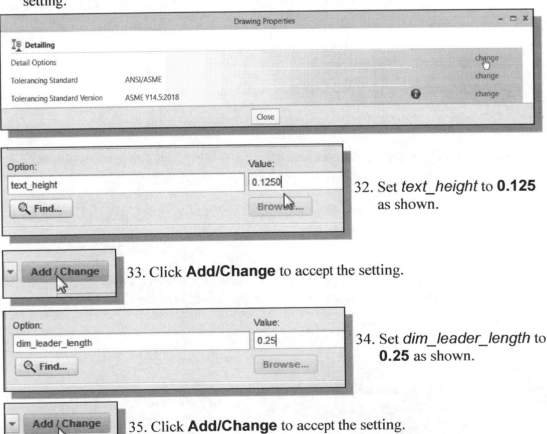

32. Set *text_height* to **0.125** as shown.

33. Click **Add/Change** to accept the setting.

34. Set *dim_leader_length* to **0.25** as shown.

35. Click **Add/Change** to accept the setting.

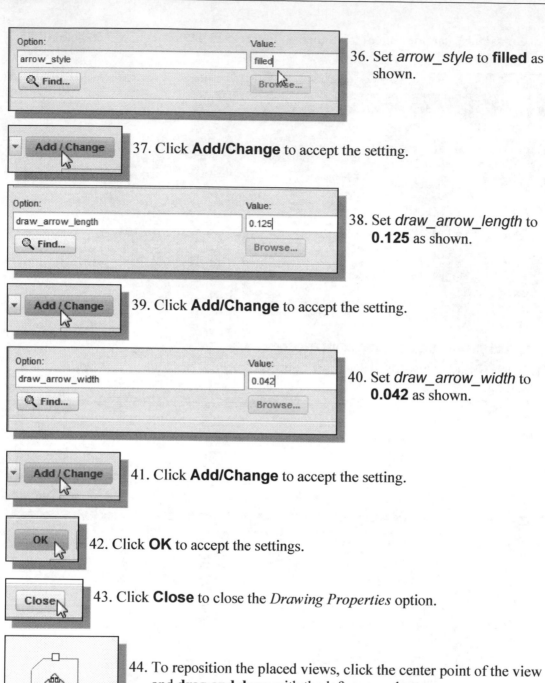

36. Set *arrow_style* to **filled** as shown.

37. Click **Add/Change** to accept the setting.

38. Set *draw_arrow_length* to **0.125** as shown.

39. Click **Add/Change** to accept the setting.

40. Set *draw_arrow_width* to **0.042** as shown.

41. Click **Add/Change** to accept the setting.

42. Click **OK** to accept the settings.

43. Click **Close** to close the *Drawing Properties* option.

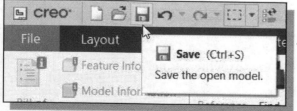

44. To reposition the placed views, click the center point of the view and **drag and drop** with the left-mouse-button.

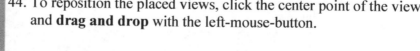

45. Click **Save** in the *Quick Access* toolbar.

46. Place a copy of the *A-H-3-Views.drw* file in the Creo 9 templates folder, or a folder of your choice.

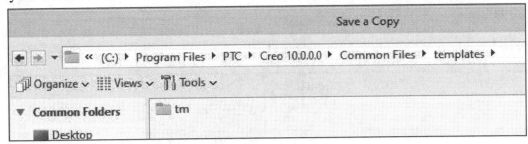

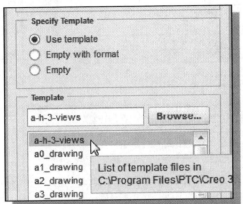

➢ The *A-H-3-Views* template can now be selected in the *template list* when a **new drawing** is created.

• Note that the three orthographic views are based on the view angle defined with the FRONT view. The FRONT view needs to be pre-defined in the model mode for the template to work properly. The **Reorient** command can be used to re-define the FRONT view if necessary.

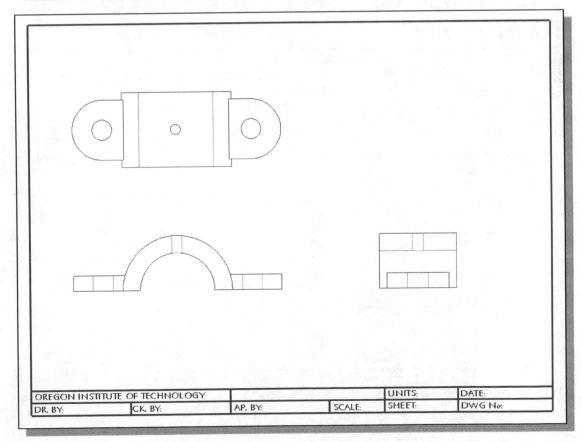

Additional Title Blocks

Drawing Paper and Border Sizes

The standard drawing paper sizes are as shown in the below tables. The edges of the title block border are generally 0.5 ~ 1 inches or 10~20 mm from the edges of the paper.

American National Standard	Suggested Border Size
A – 8.5" X 11.0"	A – 7.75" X 10.25"
B – 11.0" X 17.0"	B – 10.0" X 16.0"
C – 17.0" X 22.0"	C – 16.0" X 21.0"
D – 22.0" X 34.0"	D – 21.0" X 33.0"
E – 34.0" X 44.0"	E – 33.0" X 43.0"

International Standard	Suggested Border Size
A4 – 210 mm X 297 mm	A4 – 190 mm X 276 mm
A3 – 297 mm X 420 mm	A3 – 275 mm X 400 mm
A2 – 420 mm X 594 mm	A2 – 400 mm X 574 mm
A1 – 594 mm X 841 mm	A1 – 574 mm X 820 mm
A0 – 841 mm X 1189 mm	A0 – 820 mm X 1168 mm

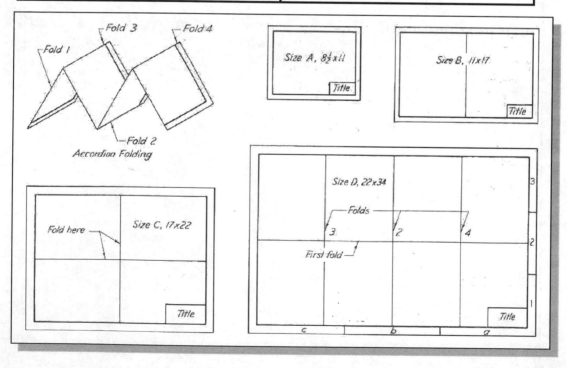

- **English Title Block** (For A size paper, dimensions are in inches.)

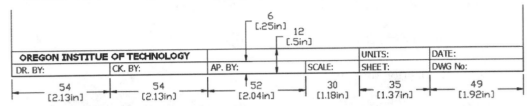

- **Metric Title Block** (For A4 size paper, dimensions are in mm.)

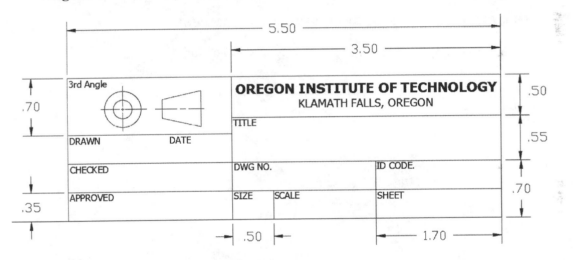

- **English Title Block** (Dimensions are in inches.)

- **Metric Title Block** (Dimensions are in mm.)

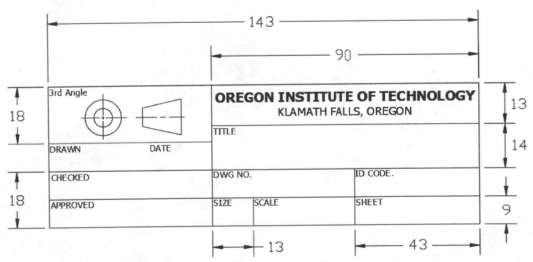

Review Questions:

1. What does *Creo Parametric's bi-directional associative functionality* allow us to do?

2. What essential element is common to revolved features?

3. How do you move a view on the *drawing sheet*?

4. How do you move a dimension from one view to another?

5. What is the difference between *erasing* and *deleting* a dimension?

6. How do you reposition dimensions?

7. What are the required elements in order to generate a sectional view?

8. What is the first feature in a pattern called in *Creo Parametric*?

9. What is a "radial hole" in *Creo Parametric*?

10. Create sketches showing the steps you plan to use to create the two models shown on the next page:

Ex.1)

Ex.2)

Exercises: (All dimensions are in inches.)

1. Shaft Support

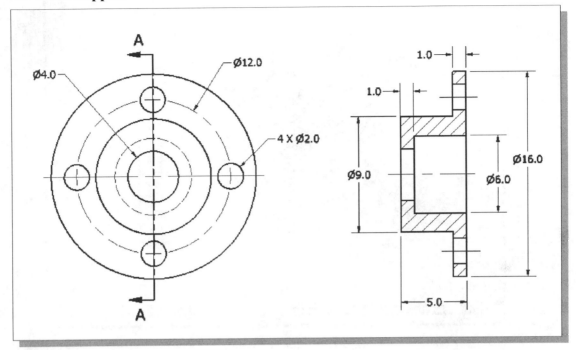

2. Ratchet Plate (Thickness: 0.25 inch)

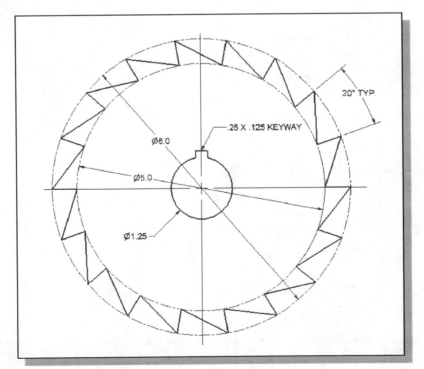

3. Geneva Wheel

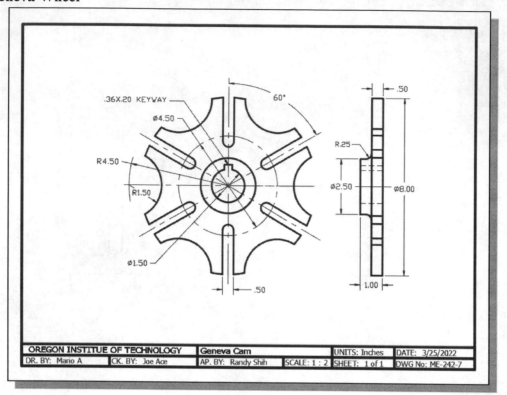

OREGON INSTITUE OF TECHNOLOGY		Geneva Cam		UNITS: Inches	DATE: 3/25/2022
DR. BY: Mario A	CK. BY: Joe Ace	AP. BY: Randy Shih	SCALE: 1 : 2	SHEET: 1 of 1	DWG No: ME-242-7

4. Support Mount

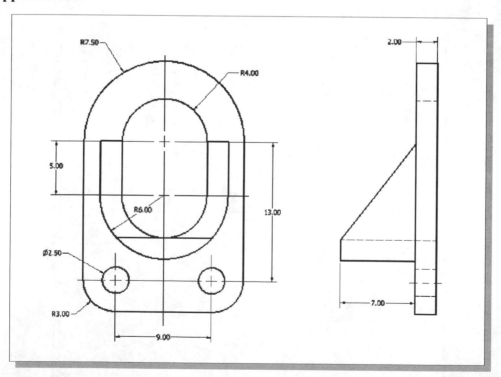

5. Hub

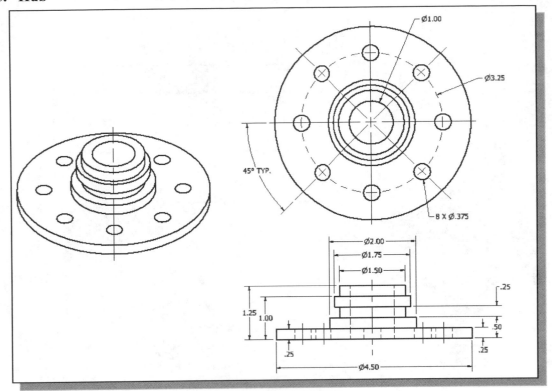

6. Circular Spacer

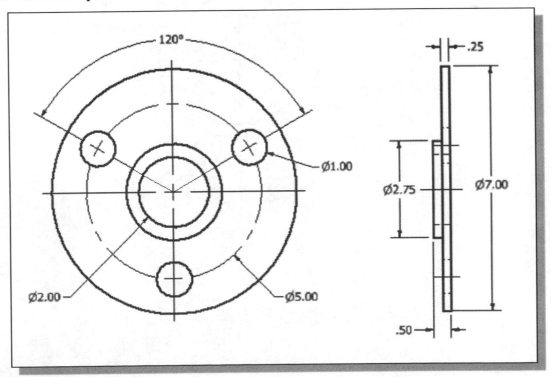

7. Bearing Seat (Create a section view in the associated multi-view drawing.)

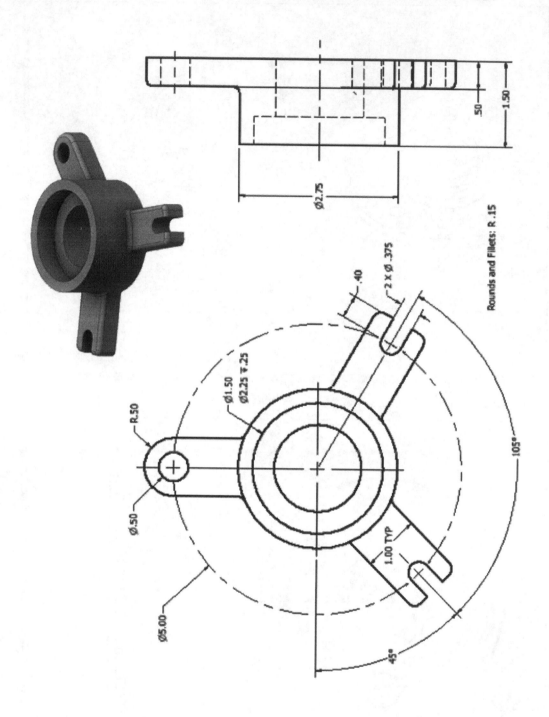

Rounds and Fillets: R .15

Ø2.75

.50

1.50

.40

2 X Ø .375

105°

R.50

Ø1.50

Ø2.25 ▼ .25

Ø.50

Ø5.00

1.00 TYP

45°

Chapter 9
Three-Dimensional Construction Tools

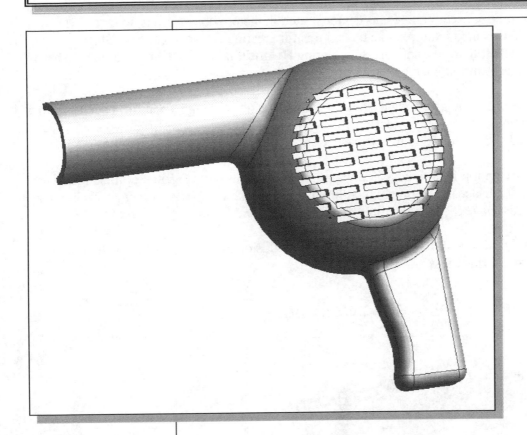

Learning Objectives

- ♦ **Understand the Parent/Child relations between features**
- ♦ **Create Swept Features**
- ♦ **Set up Multiple Work planes**
- ♦ **Create Lofted Features**
- ♦ **Use the Shell Command**
- ♦ **Create 3D Rounds & Fillets**

Introduction

Creo Parametric provides an assortment of three-dimensional construction tools to make the creation of solid models easier and more efficient. **Extrude** and **Revolve** are the two most commonly used methods, as demonstrated in the previous lessons, to create 3D models. In this next example, the procedures for creating **Sweep** features, **Blend** features, **Shell** features, and also for creating **Rounds** and **Fillets** are examined. These features are common characteristics of molded designs.

The **Sweep** option is defined as moving a cross-section through a path in space to form a three-dimensional object. To define a *sweep feature* in *Creo Parametric*, two sections are required: the trajectory and the cross-section.

The **Blend** option is defined as joining several planar sections together at the edges, with transitional surfaces, to form a three-dimensional object. In *Creo Parametric*, each planar section must have the same number of segments/vertices.

The **Shell** option is defined as hollowing out the inside of a solid, leaving a shell of specified wall thickness.

A Thin-Walled Part: *Dryer Housing*

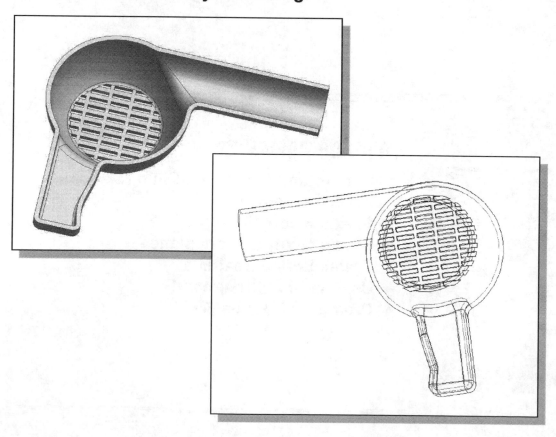

Modeling Strategy

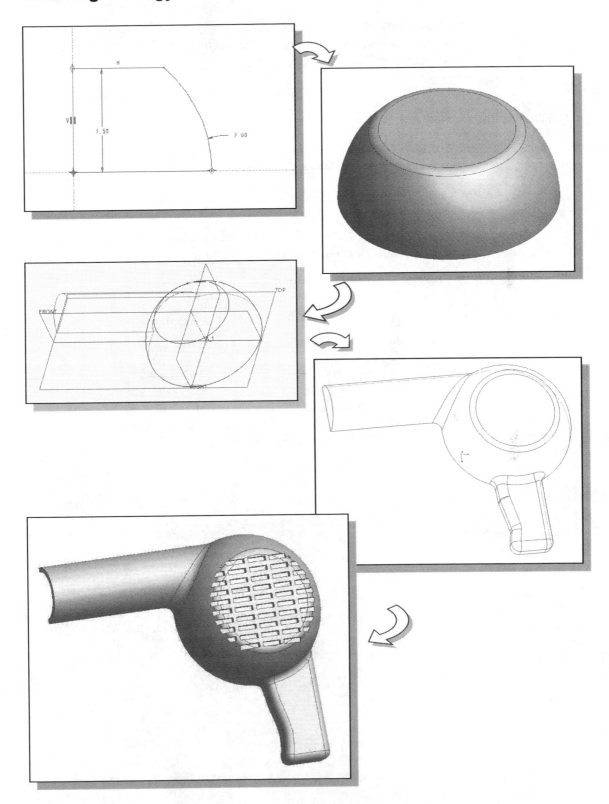

Starting *Creo Parametric*

1. Select the **Creo Parametric** option on the *Start* menu or select the **Creo Parametric** icon on the desktop to start *Creo Parametric*. The *Creo Parametric* main window will appear on the screen.

2. Click on the **New** icon, located in the *Ribbon toolbar* as shown.

3. On your own, start a new solid part file using **Dryer_Housing** as the part Name.

4. Confirm the **Use default template** option is turned *on* so that the system is using the *Creo Parametric* default (**Inch-Ibm-Second**) settings.

5. Click on the **OK** button to accept the settings.

Entity Display Configurations

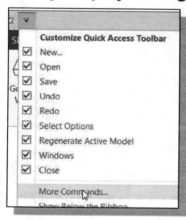

1. Click on the **triangle** icon in the *Quick Access* toolbar. The *Customize Quick Access Toolbar* option list is displayed. Click on the **More commands** option; the *Creo Parametric* options dialog box appears, which allows us to adjust various settings.

2. Choose the **Entity Display** option.

3. On your own, switch *on* the **Show datum plane tags** and the **Show datum axes tags** options as shown.

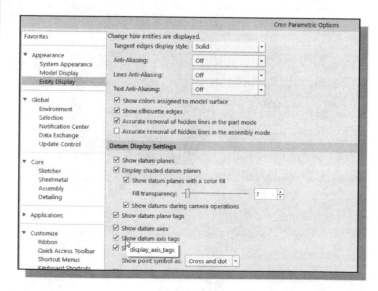

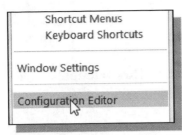

4. In the *Configuration Editor*, set the **default_dec_places** to **2** as shown. This setting controls the number of digits displayed in the dimensions. (Use the **Find** function if necessary.)

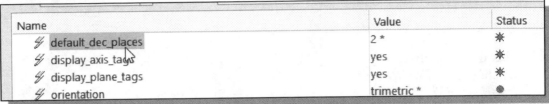

 5. Click **OK** to accept the settings.

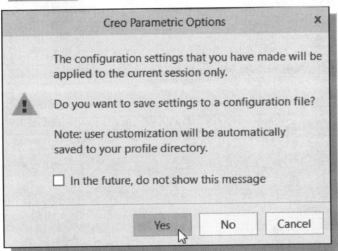

6. In the *Creo Parametric Options* dialog box, click the **Yes** button to proceed with saving the settings.

7. Click **OK** to save the settings in the config.pro file as shown.

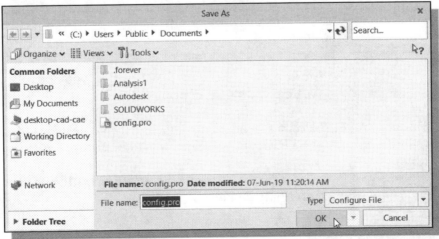

Create the Base Feature

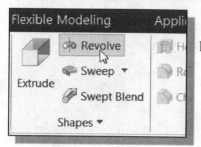

1. In the *Shapes* toolbar, select the **Revolve** tool option as shown.

2. Click the **Placement** option and choose **Define** to create a new *internal sketch*.

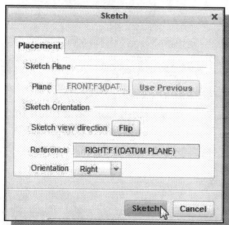

3. Set up the **FRONT** datum plane as the sketch plane with the **RIGHT** datum plane facing the **right** edge of the computer screen, and pick **Sketch** to enter the *Sketcher* mode.

❖ Note the default selection of the two datum planes, **RIGHT** and **TOP**, as the references for the 2D sketch.

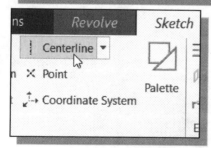

4. In the *Sketching* toolbar, select **Create Centerline**.

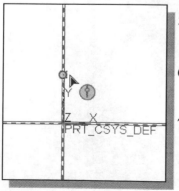

5. Create a vertical line by first selecting a point on the **vertical axis** (datum plane RIGHT).

6. Select a second point aligned to the vertical axis (datum plane RIGHT) to create a **vertical centerline**.

7. Click once with the **middle-mouse-button** to end the Centerline command.

8. On your own, create and modify the 2D section (consisting of an arc, two horizontal lines, and one vertical line) as shown below. Note that the arc *radius* is **2.0**, and the *vertical distance* is **1.5**.

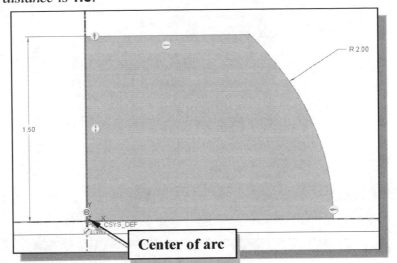

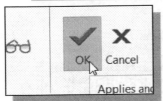

Center of arc

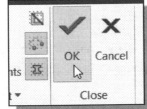

9. Click on the **OK** icon to accept the completed section.

Create the Revolved Feature

1. In the **Revolve** option menu, confirm the revolve angle is set to **360**.

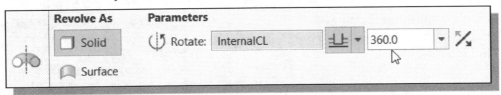

2. Pick **OK** in the *Feature Option Dashboard* to create the solid feature.

3. Click on the **Saved View List** icon and choose **Standard Orientation** to view the completed 3D solid feature in the direction of the preset standard *model* view.

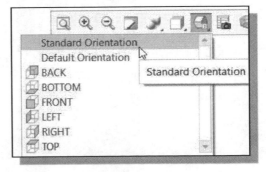

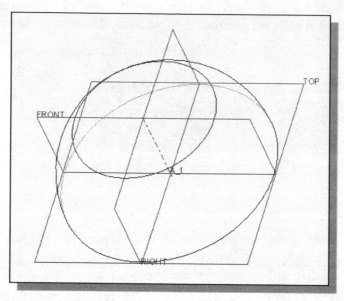

Create the second Feature

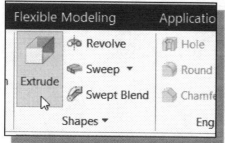

1. In the *Shapes* toolbar, select the **Extrude** tool option as shown.

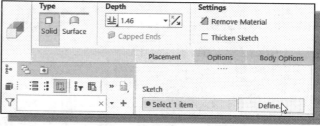

2. Click the **Placement** option and choose **Define** to begin creating a new *internal sketch*.

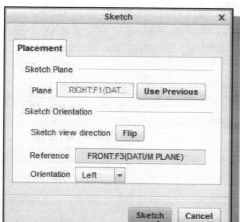

3. On your own, set up the **RIGHT** datum plane as the sketch plane with the **FRONT** datum plane facing the **left** edge of the computer screen and pick **Sketch** to enter the *Sketcher* mode.

❖ Notice the use of the two datum planes, **FRONT** and **TOP**, as the references for the 2D sketch.

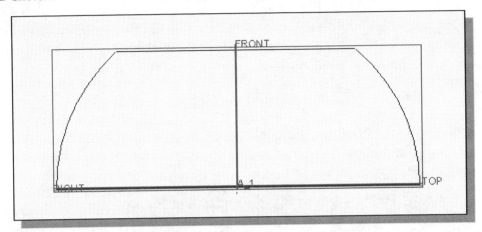

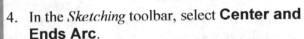

4. In the *Sketching* toolbar, select **Center and Ends Arc**.

5. Place the center of the arc to the right of the vertical reference, and place both ends of the arc aligned to datum plane **TOP**, as shown in the below figure.

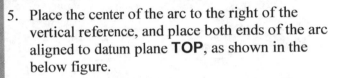

6. Pick **Line** in the *Sketching* toolbar and create a line connecting the two ends of the arc to form a closed region.

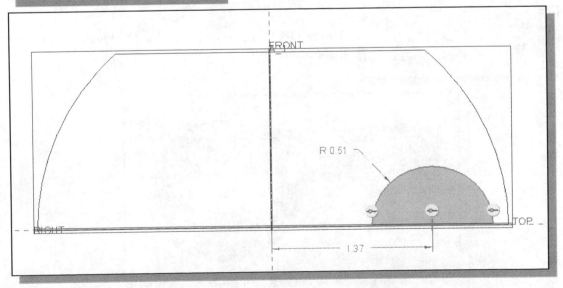

7. On your own, adjust the dimensions as shown in the below figure.

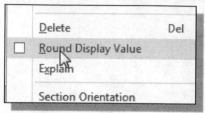

8. Bring up the *option menu* on the linear dimension and uncheck the Round Display Value to display 3 digits after the decimal point.

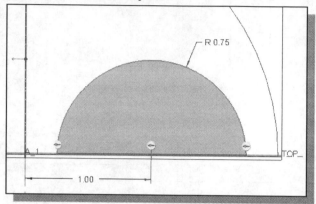

9. Click on the **OK** icon to accept the completed *extrude section*.

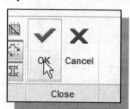

10. **Uncheck** the ***Remove material*** option and set the extrude *distance* to **5.5** and change the extrusion direction by clicking once on the **flip direction** button.

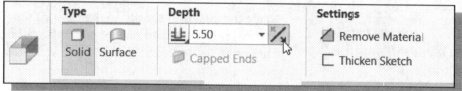

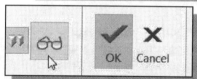

11. Pick **Preview** in the *Feature Option Dashboard* to examine the solid feature.

12. On your own, use the *Dynamic Viewing* functions to view the 3D model.

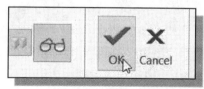

13. Pick **OK** in the *Feature Option Dashboard* to create the solid feature.

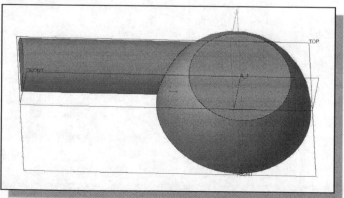

Create a Blend Feature

A *blend* feature is a series of two-dimensional sections that are joined together at the edges with transitional surfaces to form a continuous solid.

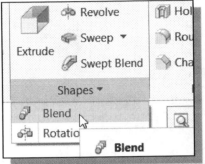

1. In the **Shapes toolbar**, select **Shapes→Blend**.

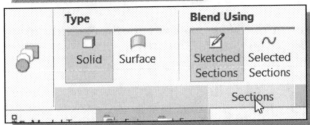

2. In the Dashboard, click on the **Sections** tab to display the available options.

3. In the Sections tab, select **Define** to start a new sketch.

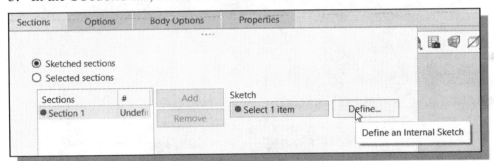

4. Pick datum plane **FRONT** as the sketching plane and set the direction arrow, by clicking on it, so that it is pointing toward the back as shown in the figure.

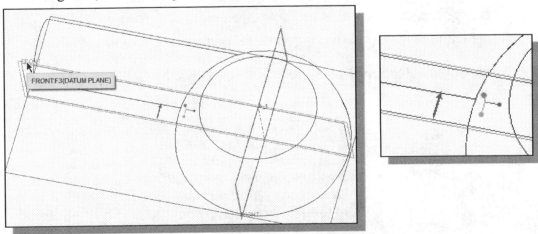

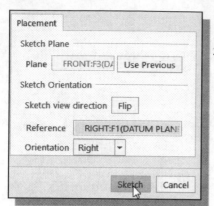

5. Click **Sketch** to accept the **RIGHT** plane to be used as the **Right side** orientation reference as shown.

Create 2D Blend Sections

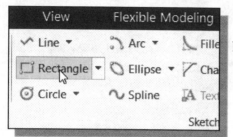

1. In the *Sketching* toolbar, click the **Rectangle** icon with the left-mouse-button.

2. Create a rectangle as shown; place the second corner of the rectangle on the horizontal axis.

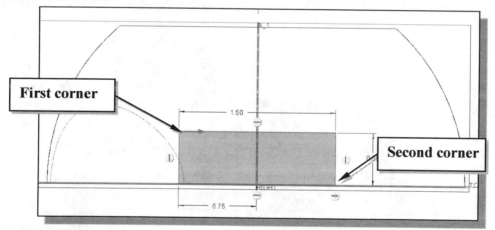

First corner

1.50

Second corner

0.75

3. Use the **Dimension** option to position the rectangle centered about the vertical axis.

4. Use **Modify** and **Regenerate** in the *Sketching* menu to adjust the size of the rectangle to **1.50 x .50**.

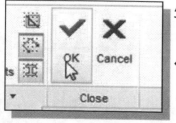

5. Click **OK** to accept the current sketch as the first section for the *Blend* feature.

❖   We have completed the first 2D section. We will next create the second 2D section. A *blend* feature is **a series of two-dimensional sections** that are joined together.

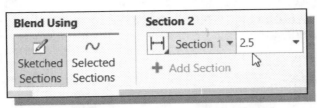

6. In the dashboard area, enter **2.5** as the offset distance for the second 2D section.

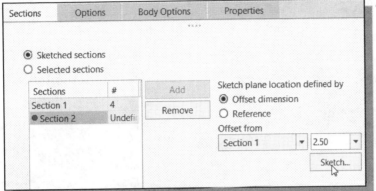

7. In the **Sections** tab, select **Sketch** to start a new sketch.

• The color of the previously sketched rectangle turned gray signifying that *Creo Parametric Sketcher* is ready to create the second 2D section.

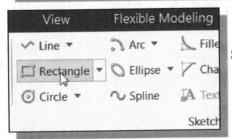

8. In the *Sketching* toolbar, select **Rectangle**.

9. Place the first corner of the rectangle on the horizontal axis and align the second corner as shown. The arrowhead indicates the direction of the blend. By placing the rectangle the same way as we did the first section, the blend will be aligned at the corresponding corners.

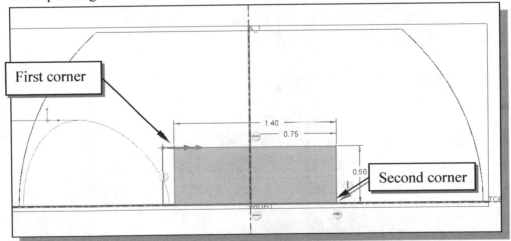

10. On your own, create and modify the three dimensions as shown.

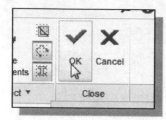

11. Click **OK** to accept the current sketch as the second section for the *Blend* feature.

12. In the **Sections** tab, click **Add** to add another section to the *Blend* feature.

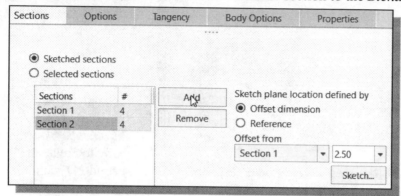

13. Enter **1.0** as the offset distance for the next 2D section.

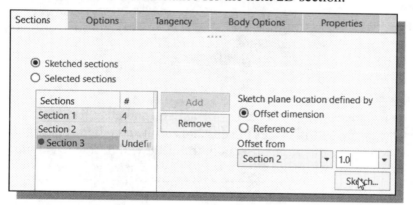

14. Click **Sketch** to enter the 2D sketcher mode.

15. On your own, create the 2D section as shown. (Start at the top left corner first.)

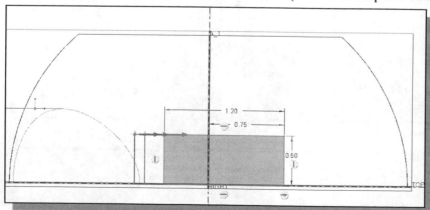

16. On your own, use the **Add** option to add another 2D section with offset distance set to **0.75** as shown.

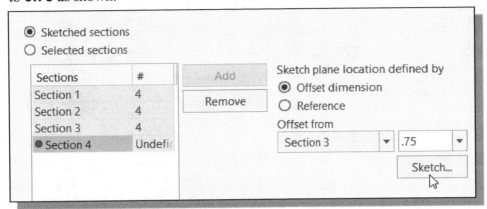

17. On your own, create the 2D section as shown.

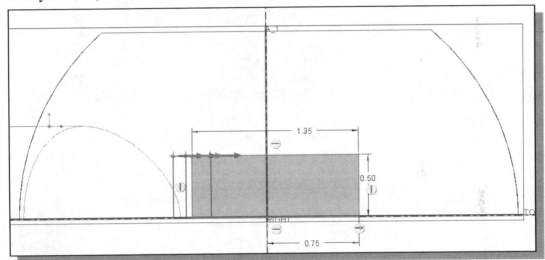

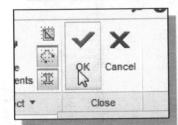

18. Click **OK** to accept the current sketch as the last section for the *Blend* feature.

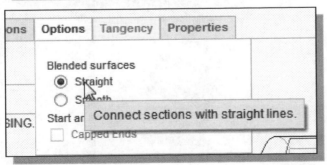

19. In the *Options tab*, set the **Blended Surfaces** option to **Straight** as shown.

• Note that only *one* loop per subsection is allowed for the **Blend** operation.

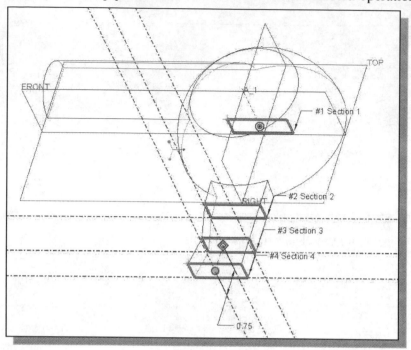

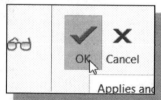 20. Click **OK** to create the solid feature.

21. On your own, use the *Dynamic Viewing* functions to view the completed 3D solid feature.

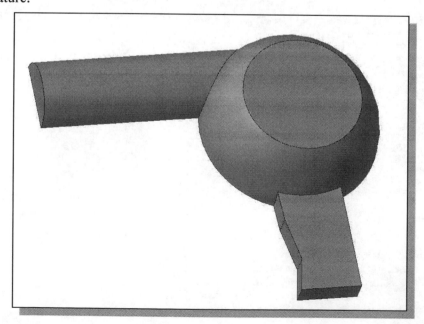

Create 3D Rounds and Fillets

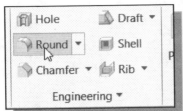

1. In the *Engineering* toolbar, select **Round** tool.

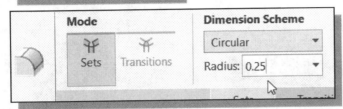

2. In the *Round Dashboard*, set the radius to **0.25** as shown.

3. Pick the top circle, the intersection curve, and all the edges connected to the top surface of the handle: the highlighted edges shown in the figure. (Also refer to the resulting features shown on the next page.)

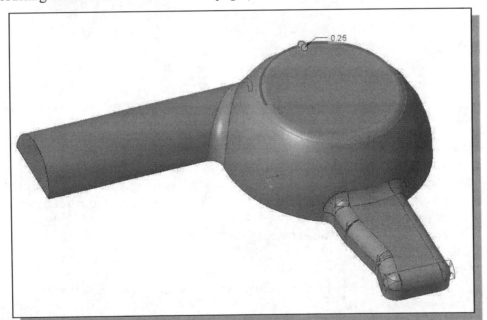

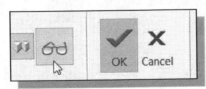

4. On your own, use the **Preview** option in the feature option menu to confirm the feature looks the same as shown on the next page.

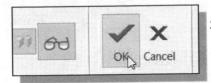

5. Pick **OK** in the feature option menu to create blended feature.

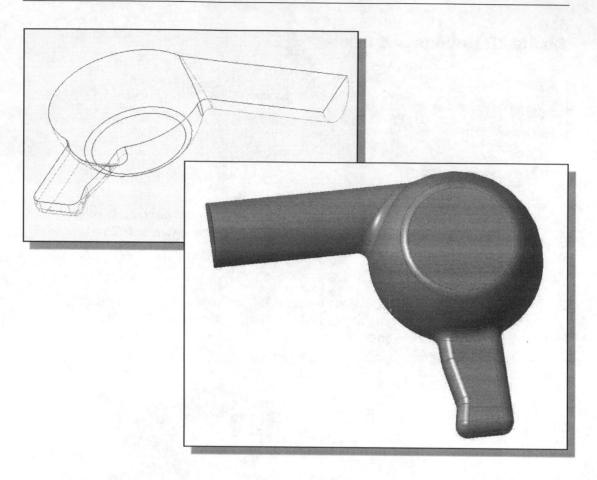

Create a Shell Feature

The **Shell** operation hollows out the inside of the solid, leaving a shell of a specified wall thickness. We can specify a surface or surfaces to be removed with the **Shell** operation. If no surfaces were selected to be removed, a "closed" shell is created with the whole inside of the part hollowed out and no access to the hollow.

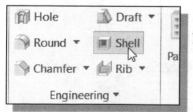

1. In the *Engineering* toolbar, select **Shell** tool.

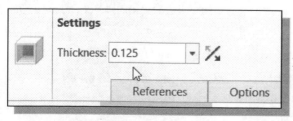

2. In the feature option menu, enter **0.125** as the wall thickness.

3. The message "*Select*" is displayed in the message area. Use the *Dynamic Rotation* function to rotate the model and pick the **two surfaces**--the *surfaces to be removed from the part*. (Hold down the [**Ctrl**] key while selecting.)

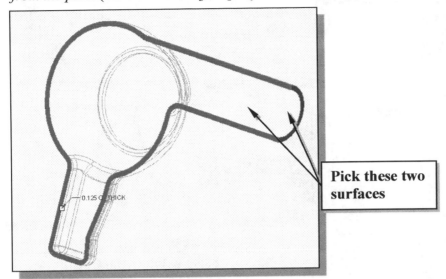

Pick these two surfaces

4. On your own, use the **Preview** option in the feature option menu to examine the feature.

5. Pick **OK** in the feature option menu to create the shell feature.

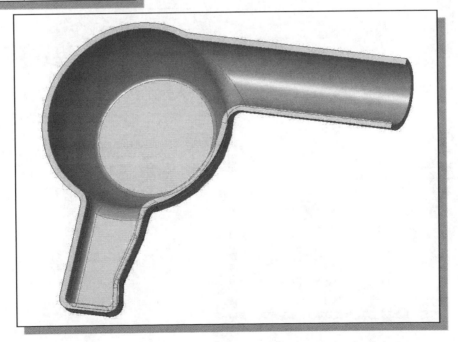

Create a Pattern Leader

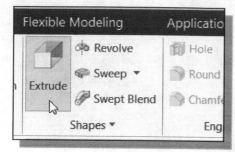

1. In the *Shapes* toolbar, select the **Extrude** tool option as shown.

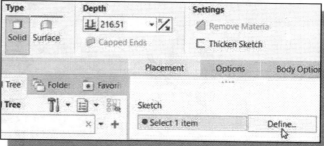

2. Click the **Placement** option and choose **Define** to begin creating a new *internal sketch*.

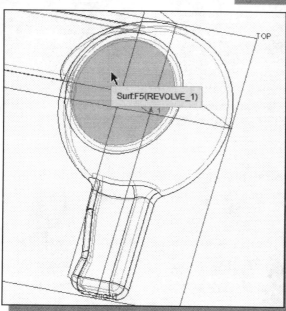

3. Pick the **top surface** of the base feature as the sketching plane, with the arrow pointing towards the base of the solid model.

4. In the *Section Placement* window, set the datum plane **FRONT** as the **bottom** side reference plane and click **Sketch** to enter the *Sketcher* mode.

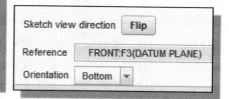

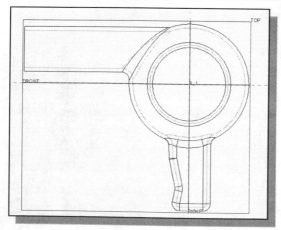

❖ Notice the two datum planes, FRONT and RIGHT, are set as the default references for the 2D sketch.

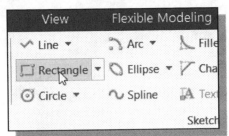

5. In the *Sketching* toolbar, select **Rectangle**.

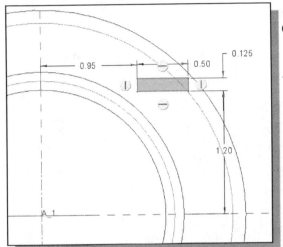

6. Create a rectangle toward the upper-right side of the solid model as shown.

7. On your own, modify the dimensions as shown in the figure. Rectangle size: **0.5 x 0.125**, location: **1.2** away from the horizontal axis and **0.95** away from the vertical axis.

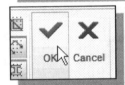

8. Click on the **OK** icon to accept the completed sweep section.

9. Toggle *on* the **Remove Material** option and enter **0.13** as the depth of the cut.

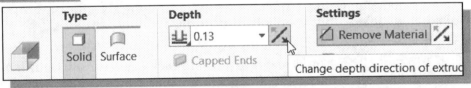

10. Click **Flip direction** once to flip the cut direction to cut into the solid model.

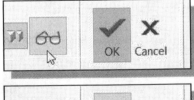

11. On your own, use the **Preview** option in the feature option menu to examine the feature.

- Notice in the message area, the message *"WARNING: EXTRUDE_2 is entirely outside the model; model unchanged."* is displayed.

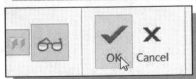

12. Pick **OK** in the feature option menu to create the shell feature.

- Even though the feature is entirely outside the model, it is created successfully. We will next duplicate this feature to form the necessary pattern on the model.

Create a Rectangular Pattern

1. In the *Editing* toolbar, select **Pattern**. (Note that the **Cut** feature is currently highlighted in the *Model Tree* area, which means it is pre-selected.)

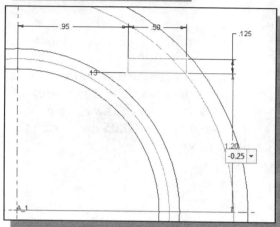

2. Pick the vertical location dimension (**1.20**).

3. In the *message area*, enter **-0.25** to place the copies of the feature below the pattern leader.

4. In the feature option menu area, enter **11** as the number of copies (including the original) needed in the vertical direction.

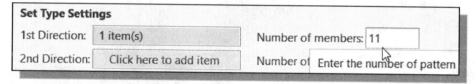

Set Type Settings				
1st Direction:	1 item(s)		Number of members:	11
2nd Direction:	Click here to add item		Number of	Enter the number of pattern

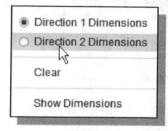

- ● Direction 1 Dimensions
- ○ Direction 2 Dimensions
- Clear
- Show Dimensions

5. Inside the *display area*, click once with the **right-mouse-button** to bring up the option menu.

6. Select **Direction 2 Dimensions** in the option list.

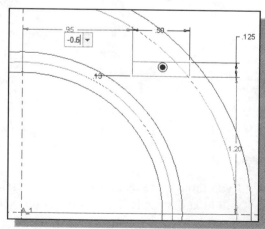

7. Pick the horizontal location dimension (**0.95**).

8. In the message area, enter **-0.60** to place the copies of the feature to the left of the pattern leader.

9. In the feature option menu area, enter **5** as the number of copies (including the original) needed in the second direction.

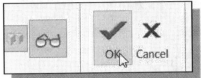

10. Click **OK** to create the pattern.

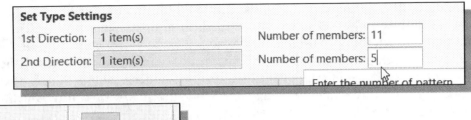

- *Creo Parametric* will now duplicate the *cut* feature and generate the rectangular cut pattern.

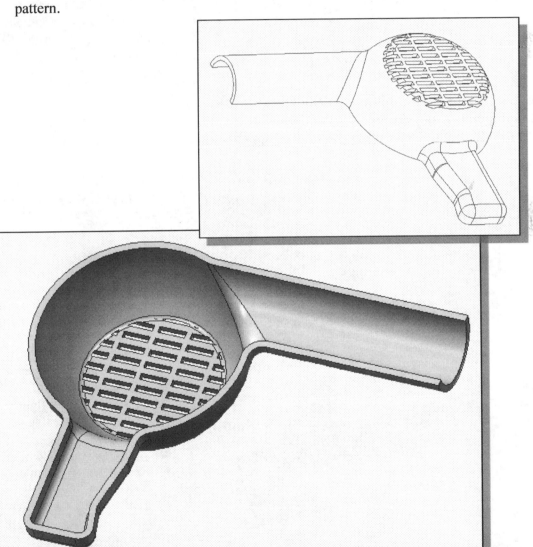

Create a Swept Feature

A *sweep* operation is defined as moving a **planar section** through a **path** in space to form a three-dimensional object. The path of the sweep can be a straight line or a curve. The **Extrude** operation, which we have used in the previous lessons, is a specific type of sweep. The Extrude operation is also known as the *linear sweep* operation, in which the sweep control path is always a line perpendicular to the two-dimensional section. Linear sweeps of unchanging shape result in what are generally called *prismatic solids*, which means solids with a constant cross-section from end to end.

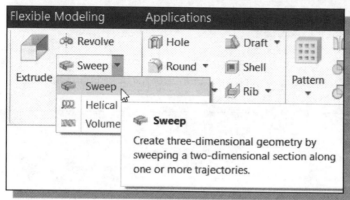

1. In the *Engineering* toolbar, select **Sweep** as shown.

➢ Note the Helical Sweep command is also available to create helical features for threads and coil springs.

➢ **Define the Sweep Trajectory**
The *sweep trajectory* is the *sweep path* that the *sweep section* is moved along.

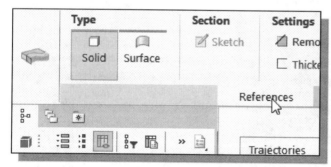

2. Click the **References** tab in the dashboard.

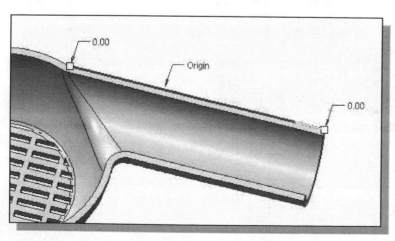

3. Select the **top edge** of the model as shown in the figure below. Note that an arrowhead is displayed indicating the sweep direction.

4. Click on the **arrow** to switch to the right end as shown.

➢ Note that for a single path constant cross section sweep, all entities forming the path need to be within the same chain.

5. Click **Details** to review and edit the chain properties.

6. Hold down the **[Ctrl]** key and select the adjacent edge as shown. Note the selected edge appears in the references list.

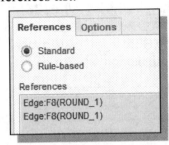

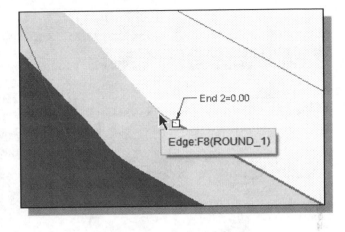

7. Hold down the **[Ctrl]** key and continue to select the adjacent outer edges of the base as shown in the figure.

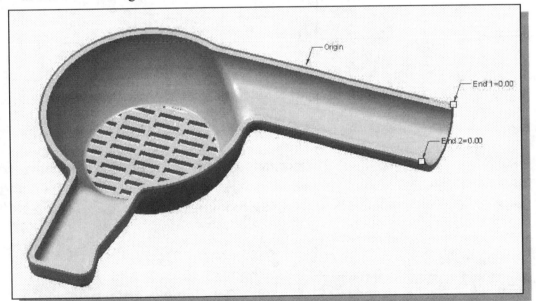

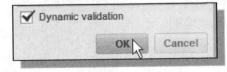

8. Click **OK** to close the chain properties dialog box.

➢ Now the sweep path has been set up, the other element is to set up a planar section. By default, the sketch plane of the section is aligned to the origin of the starting point of the sweep path.

➢ **Define the Sweep Section**

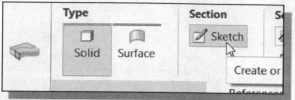

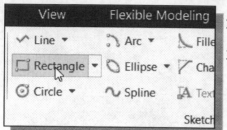

1. Click **Create or edit sweep section** in the dashboard.

2. In the *Sketching* toolbar, select **Rectangle**.

3. Create a rectangle, **0.125 × 0.0625**, aligned to the top corner of the model as shown in the figure below. (Hint: First rotate dynamically to confirm the orientation of the model.)

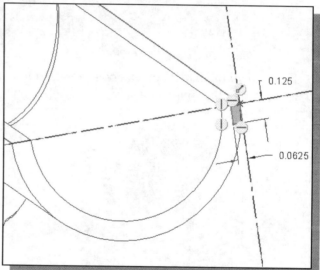

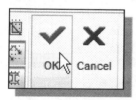

4. Click **OK** to accept the completed sweep section.

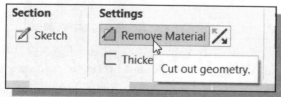

5. Click **Remove material** to create a *cut* feature.

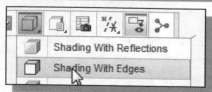

6. In the *Display Control* toolbar, adjust the display to **Shading with Edges**.

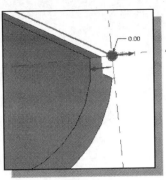

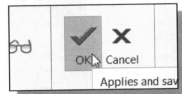

7. Confirm the arrow points towards the center of the small rectangle; the arrow indicates the *material side* of the feature.

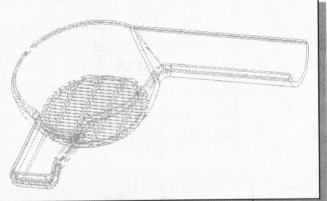

8. Notice that all elements are defined in the feature dialog window; click **OK** to create the solid feature.

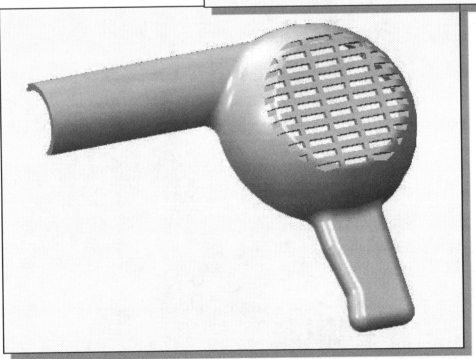

Review Questions:

1. Describe two different methods to *modify* dimensions.

2. Keeping the *Model History Tree* in mind, what is the difference between *Cut with a Pattern* and *Cut Each One Individually*?

3. What are the differences between **Sweep** and **Blend**?

4. What are the differences between **Sweep** and **Extrude**?

5. How do we modify the pattern parameters after the model is built?

6. Describe the elements required in creating a **Blend** feature.

7. Create sketches showing the steps you plan to use to create the model shown on the next page:

Exercises:

1. **Motor Housing** (Dimensions are in inches.)

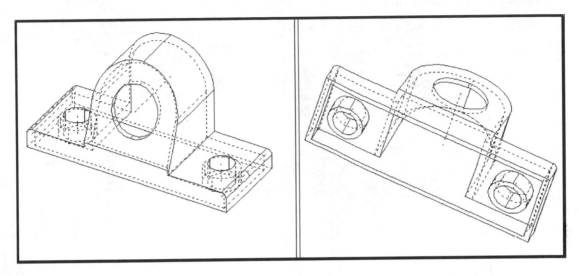

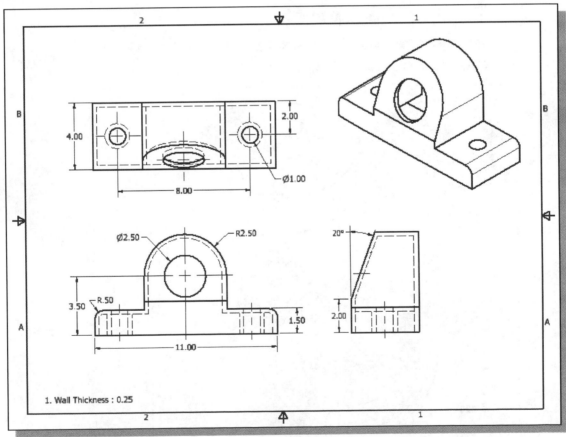

1. Wall Thickness : 0.25

2. Using the same dimensions given in the tutorial, construct the other half of the dryer housing. Plan ahead, and consider how you would create the matching half of the design. (Save both parts so that you can create an assembly model once you have completed Chapter 12.)

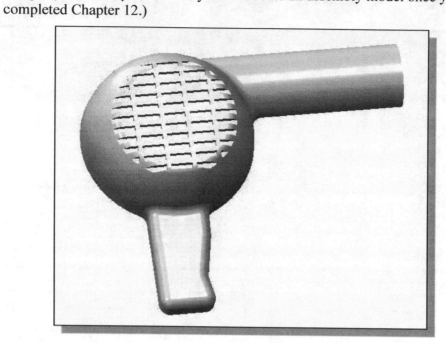

3. **Piston Cap** (Dimensions are in inches.)

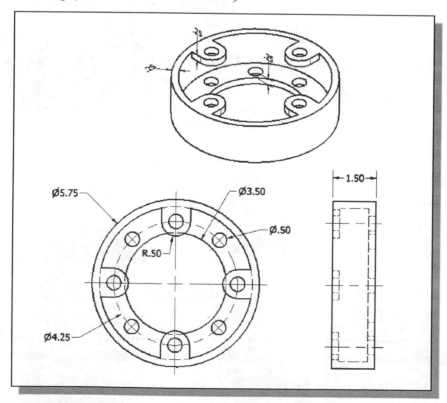

4. Jig Base (Dimensions are in millimeters.)

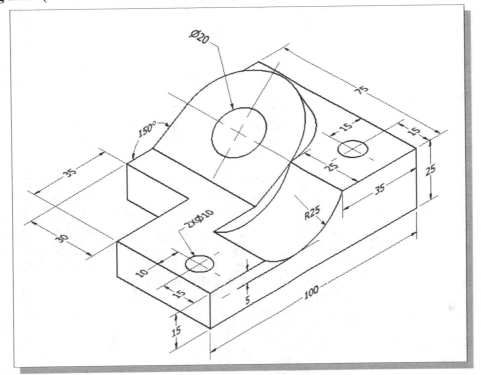

5. Anchor Base (Dimensions are in inches.)

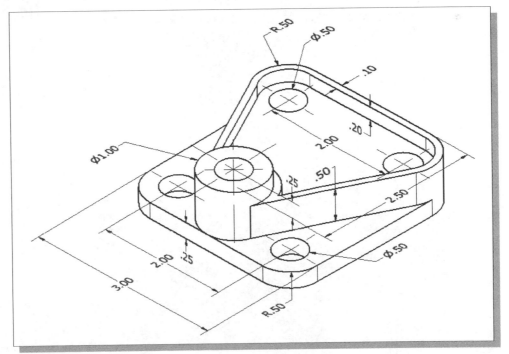

6. Slide Support (Rounds: 0.25 & 0.5)

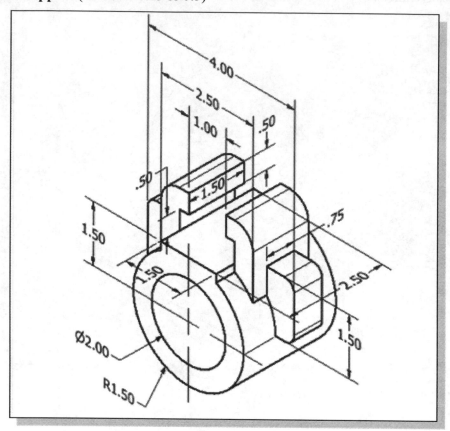

Chapter 10
Basic Sheet Metal Designs

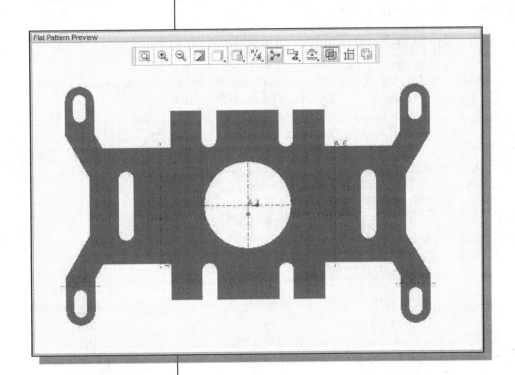

Learning Objectives

♦ **Understand the Sheet Metal Manufacturing Processes**

♦ **Understand the Creo Sheet Metal Modeling Methodology**

♦ **Create Parts in the Sheet Metal Modeler Mode**

♦ **Utilize the Sheet Metal Tools to Create Bends and Flanges**

♦ **Create Flat Pattern Layouts**

Sheet Metal Processes

Sheet metal is one of the most commonly used materials in our everyday life. Sheet metal is simply a thin and flat piece of metal, which can be cut and bent into a variety of different shapes. The thicknesses of sheet metal can vary significantly, but the thickness is generally between 0.006" and 0.250".

Sheet metal is generally produced by reducing the thickness of a work piece by compressive forces applied through a set of rolls. This process is known as rolling and has been around since 1500 AD. Sheet metal is identified by the thickness, or gauge, of the metal and is generally available as flat pieces or in coils. The gauge of sheet metal (see Appendix A) ranges from 30 gauge to about 6 gauge. The higher the gauge number, the thinner the metal is. Aluminum, brass, copper, cold rolled steel, tin, nickel and titanium are some of the more commonly available sheet metal materials. Typical sheet metal applications are seen in cars, boats, airplanes, casing for electronic devices and many other things.

The main feature of sheet metal is its ability to be formed and shaped by a variety of processes, such as **bending** and **cutting**. Different processes can be used to achieve the desired shape and form. Some of the more commonly used sheet metal processes include:

Drawing
Drawing forms sheet metal into parts by using a punch, where the punch presses a sheet metal blank into a die cavity. This process is generally used to create shallow or deep parts with relatively simple shapes. **Soft punches** can also be utilized to create more arbitrary shapes. **Deep drawing** is generally done by making multiple steps; this process is known as *draw reductions*.

Stretch forming
Stretch forming is a process where the sheet metal is clamped around its edges and stretched over a die. This process is mainly used for the manufacturing of large parts with shallow contours, such as aircraft wings, or automotive door and window panels.

Spinning
Spinning is the process used to make axis-symmetric parts by applying a work piece to a rotating mandrel with the help of rollers. *Spinning* is commonly used to make cylindrical shapes, such as missile nose cones and satellite dishes.

Stamping
Stamping is the general term used to describe a variety of operations, such as bending, flanging, punching, embossing, and coining. The main advantage of stamping is its speed; designs containing simple or complex shapes can be formed at relatively high production rates.

Flanging
Flanging is a process used to strengthen different sections of a sheet metal part and also to form various shapes. This process is commonly used for a variety of parts, for example, aluminum cans for soft drinks.

Bending

Bending is a process by which sheet metal can be deformed by plastically deforming the material and changing its shape. The material is stressed beyond the yield strength but below the ultimate tensile strength. With this process, the surface area of the material does not change much. *Bending* usually refers to deformation about one axis.

Bending is a flexible process by which many different shapes can be produced. Standard die sets are used to produce a wide variety of shapes. The material is placed on the die and positioned in place with *stops* and *gauges*. The material is held in place with *hold-downs*. The upper part of the press, the ram, with the appropriately shaped punch descends and forms the v-shaped bend.

Bending is usually done using ***press brakes***. The lower die of the press contains a V-shaped groove. The upper part of the press contains a punch that will press the sheet metal down into the v-shaped die, causing it to bend.

The most commonly used modern *Bending* method is the ***air bending*** method, where a sharper die angle is used; for example, an 85 degree angle is used for a 90 degree bend. *Air Bending* is done with the punch touching the work piece, and the work piece not bottoming in the lower die. By controlling the push stroke of the upper punch, the metal is pushed down to the required bend angle. The groove width of the lower die is typically 8 to 10 times the thickness of the metal to be bent. The *press brake* can also be computer controlled to allow the making of a series of bends to assure a high degree of accuracy in manufactured parts.

Cutting

Cutting sheet metal can be done in various ways, from using a variety of hand tools to very large powered shears. Today, computer-controlled cutting is also available for very precise cutting. Most modern computer-controlled sheet metal cutting operations are ***CNC laser cutting*** and ***CNC punch press***.

CNC laser cutting is done by moving the laser beam over the surface of the sheet metal. The sheet metal is heated and then burnt by the laser beam. The quality of the edge can be extremely smooth. *CNC punching* is performed by moving the sheet metal between the computer-controlled punch. The top punch mates with the bottom die, cutting a simple shape, such as a square, circle, or hexagon from the sheet. An area can be cut out by making several hundred small square cuts around the perimeter. A *CNC punch* is less flexible than a laser for cutting compound shapes, but it is faster for repetitive shapes. A typical *CNC punch* has a choice of up to 60 tools in a ***turret***. A modern *CNC punch* can run as fast as 600 blows per minute. A *CNC punch* or a *CNC laser* machine can typically cut a blank sheet into the desired shapes in less than 15 seconds, with very high precision.

Sheet Metal Modeling

In reality, a sheet metal part is made from a piece of flat metal sheet of uniform thickness by cutting out a flat pattern and then folding it into the desired shape. To construct a computer sheet metal part, we can (1) simulate the actual production methods and start with a flat pattern layout to make the model; (2) use the building block approach which concentrates on the different sections of the formed 3D design; or (3) construct a solid model first, then convert it into a sheet metal model. All three methods are applicable in modern parametric modeling software such as *Creo Parametric*.

Since the actual sheet metal manufacturing process requires a flat pattern layout, the accurate generation of the flat pattern layout in the computer modeling software is critical. The conversion between the 3D formed designs and 2D flat pattern layouts requires the use of the correct *K-Factor*, which can be used to determine the required *Bend Allowance*.

Bend allowance is the term used to describe how much material is needed between two panels to accommodate a given bend. Determining bend allowance is commonly referred to as *Bend Development*.

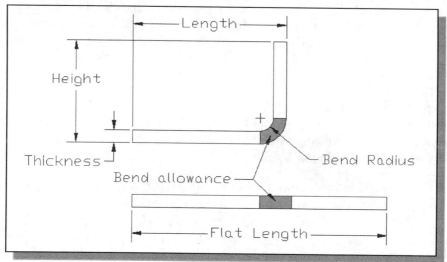

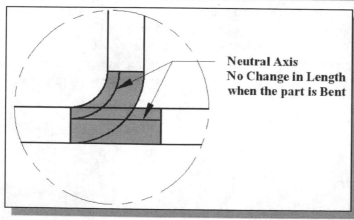

**Neutral Axis
No Change in Length
when the part is Bent**

In sheet metal, the *Neutral Axis* is defined as the location where there is no change in length when the part is bent.

On the inside of the bend, above the neutral axis in the figure, the material is in compression, where the area below the neutral axis is in tension.

K-Factor and Y-Factor

The location of the neutral axis in a bend is used to calculate the *k-factor*. The K-Factor is the ratio of the neutral axis to the material thickness. Since the amount of inside compression is always less than the outside tension, the k-factor can never exceed **0.50** in practical use. To the other extreme, a reasonable assumption is that the k-factor cannot be less than **0.25**.

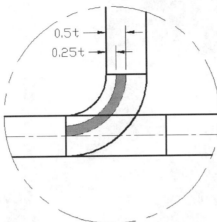

Several factors can change the k-factor, such as the type of bending (free vs. constrained), tool geometry, rate of bend, material (Mild Steel, Cold Rolled Steel, Aluminum, etc.), and even grain direction. With some grades of aluminum, the age of the material can also be a factor.

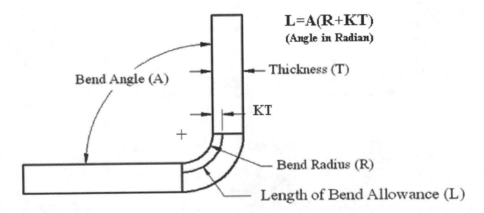

Sheet metal fabricators will typically have developed a k-factor table (usually through trial and error) to use. *Creo Parametric* is set up to allow the k-factor to be used to create fairly accurate and reliable flat patterns.

The Y-Factor is simply a variable based on the more commonly used K-Factor. It is derived by taking half of the K-Factor multiplied by pi.

Y-Factor:
$$Y = \frac{K \cdot \pi}{2}$$

The Actuator Bracket – 16 Gage Sheet Metal Design

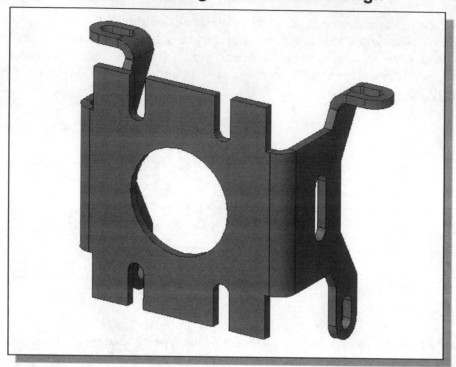

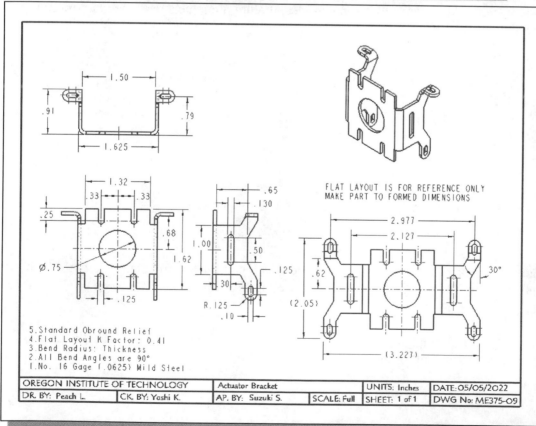

FLAT LAYOUT IS FOR REFERENCE ONLY
MAKE PART TO FORMED DIMENSIONS

5. Standard Obround Relief
4. Flat Layout K Factor: 0.41
3. Bend Radius: Thickness
2. All Bend Angles are 90°
1. No. 16 Gage (.0625) Mild Steel

OREGON INSTITUTE OF TECHNOLOGY		Actuator Bracket		UNITS: Inches	DATE: 05/05/2022
DR. BY: Peach L.	CK. BY: Yoshi K.	AP. BY: Suzuki S.	SCALE: Full	SHEET: 1 of 1	DWG No: ME375-09

Starting Creo Parametric

1. Select the **Creo Parametric** option on the *Start* menu or select the **Creo Parametric** icon on the desktop to start *Creo Parametric*. The *Creo Parametric* main window will appear on the screen.

2. Select the **New File** icon with a single click of the left-mouse-button in the *Ribbon* toolbar.

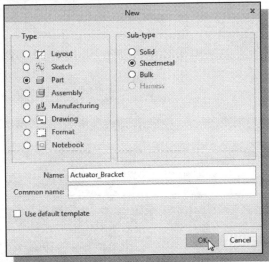

3. Select **Part** tab in the *Type* list, and set the *Sub Type* list to **SheetMetal** as shown.

4. Enter **Actuator_Bracket** as the new design name.

5. Uncheck the **Use default template** option.

6. Click on the **OK** button in the *New File* dialog box to accept the selected settings and proceed to the next step.

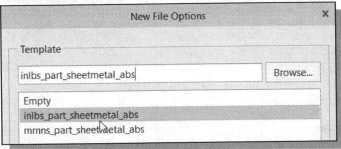

7. In the *New File Options* window, choose the **inlbs_part_sheetmetal_abs** template for our design.

8. Click **OK** to create the new design.

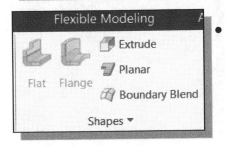

- In the *Ribbon* toolbar area, sheet metal related commands are displayed. In general, the starting point of a sheet metal design is either using the **Extrude** command or the **Planar** command to create the first wall.

Examine and Set Sheet Metal Defaults

In general, sheet metal models use a set of default properties, such as **K-Factor**, **Bend Radius**, **Corner Relief**, and **Edge Treatment** that are defined in the *Creo Model Properties* option.

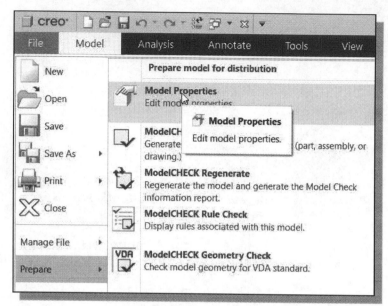

1. In the *Ribbon toolbar* area, select **File → Prepare → Model Properties** with the left-mouse-button.

2. In the *Sheet metal* section, click **Change** in the first item as shown.

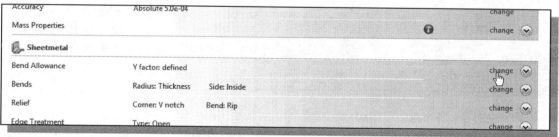

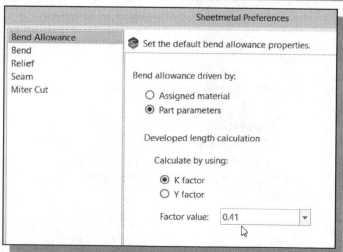

3. Choose to use **K factor** and enter **0.41** as the *K Factor Value* as shown.

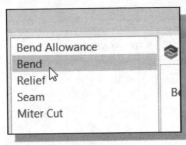

4. Select **Bend** to view the settings related to sheet metal Bends.

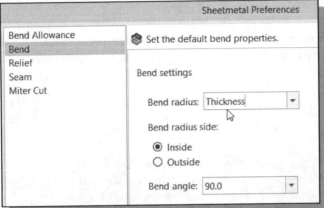

5. Set the *Bend Radius* to the sheet metal thickness, **thickness**.

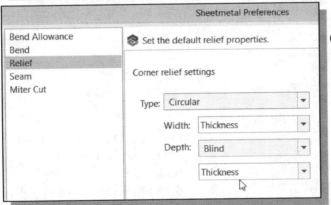

6. In the *Reliefs* section, set the *Corner relief type* to **Circular**, *Width* to **Thickness**, *Depth* to **Blind** and **Thickness** as shown.

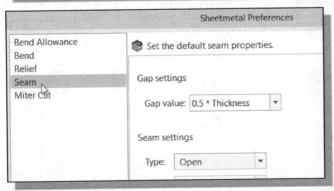

7. On your own, examine the other options in the different sections. Note that these settings can also be adjusted while features are being created.

8. Click **OK** to accept the settings and exit the *sheet metal properties* option. Also click **Close** to exit the *Model Properties* menu.

Create the Base Face Feature of the Design

The main section of a sheet metal design is generally treated as the stationary portion of the design, to which all the other sections are added to form the final design. The main section is also typically the starting point of sheet metal modeling, and thus the base feature in parametric modeling.

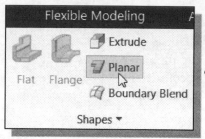

1. In the *Shapes* toolbar select the **Planar** command by left-clicking once on the icon.

- The **Planar** command is similar to the Extrude command in solid model; we create a 2D sketch and the extrusion distance is preset to the thickness of the sheet metal.

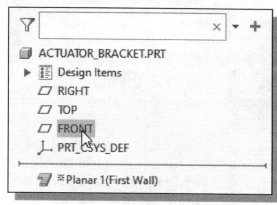

2. Select the **Front Plane**, in the *Model Tree* window, of the sheet metal part Part1 to align the sketch plane.

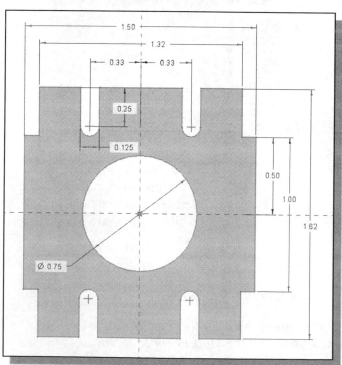

3. On your own, construct the 2D sketch and apply the proper constraints and dimensions as shown. (Hint: all line segments are either horizontal or vertical.)

❖ Note the design is symmetrical about the horizontal and vertical axes.

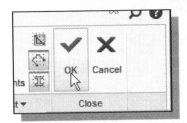

4. In the *Close toolbar*, select **OK** to end the *2D Sketch* mode.

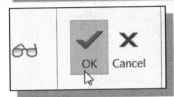

5. In the *dashboard*, set the thickness to **0.0625** as shown.

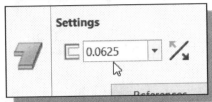

6. In the *Planar Option Dashboard*, click on the **Accept** button to proceed with the feature creation.

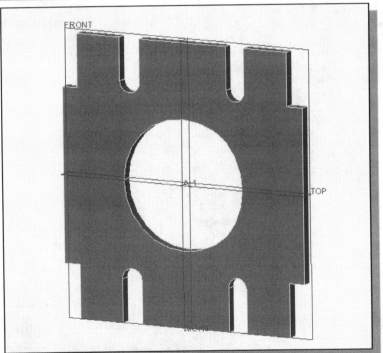

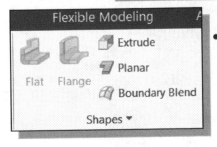

- The **Planar** command allows us to create the profile of the flat wall; the **Extrude** command provides us the *Edge view* option which can be used to create multiple flat surfaces of the sheet metal design.

Using the Flat Command

With sheet metal design, the **Flat** command will attach a flat face, which has a bend that connects to an existing straight edge of the existing sheet metal model. Flat wall features are added by selecting one edge and by specifying a set of options which determine the size and shape of the material added. In Creo, the **Flange** command is similar to the *Flat* command, but it can be used to build more complex attached walls with additional settings and options.

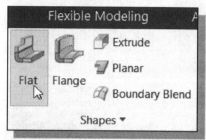

1. In the *Sheet Metal Features* panel, select **Flat** to activate the command.

❖ The **Flat** command requires the selection of an existing straight edge to create a flat feature, while the **Flange** command allows the creation of 3D surfaces that are not flat.

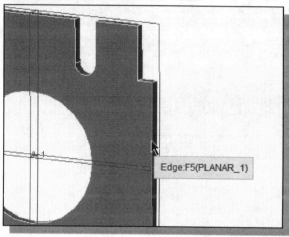

2. Select the **right vertical edge** on the front face to add a flat wall as shown.

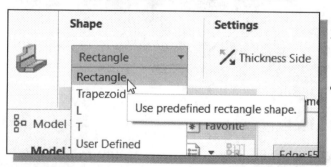

3. Set the *Shape* of the attached wall to **Rectangle** as shown.

• Notice the different pre-defined shapes available in the list.

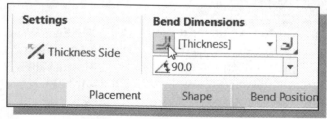

4. Confirm the *Bend Radius* to the default setting, matching the sheet metal thickness, **thickness**.

5. Set the radius dimension to be on the inside surface of the bend.

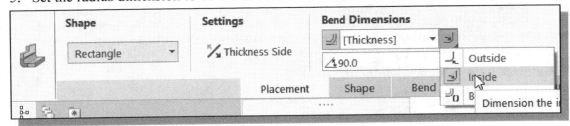

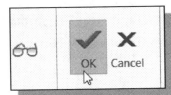

6. Set the *Bend angle* to **90** degrees and click once on the **Flip Direction** icon to switch the bend edge to align to the inner edge. Thus the position of the bend is offset by the thickness of the sheet metal.

7. In the *graphics area*, adjust the wall length to **0.5625** inches as shown in the figure. Note the distance is measured to the outside edge of the first wall.

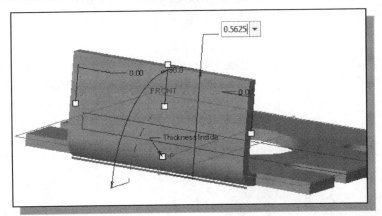

8. In the *Dashboard* area, click **OK** to accept the settings and proceed to create the feature.

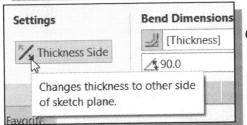

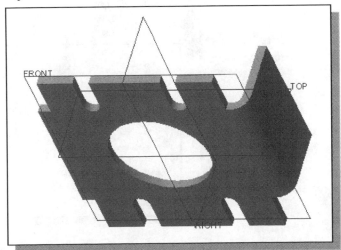

Add another Planar Feature

The main section of a sheet metal design is generally treated as the stationary portion of the design, to which all the other sections are added to form the final design. The main section is also typically the starting point of sheet metal modeling, and thus the base feature in parametric modeling.

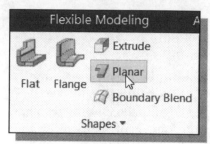

1. In the *Shapes* toolbar select the **Planar** command by left-clicking once on the icon.

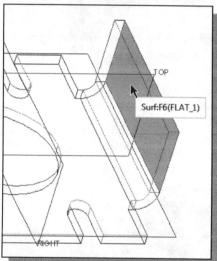

2. Select the **Outside surface** of the last feature to align the sketch plane, and use the **Front** plane to face the **bottom** edge of the screen for the orientation as shown in the below figure.

- The *Planar* command can be used to create 2D sketches that contain multiple regions. The *shape* option in the *Flat* command does not allow multiple regions.

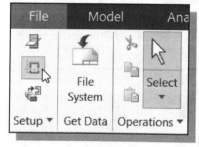

3. Use the **Reference** command and add the three outside edges of the last feature we just created as sketching references.

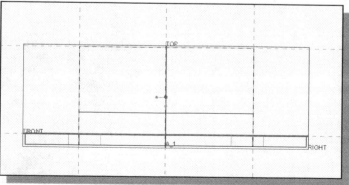

4. On your own, construct the 2D sketch and apply the proper constraints and dimensions as shown. (Hint: Use the **Equal** and **Horizontal align** constraints to align the symmetrical geometry.)

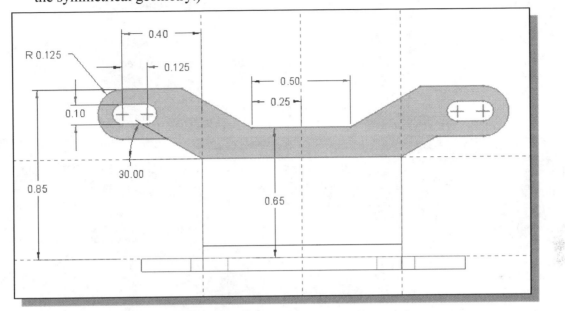

❖ Note the design is symmetrical about the center axis and there are two sets of parallel inclined lines at 30 degrees angle.

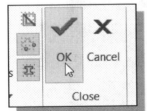

5. Click **OK** to end the *2D Sketch* mode and return to the feature dashboard.

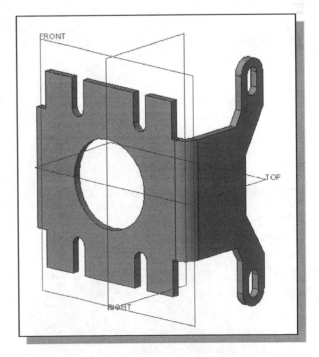

6. Click **OK** to create the feature.

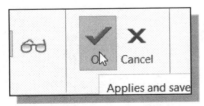

Add a Cut Feature

The main section of a sheet metal design is generally treated as the stationary portion of the design, to which all the other sections are added to form the final design.

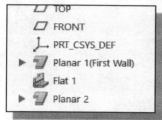

1. De-select the last feature in the model tree.

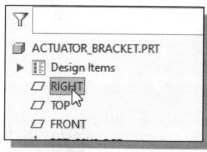

2. In the *Shapes* toolbar select the **Extruded Cut** command by left-clicking once on the icon.

3. In the *Placement tab*, select **Define** to set up the sketching plane for the feature.

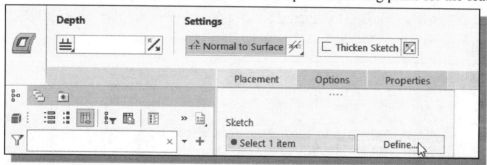

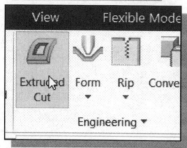

4. Select the **Right Plane** of the sheet metal *part* to align the sketch plane.

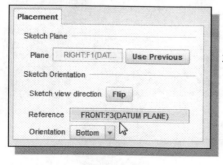

5. On your own, set the **Front** plane to face the **Bottom** edge of the display screen.

6. Click **Sketch** to enter the *2D sketch* mode.

7. On your own, construct the 2D sketch and apply the proper constraints and dimensions as shown.

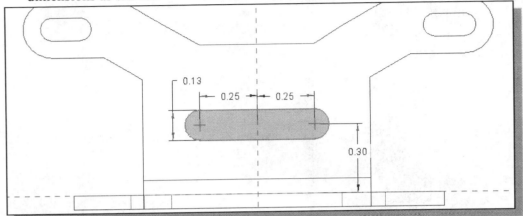

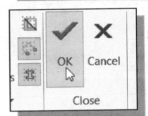 8. In the *Close* toolbar, click **OK** to exit the 2D sketch mode.

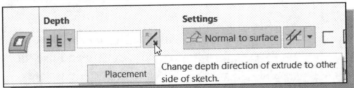 9. In the *dashboard*, set the extrusion option to **Thru-All** as the extrusion distance.

10. Click on the **Flip direction** button to flip the extrusion direction and create the cut feature as shown.

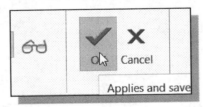

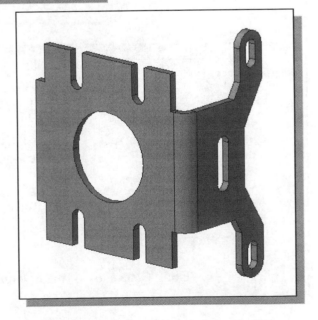

Create a Bend Feature

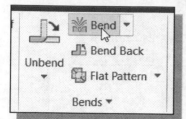

1. In the *Bends* toolbar select the **Bend** command by left-clicking once on the icon.

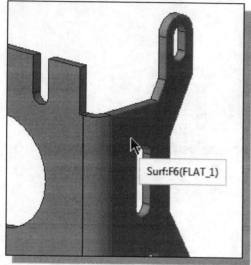

Surf:F6(FLAT_1)

2. Select the **outside surface** of the *Flat* feature to set the bend surface as shown.

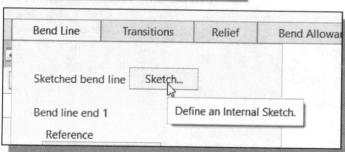

Bend Line	Transitions	Relief	Bend Allowa

Sketched bend line   Sketch...

Define an Internal Sketch.

Bend line end 1

Reference

3. In the *dashboard*, click the **Bend Line** tab and use the **Sketch** option to create a bend line.

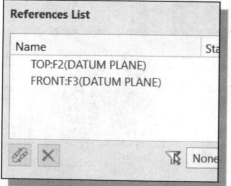

References List

Name	Sta
TOP:F2(DATUM PLANE)	
FRONT:F3(DATUM PLANE)	

None

4. On your own, choose to use the **Front** plane and the **Top** plane as the two references.

5. Click **Update** to confirm the reference selections.

Update the sketch.

Close

6. Click **Close** to exit the *References* command.

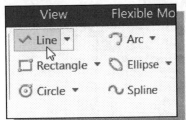

7. In the *Sketching* toolbar, select the **Line** command as shown.

8. On your own, construct a vertical line and apply the three dimensions to the two references, as shown. Note the length of the bend line needs to go past the length of the bend portion.

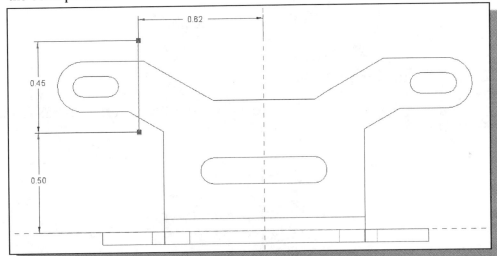

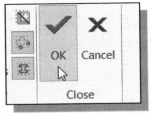

9. In the *Close* toolbar, click **OK** to exit the 2D sketch mode.

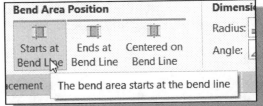

10. Set the bend edge to **align on the inside edge** as shown.

11. Click on the **Change Direction** arrow to flip the bend direction of the bend.

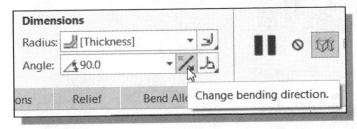

- Note the Fixed Side button can be used to set which side is rotated; set the bend as shown before proceeding to the next step.

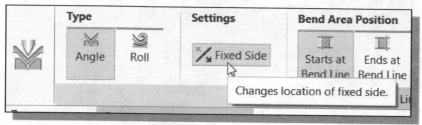

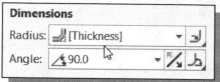

12. Set the *Bend Radius* to **Thickness**, and measurement is to the inside as shown.

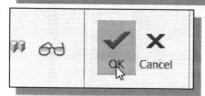

13. In the *dashboard,* click **OK** to create the bend feature as shown.

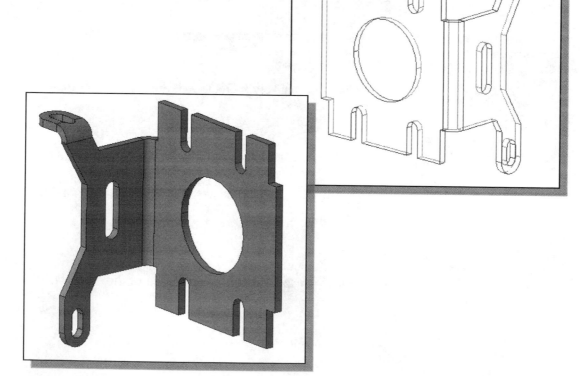

> The ***Bend*** feature requires the use of a ***bend line***, or an edge on the model, which simulates the edge where the bend will occur.

Create a Mirrored Feature

In *Creo Parametric*, sheet metal features can be mirrored just like regular solid features. The mirrored features are parametrically linked to the original parametric definitions.

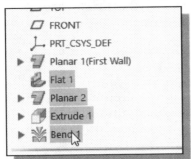

1. Pre-select the last four features, **Flat 1**, **Planar 2**, **Extrude 1** and **Bend 1**, by holding down the [CTRL] key and clicking with the left-mouse-button.

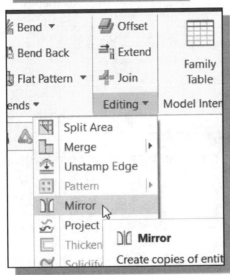

2. In the *Editing* toolbar pull-down list, select the **Mirror Feature** command by clicking the left-mouse-button on the icon as shown.

3. Activate the **Mirror Plane Selection** by clicking inside the edit box.

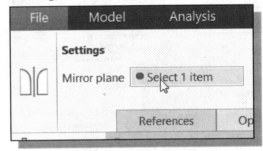

4. Select the **Right Plane** in the *Model Tree* window.

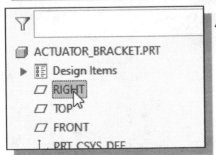

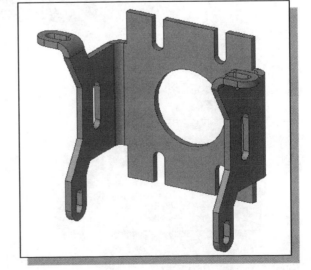

5. Click **OK** to create the *Mirrored* features.

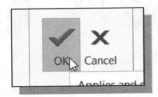

Confirm the Model Dimensions

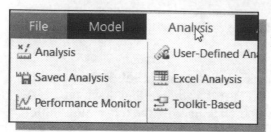

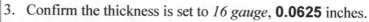

1. In the *Ribbon toolbar*, switch to the **Analysis** as shown.

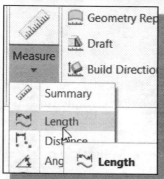

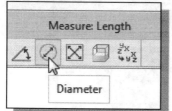

2. Activate the **Measure Length** command as shown.

3. Confirm the thickness is set to *16 gauge*, **0.0625** inches.

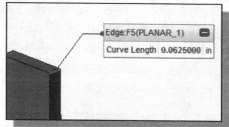

4. Switch to the *Diameter* option and confirm the bend radius is **thickness** as shown.

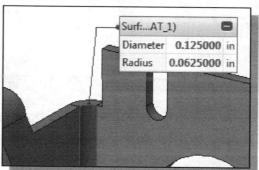

5. On your own, use the *Measure Distance* option and confirm the inside width of the model is **1.50**.

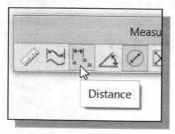

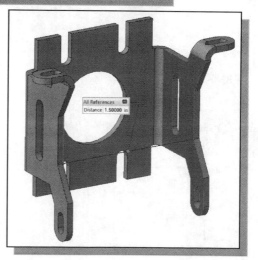

Create the Associated Flat Pattern

A sheet metal flat pattern is the shape of the sheet metal part before it is formed. A flat pattern is required to create drawings for manufacturing. The flat pattern shows the shape of the sheet metal part before it is formed showing all the bend lines, bend zones, punch locations, and the shape of the entire part with all bends flattened and bend factors considered.

The *Creo* **Flat Pattern** command calculates the material and layout required to flatten a 3D sheet metal model. Once the flat pattern is created, the part *browser* window displays a *Flat Pattern* item, and the flattened state of the model is displayed whenever this item is active. The flat pattern is typically created normal to the sketched base face feature, but this can be edited if necessary.

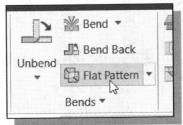

1. In the *Bends* toolbar, select the **Flat Pattern** command by clicking the left-mouse-button on the icon.

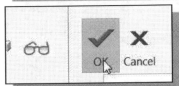

2. Click **OK** to accept the use of the base feature as the base of the flat pattern.

❖ *Creo Parametric* calculates the material and layout required to flatten the 3D sheet metal model and displays the flat pattern as shown.

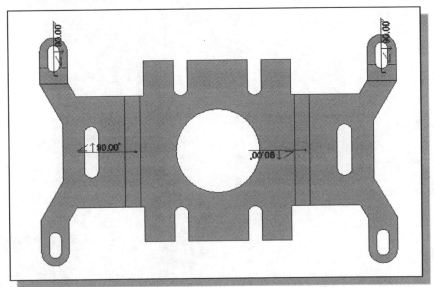

❖ Note that the displayed angles identify the locations and sizes of the bends.

Confirm the Flattened Length

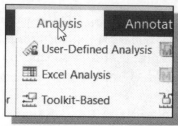

1. In the analysis tab, activate the **Measure Distance** option.

2. Confirm the center to the right edge width of the model is **1.41343** as shown.

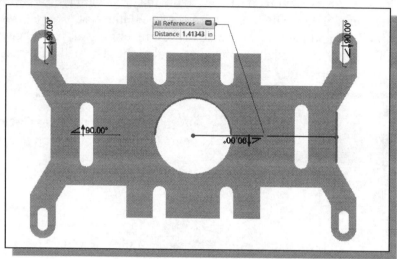

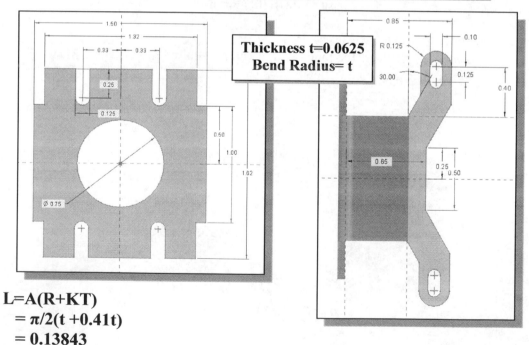

Thickness t=0.0625
Bend Radius= t

$$L=A(R+KT)$$
$$= \pi/2(t +0.41t)$$
$$= 0.13843$$

Flattened Length = W+H+L
$$=(1.50/2-t)+(0.65-t)+0.13843=1.41343$$

Create an Instance of the Flat Pattern

Creo provides several options to include the flat pattern of a sheet metal design in a 2D drawing. One option is to create an **instance** of the flat pattern; this will create a copy of the design with specified variations to the original.

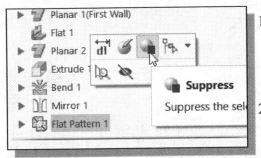

1. In the *Model Tree* window, right-click on **Flat Pattern 1** to bring up the *option list* and select **Suppress** to switch back to the 3D folded model.

2. Click **OK** to proceed with the suppression of the flat pattern.

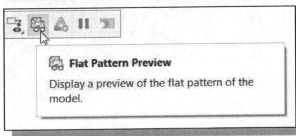

3. In the *Display View* toolbar, click on the **Flat Pattern Preview** icon to show the flat pattern in a preview window.

4. In the preview window, click on the **Create Instance** icon to create an instance of the flat pattern. (Hint: Resize the window if the icon isn't showing.)

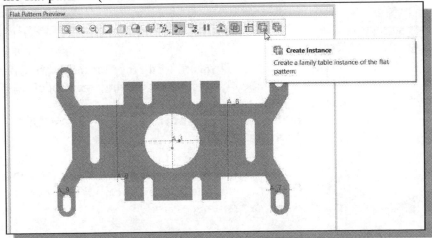

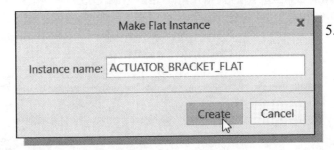

5. In the *New Instance* window, Creo provides an instance name for the design. We will accept the default name. Click **Create** to accept the setting and create the instance.

Create a 2D Sheet Metal Drawing

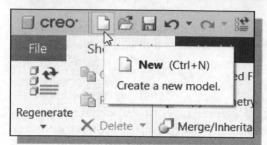

1. Pick the **New** icon in the toolbar. (We can also use the key combination **Ctrl-N** to start a new object.)

2. In the *New* form, select **Drawing** under the **Type** list.

3. Toggle *off* the **Use default template** option.

4. Choose *Use drawing model file name* so the drawing file name is the same as the model file name.

5. Pick **OK** to start a new drawing file.

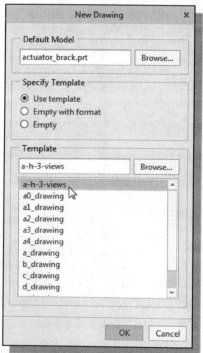

❖ In the *New Drawing* form, the currently active part is automatically selected as the drawing model.

6. In the **Specify Template** option, choose **Use Template**.

7. Pick **a-h-3views** in the **Template** list. (Use the **Browse** button to locate a specific *Creo Parametric template file* if necessary.)

8. Pick **OK** to proceed with the drawing template layout.

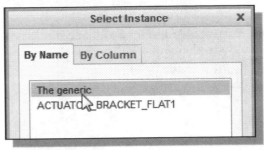

9. In the *Select Instance* window, confirm the **Generic** instance is highlighted as shown.

- Note the flat pattern instance is also available, which will be added in the next section.

10. Click Open to use the selected instance and start the 2D drawing.

- The **a-h-3views** template is set to using the three standard views (**TOP**, **FRONT** and **RIGHT**) with the *View Display* option set to the **Hidden line** option as shown in the figure.

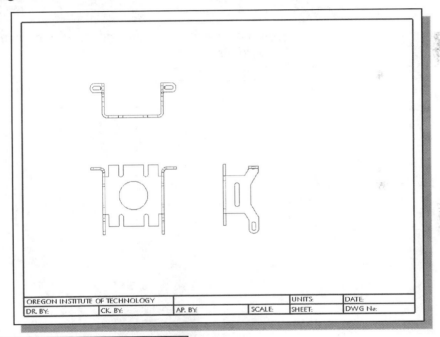

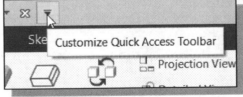

11. Click on the **Down arrow** icon in the quick Access toolbar and choose **More Command**.

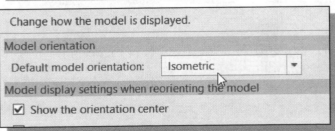

12. On your own, set the *Default orientation* to **Isometric**.

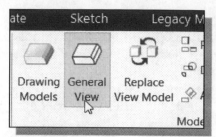

13. Click on the **General View** icon and add an **Isometric view** to the upper right corner of the title block as shown in the below figure.

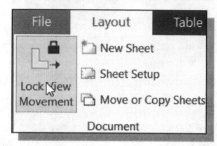

14. Turn **off** the *Lock View Movement* option by clicking once on the icon as shown.

15. On your own, use drag & drop with the left-mouse-button and reposition the four views roughly as shown in the below figure.

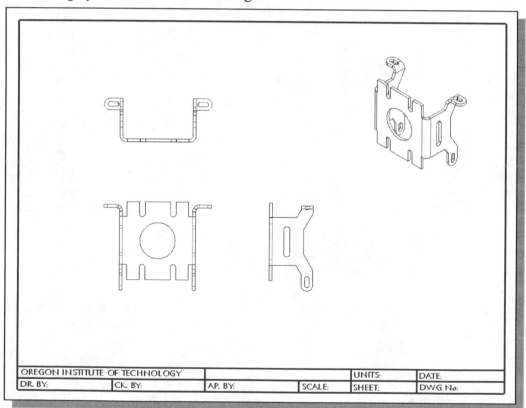

• The four views are created using the original copy of the sheet metal design. These views show the 3D sheet metal design, which identifies the expected final product. The Flat pattern of the design will be used as an aid for creating the 3D design.

Include a 2D View of the Flat Pattern

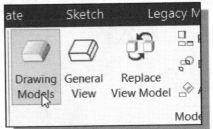

1. Click on the **Drawing Models** icon to start the process of adding a 2D view of the flat pattern of the design.

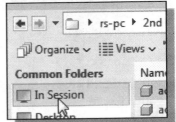

2. In the *Menu Manager* window, click **Add Model**.

3. In the *Open Model* window, select **In Session**.

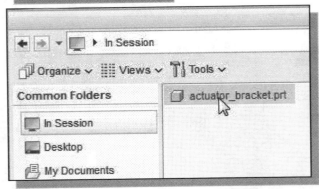

4. In the file list, choose the **Actuator_Bracket** part as shown.

5. Click **Open** to accept the selection.

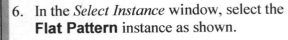

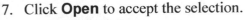

6. In the *Select Instance* window, select the **Flat Pattern** instance as shown.

7. Click **Open** to accept the selection.

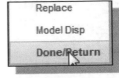

8. In the *Menu Manager* window, click **Done/Return** to end the *Drawing Model* command.

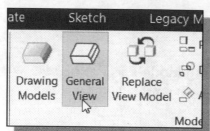

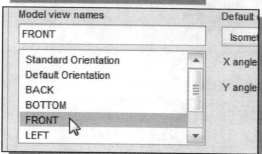

9. Click on the **General View** icon.

10. Accept the **No Combined State** option.

11. Select a location below the isometric view.

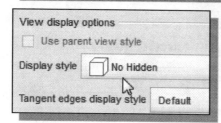

12. Select **Front View** as the view orientation.

13. On your own, adjust the display style of the view to **No Hidden** as shown.

- Note the dimensions showing the bend angles and the overall size of the flat pattern are automatically displayed.

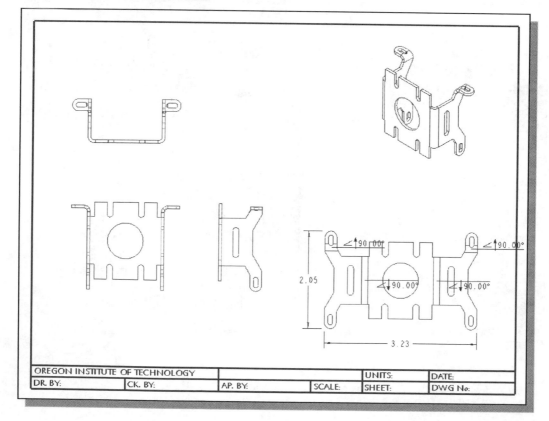

Display Center Axes and Dimensions

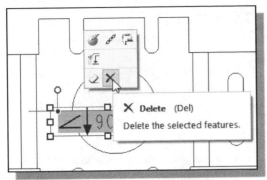

1. Pick the bend angle dimension shown in the center of the flat pattern and choose **Delete** to remove the angle dimension.

2. On your own, delete all of the bend angle dimensions shown in the flat pattern.

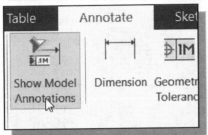

3. In the **Annotate** tab of the *Ribbon* toolbar, select the **Show Model Annotations** command.

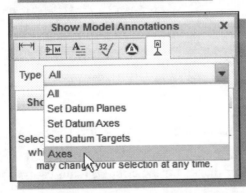

4. Select the **Axes and Set Datum objects** in the *Show Model Annotations* window as shown.

5. Choose **Axes** in the Type option list as shown.

6. Click on the base feature on the flat pattern to add the center lines as shown.

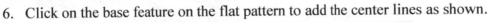

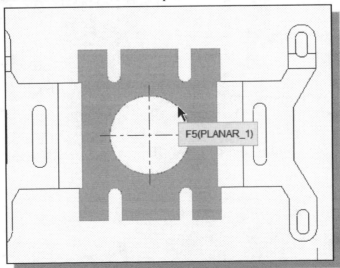

7. On your own, add the center lines to the other views.

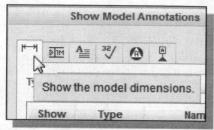

8. Select the **Show Model Dimensions** in the *Show Model Annotations* window as shown.

9. Click on the base feature in the front view to show the model dimensions of the base feature.

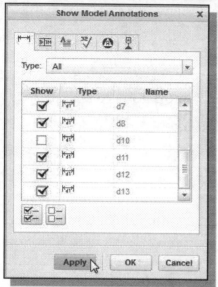

10. On your own, select the dimensions to show in the view.

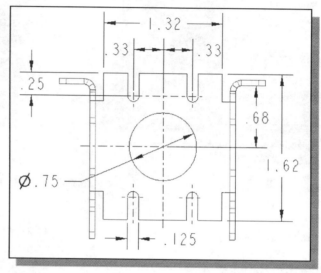

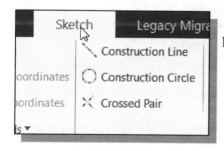

11. On your own, use the sketch toolbar to create and adjust the center marks for the circular features in the side view.

12. On your own, repeat the above steps to the side view as shown.

13. Complete the drawing and print a copy of the 2D drawing.

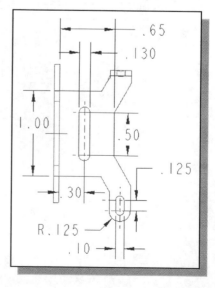

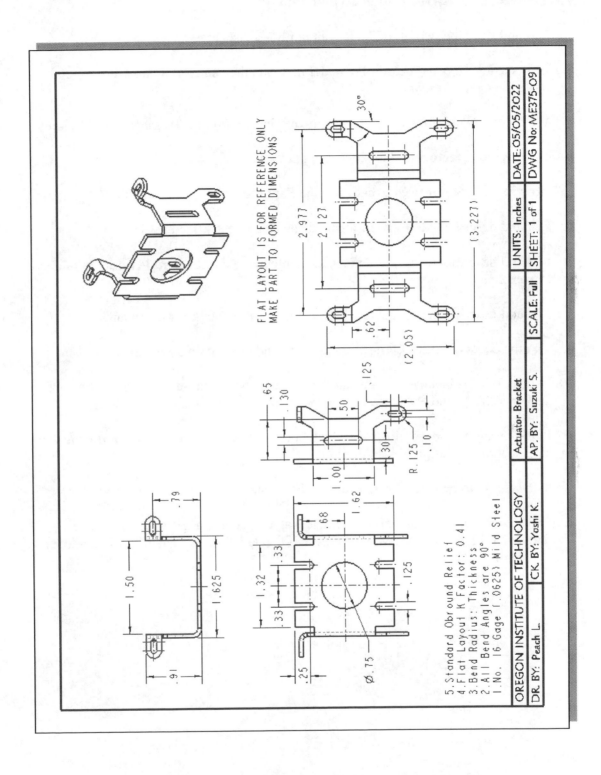

FLAT LAYOUT IS FOR REFERENCE ONLY
MAKE PART TO FORMED DIMENSIONS

30°

2.977
2.127
(3.227)
.62
(2.05)
.125

.65
.130
.50
.30
1.00
R.125
.10

1.62
.68
1.32
.33
.33
.125
Ø.75
.25

.79
1.50
1.625
.91

5. Standard Obround Relief
4. Flat Layout K Factor: 0.41
3. Bend Radius: Thickness
2. All Bend Angles are 90°
1. No. 16 Gage (.0625) Mild Steel

| OREGON INSTITUTE OF TECHNOLOGY | Actuator Bracket | UNITS: Inches | DATE: 05/05/2022 |
| DR. BY: Peach L.   CK. BY: Yoshi K. | AP. BY: Suzuki S. | SCALE: Full   SHEET: 1 of 1 | DWG No: ME375-09 |

Review Questions: (Time: 30 minutes)

1. List and describe two of the more commonly used sheet metal processes.

2. Is it possible to construct a solid model first, and then convert it into a sheet metal model in *Creo Parametric*?

3. Which command do we issue to display the flat pattern of a 3D sheet metal design?

4. What is the **k-factor** used in sheet metal processes?

5. How is the **k-factor** used to calculate the flattened length in sheet metal flat patterns?

6. List and describe two of the factors that can change the k-factor value.

7. List and describe two of the settings available in the **Sheet Metal Defaults** in *Creo Parametric*.

8. List and describe the differences between the **Flat** and **Planar** commands.

9. Can the **Show Model Dimensions** command be used on a *flat pattern* view?

10. In the *Creo Parametric* sheet metal module, can the feature-duplicating commands, such as **Mirror** and **Pattern**, be used on sheet metal features?

11. In the *Drawing View* dialog box, which option allows us to display the *flat pattern* of the sheet metal design?

12. Is the *flat pattern* item always available in the sheet metal part *browser* window?

13. Can we create a sheet metal feature that is at a 30 degree angle to the base face feature? Which command would you use if the new feature contains fairly complex 2D geometry?

Exercises: (Time: 125 minutes)
(Create the 3D model and the associated 2D drawings. All dimensions are in inches.) (Hint: Create the four 0.125 cut corners on the main face.)

1. Cooling Fan Cover

1. No. 16 Gauge (0.0625) Mild Steel
2. Flat Layout K-Factor: 0.44
3. Bend Radius: Thickness
4. No Bend Relief
5. All Bend Angles are 90 degrees

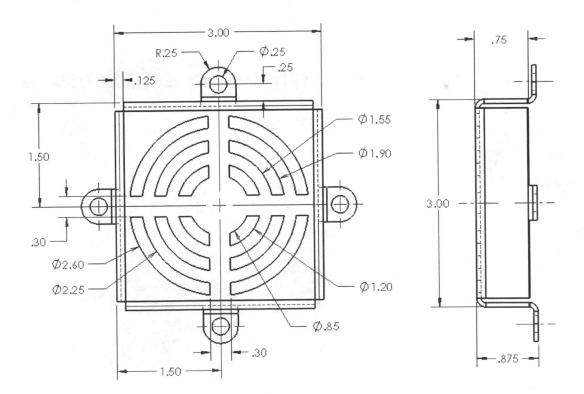

2. Sheet Metal - Rectangle to Square Transition (Create and convert a Solid model to a Sheet Metal model.)

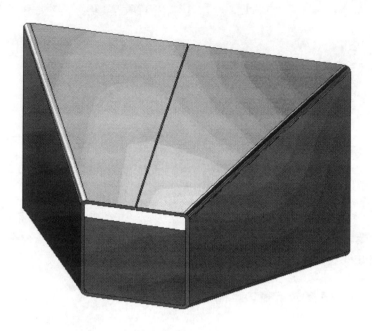

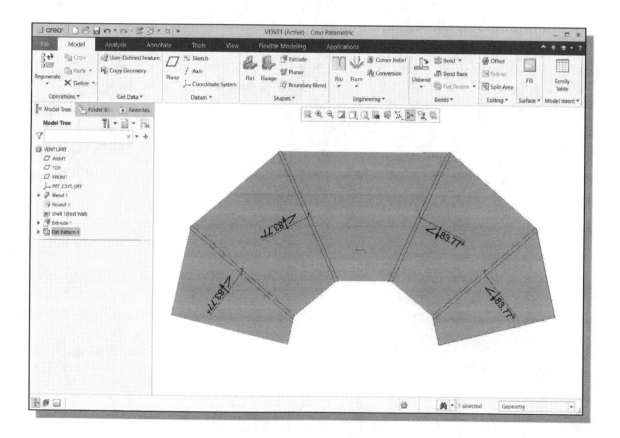

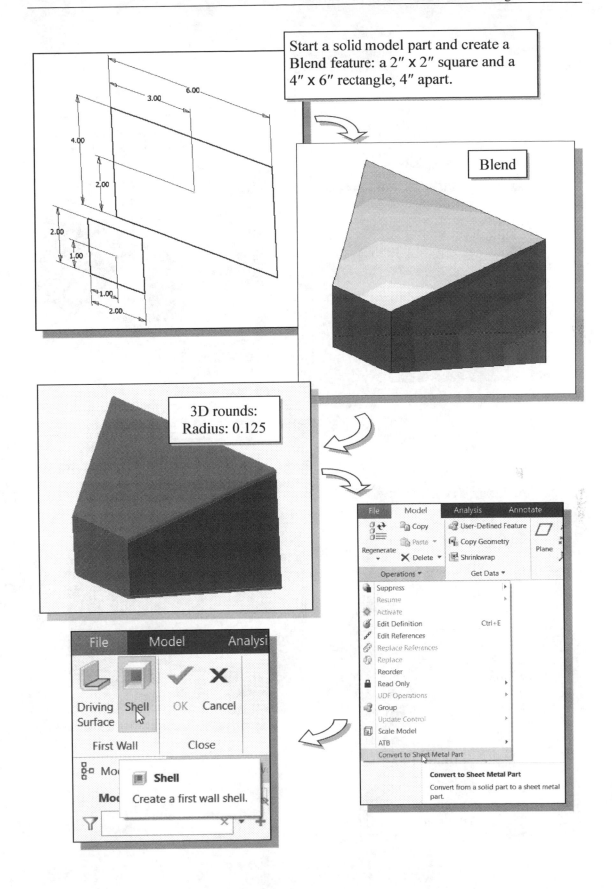

Start a solid model part and create a Blend feature: a 2″ x 2″ square and a 4″ x 6″ rectangle, 4″ apart.

Blend

3D rounds:
Radius: 0.125

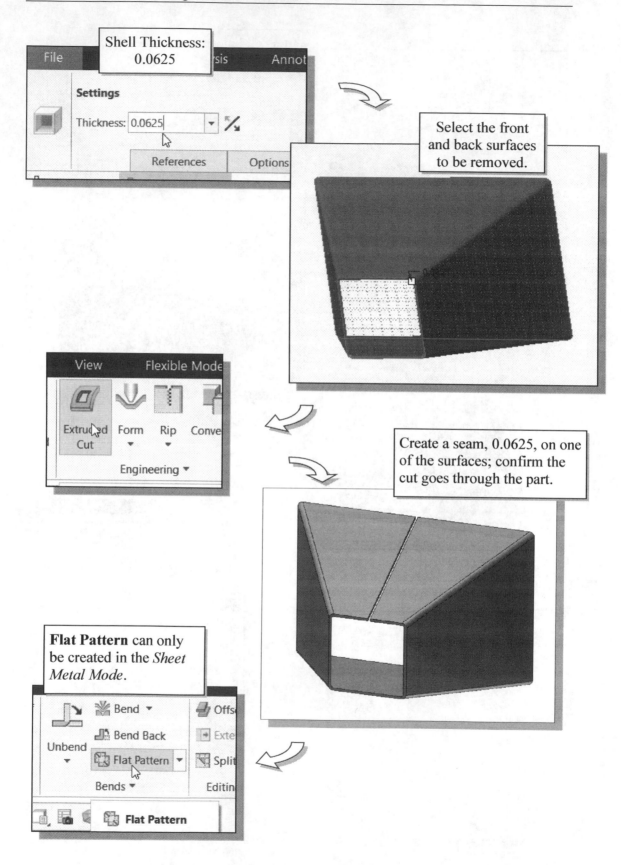

Shell Thickness: 0.0625

Settings

Thickness: 0.0625

References   Options

Select the front and back surfaces to be removed.

View   Flexible Mode

Extruded Cut   Form   Rip   Conve

Engineering ▼

Create a seam, 0.0625, on one of the surfaces; confirm the cut goes through the part.

Flat Pattern can only be created in the *Sheet Metal Mode*.

Unbend ▼

Bend ▼
Bend Back
Flat Pattern
Bends ▼

Offs
Exte
Split
Editin

Flat Pattern

Chapter 11
Advanced Modeling Techniques

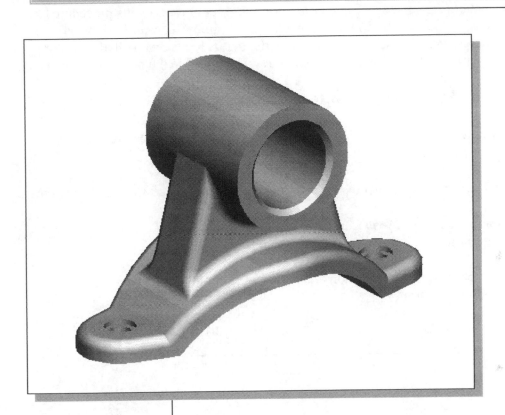

Learning Objectives

- ♦ **Perform the Boolean – Intersect Operation**
- ♦ **Apply Draft Angles to surfaces**
- ♦ **Create complex 3D Rounds and Fillets**
- ♦ **Use the Feature Copy command**
- ♦ **Use the CHAMFER command**
- ♦ **Adjust the Model Color**
- ♦ **Using Render Studio to create photorealistic images**

Introduction

This chapter is intended to provide a summary of the basic construction techniques presented in the previous chapters. We will also demonstrate the more advanced construction techniques such as using the *Boolean-Intersect operation*, the procedure to create *drafted surfaces*, *sketched holes*, *chamfers*, and *complex 3D rounds*; these are common characteristics to casting and molded parts. In this lesson, we will also explore some of the variations of using the basic techniques introduced in the previous chapters.

Summary of Modeling Considerations

- **Design Intent** – determine the functionality of the design; select features that are central to the design.

- **Order of Features** – consider the parent/child relationships necessary for all features.

- **Dimensional and Geometric Constraints** – the way in which the constraints are applied determines how the components are updated.

- **Relations** – consider the orientation and parametric relationships required between features and in an assembly.

The Bracket Design

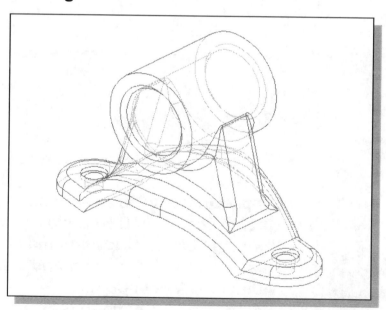

❖ Based on your knowledge of *Creo Parametric*, how many features would you use to create the design? Which feature would you choose as the **base feature** of the model? Identify the more difficult features in the design and consider the possibilities in creating the design. You are encouraged to create a similar design on your own prior to following through the tutorial.

Modeling Strategy

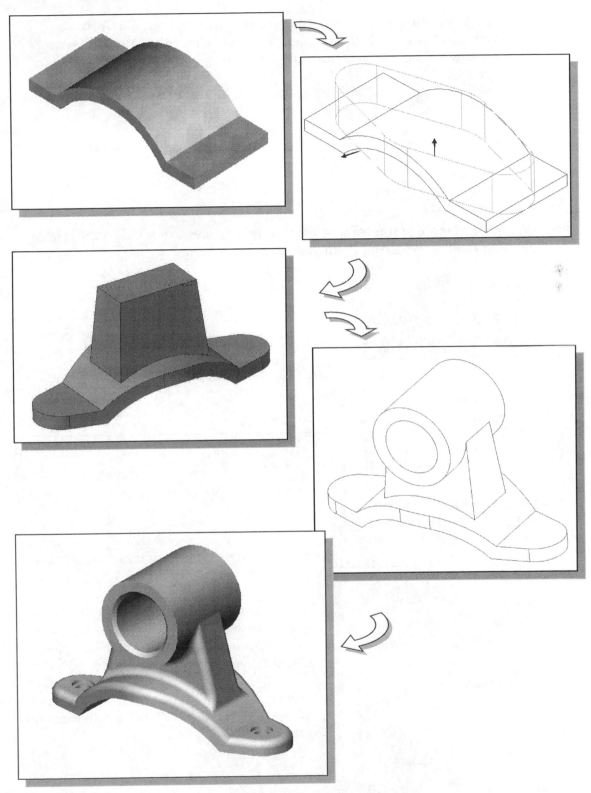

Starting Creo Parametric

1. Select the **Creo Parametric** option on the *Start* menu or select the **Creo Parametric** icon on the desktop to start *Creo Parametric*. The *Creo Parametric* main window will appear on the screen.

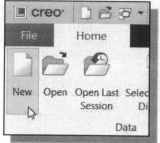

2. Click on the **New** icon, located in the *Standard* toolbar, as shown.

3. On your own, start a new solid part file using ***Bracket*** as the part **Name**.

4. Confirm the **Use default template** option is turned ***on*** so that the system is using the *Creo Parametric* default (**Inch-lbm-Second**) settings.

5. Click on the **OK** button to accept the settings.

Create the Base Feature

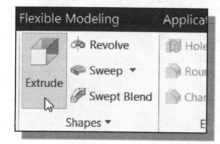

1. In the *Shapes* toolbar, select the **Extrude** tool option as shown.

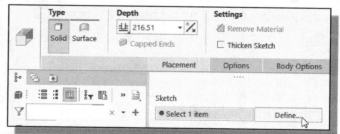

2. Click the **Placement** option and choose **Define** to create a new *internal sketch*.

3. Set up the **FRONT** datum plane as the sketch plane with the **RIGHT** datum plane facing the **right** edge of the computer screen, and pick **Sketch** to enter the *Sketcher* mode.

- Note the default selection of the two datum planes, **RIGHT** and **TOP**, as the references for the 2D sketch.

4. Create the 2D section as shown in the figure below. Note the sketch is symmetrical with respect to the vertical axis and the coincident center point of the two arcs is aligned to the vertical axis.

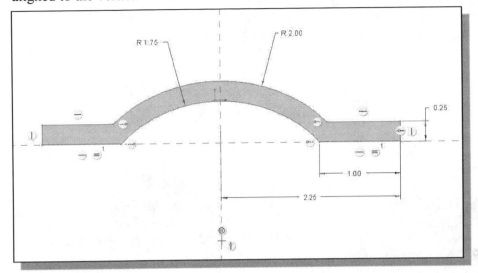

5. Complete the protrusion feature by extruding in *both directions* of the sketching plane and a *depth* of extrusion of **2.0**.

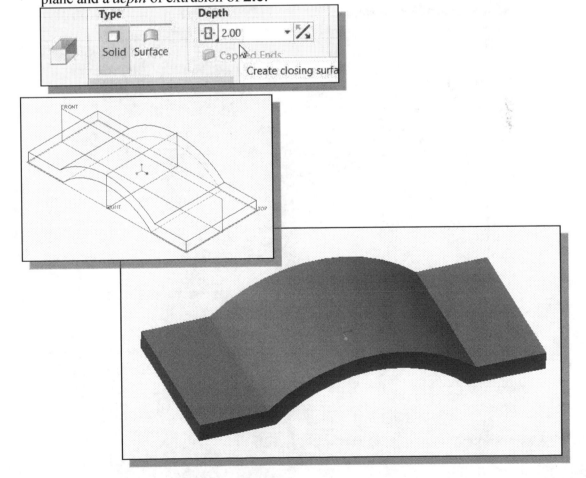

Using the CUT Option to Create a Boolean INTERSECT

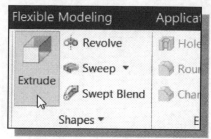

1. In the *Shapes* toolbar, select the **Extrude** tool option as shown.

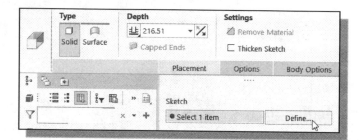

2. Click the **Placement** option and choose **Define** to begin creating a new *internal sketch*.

3. Pick the **TOP** datum plane as the sketch plane.

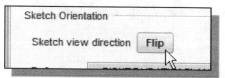

4. Click the **Flip** icon to change the arrow direction so that the arrow is pointing in the direction shown in the figure below.

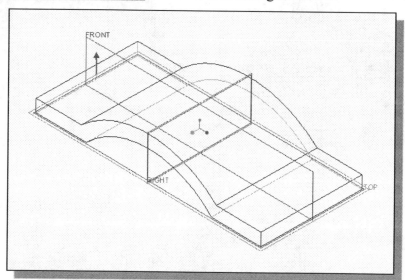

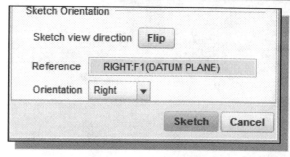

5. Confirm the **RIGHT** datum plane is facing the **right** edge of the computer screen and pick **Sketch** to enter the *Sketcher* mode.

6. On your own, add the four outer edges of the base feature as **additional references** for the 2D sketch to be created.

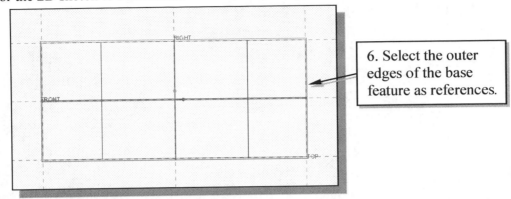

6. Select the outer edges of the base feature as references.

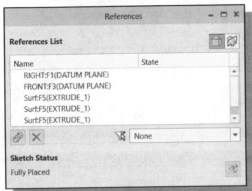

7. In the *References* window, confirm datum planes **RIGHT**, **FRONT** and the four edges we selected are listed as references for the 2D sketch. Click the **Close** button to accept the settings.

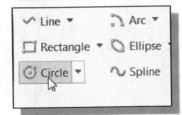

8. In the *Sketching* toolbar, select **Circle** as shown. The default option is to create a circle by specifying the center point and a point through which the circle will pass.

9. Create the three circles as shown. Note all three centers are on the horizontal axis and tangent to the outer edges of the base feature. No dimension is needed for the middle circle since the center is aligned to the origin.

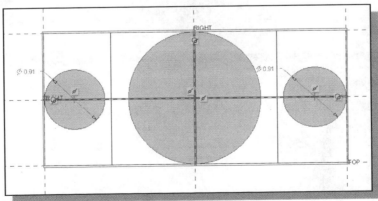

10. Add the four tangent lines and set the two smaller circles to be the same diameter as shown. (Hint: Use the Tangent and Equal constraints.)

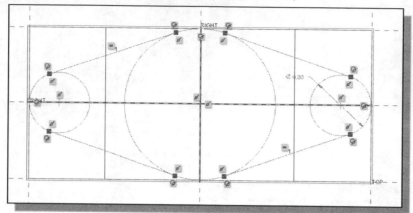

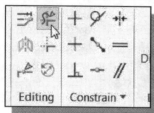

11. Select **Delete Segment** (*Dynamic Trim*) in the *Editing* toolbar. The Dynamic Trim function allows us to quickly remove unwanted portions of entities by simply clicking on them.

12. On your own, modify the sketched geometry (radius **0.5**) as shown in the figure below.

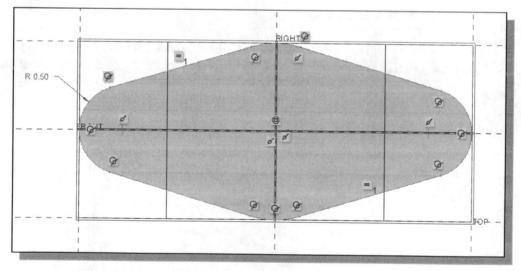

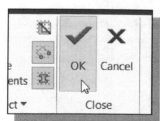

13. Click on the **OK** icon to accept the completed section.

14. In the feature option menu, switch *on* the Remove Material option and set the *extrude* option to **Thru All** as shown.

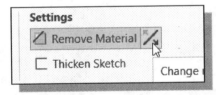

15. The first **Flip-Direction** option is used to switch the extrude *direction.* On your own, set the Remove Material direction as shown.

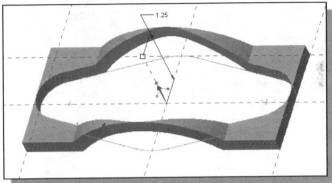

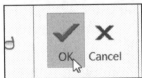

16. Click the second **Flip-Direction** option to set the Remove Material direction.

17. Pick **OK** in the feature option menu to create the blended **Intersect** feature.

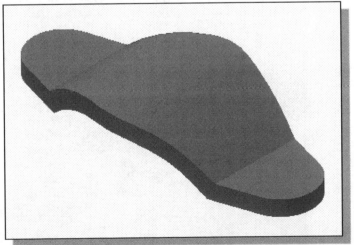

Create a Reference Plane

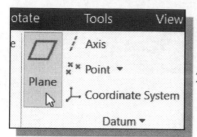

1. On your own, deselect the last feature we created.

2. In the *Datum* toolbar, select **Datum Plane** tool to start the creation of a new datum plane.

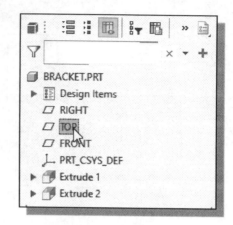

3. Pick datum plane **TOP** in the *Model Tree* window.

4. Place the datum plane at an offset distance of **2.0** from datum plane **TOP**.

5. Pick **OK** in the *Datum Plane* dialog box.

❖ Datum plane DTM1 is exactly 2.0 inches above datum plane TOP.

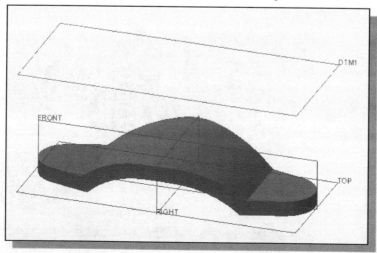

❖ In *Creo Parametric*, **datum planes** can be constructed with the following options:

- **Through**: Create a datum plane through the selected references.

- **Normal**: Create a datum plane perpendicular to the selected reference.

- **Parallel**: Create a datum plane parallel to the selected reference.

- **Offset**: Create a datum plane by specifying a distance or a specific location away from the selected reference.

- **Angle**: Create a datum plane by specifying an angle relative to the selected reference.

- **Tangent**: Create a datum plane tangent to the selected reference.

Create another Extrusion Feature

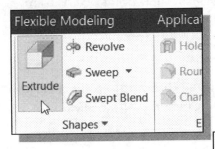

1. In the *Shapes* toolbar, select the **Extrude** tool option as shown.

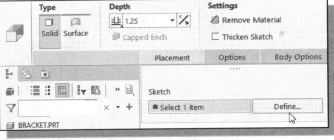

2. Click the **Placement** option and choose **Define**.

3. Pick the **DTM1** datum plane as the sketch plane.

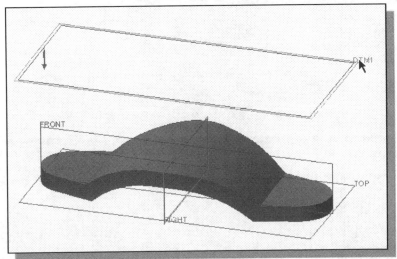

4. Confirm the view direction points downward and datum plane **RIGHT** is set as the *reference plane*, Orientation **Right**, as shown.

5. Click **Sketch** to enter the *Sketcher* mode.

6. On your own, open the *References* window and confirm datum planes **RIGHT** and **FRONT** are listed as references for the 2D sketch. Click the **Close** button to accept the settings.

7. On your own, use the **Rectangle** command and create the rectangle (**1.5 x 0.75**) as shown.

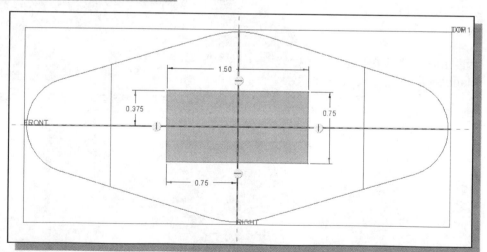

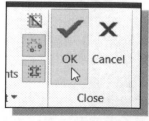

8. Click on the **OK** icon to accept the completed section.

9. Click the **Flip** icon once to change the extrude direction so that the arrow points toward the solid model as shown in the figure below.

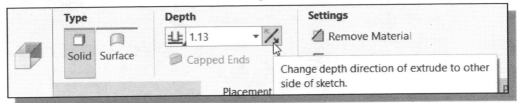

10. Set the *depth* option to **Extrude to Selected Surface** as shown.

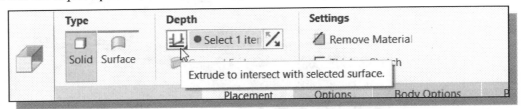

11. Pick the top surface of the solid model as shown.

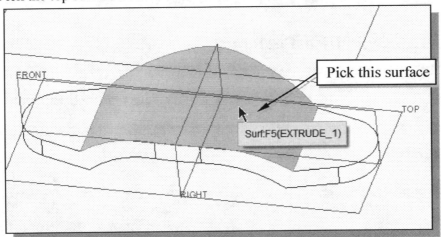

Pick this surface

Surf:F5(EXTRUDE_1)

FRONT

TOP

RIGHT

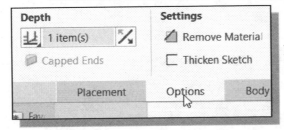

12. In the dashboard, click on the Options tab.

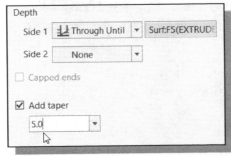

13. Click on the **Add taper** option, and set the taper angle to **5.0** degrees as shown.

- Note the additional extrusion distance option, allowing different extrusion distance on both sides, is also available.

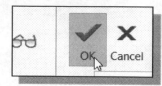

14. Pick **Done** in the feature option menu to create the feature.

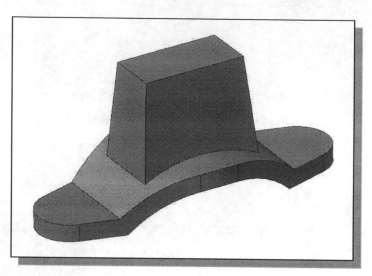

Create the Top Cylindrical Feature

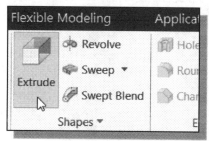

1. In the *Shapes* toolbar, select the **Extrude** tool option as shown.

2. Click the **Placement** option and choose **Define**.

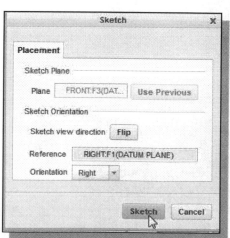

3. On your own, set up the **FRONT** datum plane as the sketch plane with the **RIGHT** datum plane facing the **right** edge of the computer screen, and pick **Sketch** to enter the *Sketcher* mode.

4. On your own, open the references window.

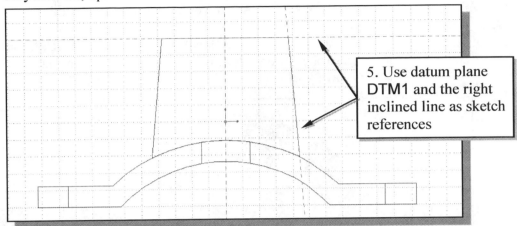

5. Use datum plane DTM1 and the right inclined line as sketch references

5. In the *References* window, remove the datum plane **TOP** from the reference list; choose datum planes **RIGHT**, **DTM1** and the right inclined line as references for the 2D sketch. Click the **Close** button to accept the settings.

6. In the *Sketching* toolbar, select **Circle** as shown. The default option is to create a circle by specifying the center point and a point through which the circle will pass.

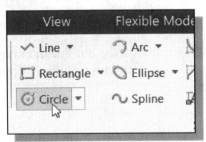

7. Create a **Circle** that is aligned to the sketch references as shown.

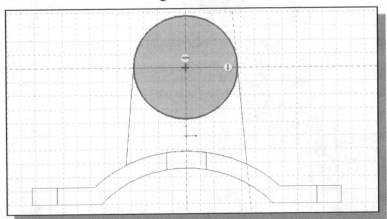

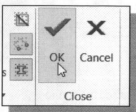

8. Click on the **OK** icon to accept the completed section.

9. Complete the cylinder (length **1.75**) using the **Both Sides** extrusion option. The complete model should appear as shown.

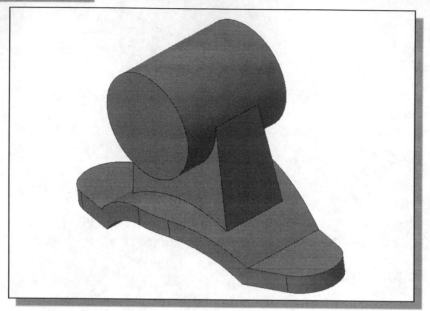

Create a Hole through the Cylinder

On your own, create a **1.0** diameter hole through the cylinder we just created. The completed solid model should appear as shown in the figure below.

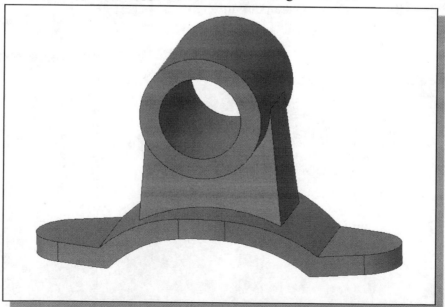

Create a Sketched Hole on the Base Feature

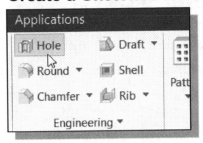

1. In the *Shapes* toolbar, select the **Hole** tool option as shown.

2. In the *Feature Option Dashboard*, select the **Sketch** option as shown.

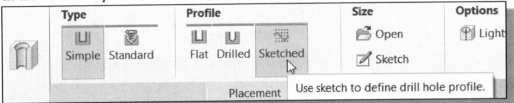

3. In the *Feature Option Dashboard*, click **Activate Sketcher**.

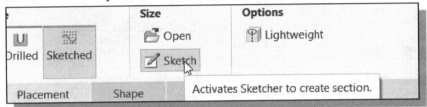

❖ The *sketched hole* option allows us to create a 2D section, which is then used to revolve about an axis and form the hole (similar to a revolved cut feature).

4. Create a rough sketch as shown (a vertical **centerline** and six regular lines) and modify the dimensions as shown in the figure below.

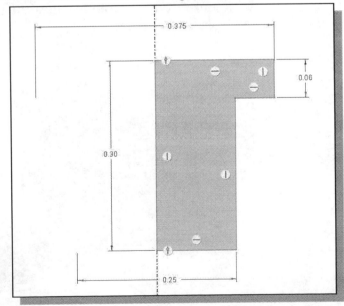

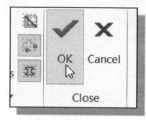

5. Click on the **OK** icon to accept the completed section.

6. Pick the top face of the base as shown as the *primary placement reference*.

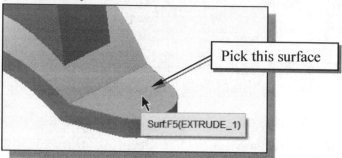

Pick this surface

Surf:F5(EXTRUDE_1)

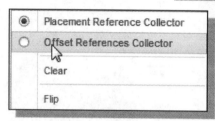

7. Inside the *graphics area*, press and hold down the **right-mouse-button** to bring up the *option menu*.

8. In the *pop-up list*, select **Offset References Collector**.

9. Inside the *Model Tree* area, press down the **[Ctrl]** key and select datum planes **RIGHT** and **FRONT** as the two *secondary references*.

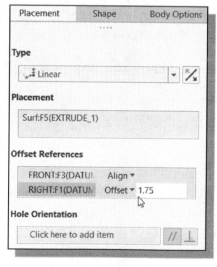

10. On your own, set up the Offset *distances* to **1.75** and **Align** as shown.

11. Pick **OK** in the **Hole** feature menu to create the placed hole feature.

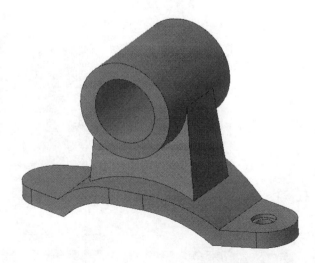

Create a Mirror Feature

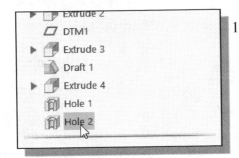

1. In the model tree window, click on the last **Hole** feature we just created.

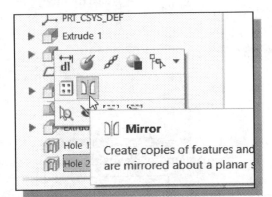

2. In the Option list toolbar, select **Mirror**.

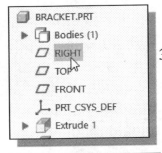

3. Pick datum plane **RIGHT** as the *reference plane* to create the mirror image feature.

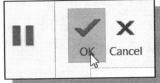

4. Click **OK** in the dashboard to accept the settings and create the mirror feature.

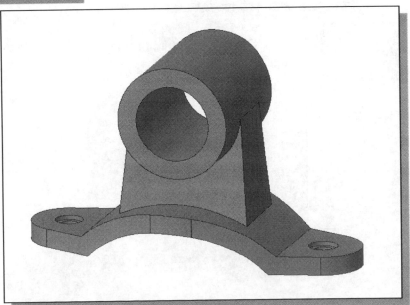

Create Chamfers

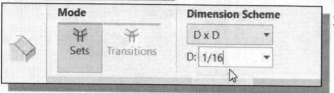

1. In the *Feature* toolbar, select the **Chamfer** tool.

2. In the *Chamfer Feature* menu, confirm the dimensioning scheme is set to D x D as shown in the figure below.

3. In the *Chamfer Feature* menu, enter **1/16** as the chamfer *dimension*.

4. Pick the front and back inside circles as shown.

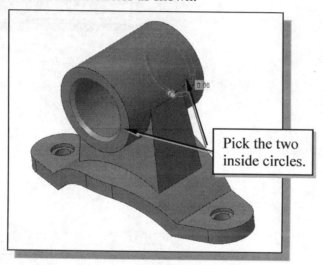

Pick the two inside circles.

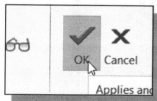

5. Pick **OK** in the feature option menu to create the chamfer feature.

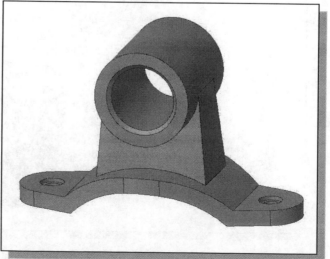

Adding Simple Rounds and Fillets – Edges

1. In the *Engineering* toolbar, select the **Round** tool.

2. In the *Round* option menu, set the *radius* to **0.125** as shown.

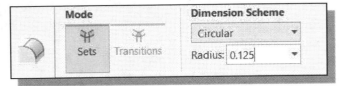

3. Pick the top edges of the base feature as shown below.

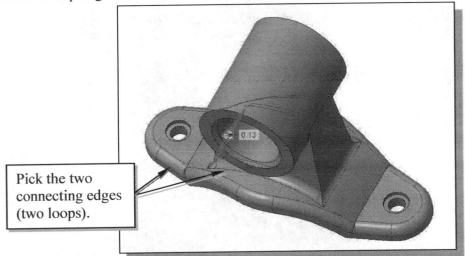

Pick the two connecting edges (two loops).

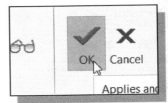

4. Pick **OK** in the feature option menu to create the round feature.

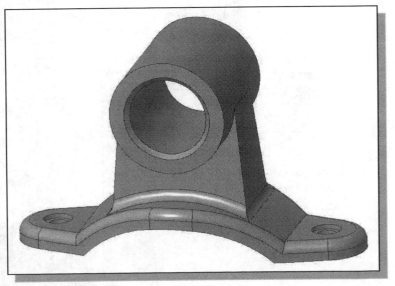

Adding Advanced Rounds and Fillets – Surfaces

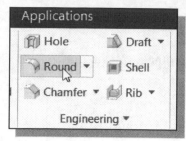

1. In the *Engineering* toolbar, select **Round** tool.

2. In the *Round* option menu, set the *radius* to **0.125** as shown.

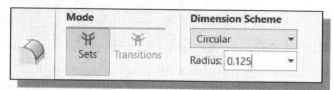

❖ We will create multiple sets of rounds in between the four vertical surfaces.

3. Pick the two surfaces as shown. (Hint: hold down the [**Ctrl**] key.)

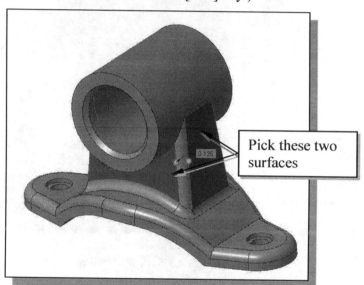

Pick these two surfaces

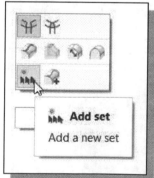

4. Inside the graphics area, press and hold down the **right-mouse-button** to bring up the option menu.

5. In the pop-up menu, select **Add set** to create another rounded edge.

6. Pick the back face and the right adjacent surface to form the second round as shown.

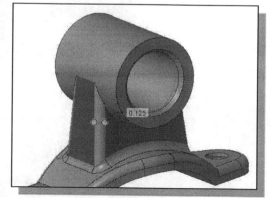

7. Repeat the above steps and create the four rounded corners as shown in the figure below.

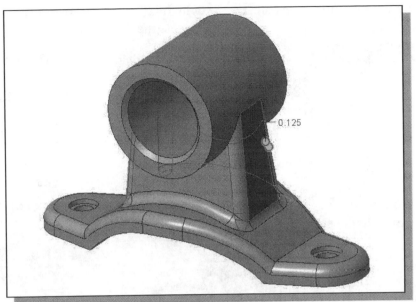

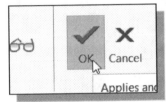

8. Pick **OK** in the feature option menu to create the feature.

Add the Last Round Feature

The last feature will be the fillets around the cylinder. Note that 3D rounds and fillets are complex geometry and it may be necessary to create them individually.

1. In the *Engineering* toolbar, select the **Round** tool.

2. In the *Round* option menu, set the *radius* to **0.125** as shown.

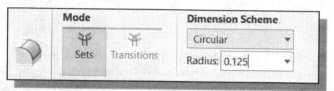

3. Pick the **two surfaces** (one cylindrical and one vertical) as shown.

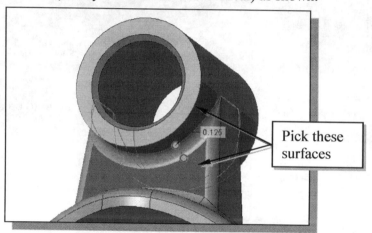

Pick these surfaces

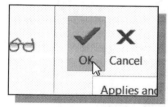

4. Pick **OK** in the feature option menu to create the feature.

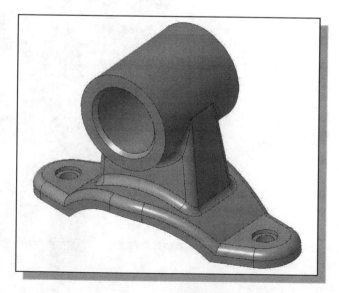

Changing Model Color

In *Creo Parametric*, changing model color is a two-step procedure: (1) define the color desired, and (2) apply the color to the model.

1. Select **Appearances** in the *Model Display* toolbar as shown.

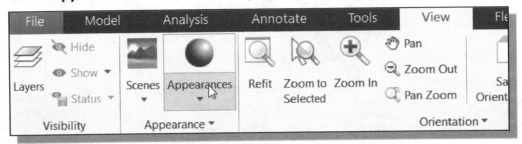

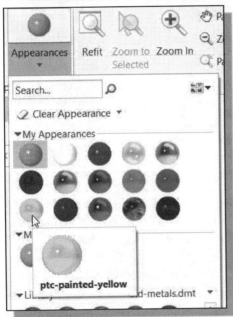

2. Note that *Creo Parametric* comes with a set of pre-defined appearances that can be used by clicking on the icons.

3. Select the **ptc-painted-yellow** appearance as shown.

4. Select the part name in the model tree.

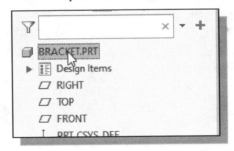

5. **Middle mouse** click once to accept the selection.

6. Select **Appearances** in the *Model Display* toolbar again.

7. Click the **Edit Model Appearances** option to define a new appearance setting.

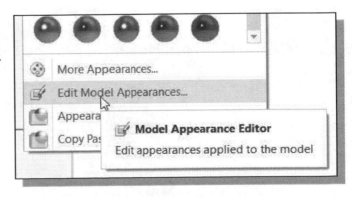

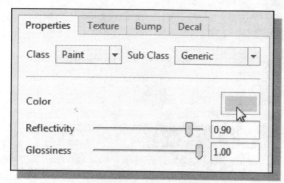

8. In the *Appearance Properties* window, click the **Color** button to open the *Color Editor*.

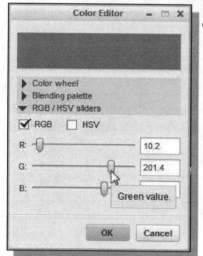

9. In the *Color Editor* window, control the amount of **red**, **green** and **blue** in the color being defined by dragging the **RGB** sliders.

10. Pick **OK**, in the *Color Editor* window, to accept the settings.

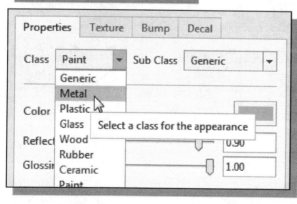

11. In the *Model Appearance Editor* window, click the **Class** list to view the available material list.

12. Choose Metal as shown.

13. Pick the **Close** icon to exit the *Appearance Editor* window.

Assign Material Property and Save the Part

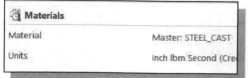

1. On your own, assign **Steel_Cast** as the model material.

2. Save the part with the file name, ***Bracket***.

- Note the bracket design will be used in the next chapter, where the procedure for creating assembly models is illustrated.

Photorealistic Renderings

Photorealistic renderings can be made of your model in Creo Parametric using **Render Studio**. The main steps for creating a rendering include applying Appearances and defining Scenes and defining a view. The view for the rendering can be one of the standard orthographic or perspective views or can be established using a camera view. **Appearances** can be used to make your model appear more realistic through the application of features including colors, material properties, transparency, illumination, and surface finish. **Scenes** can be applied and edited using the Creo Parametric library of default scenes. In this section we will activate the Render Studio module and create a rendering of the bracket design by applying a Scene in Creo Parametric.

Activating Render Studio

1. In the *Applications* tab, click the **Render Studio** button as shown.

- Note that, by default, the **Real-Time Rendering** option is activated and the rendering is done in the graphics window.

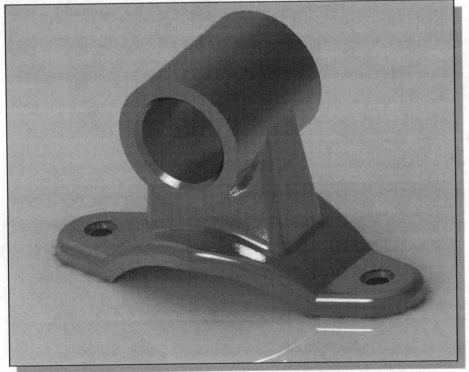

2. In the *Color Editor* window, control the amount of **red**, **green** and **blue** in the color being defined by dragging the **RGB** sliders.

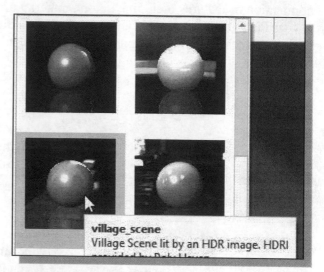

3. Choose **Village Scene** as shown. Note that Creo will import the background scene, which may take a while to complete.

4. On your own, use the dynamic viewing functions, rotate, pan, zoom to reposition the model inside the scene.

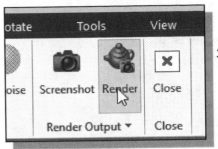

5. Choose Render to start a high-resolution image to be saved.

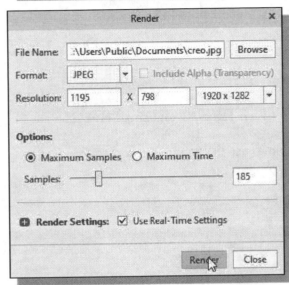

6. On your own, examine the different options available in the Render dialog box.

7. Pick **Render** in the *Color Editor* window, to accept the settings.

8. Pick the **Close** icon to exit the *Render* window.

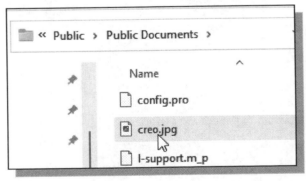

9. On your own, open the rendered image to view the saved high-resolution image.

Review Questions:

1. In *Creo Parametric*, which command can we use to adjust the tapered angle of a surface?

2. Will dimensions be updated when the geometry is modified?

3. Describe the *Boolean Intersect* operation.

4. How do you modify the directions of the arrows of a dimension?

5. What is the main difference between *straight* and *sketched* holes?

6. What is the difference between *coaxial* and *radii* holes?

7. When extruding, what is the difference between **Thru All** and **Until Next**?

8. Create sketches showing the steps you plan to use to create the two models shown on the next page:

Ex.1)

Ex.2)

Exercises:

1. **Switch Base** (Dimensions are in inches.)

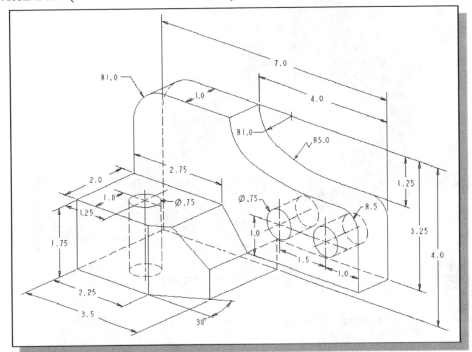

2. **Angle Support** (Dimensions are in inches.)

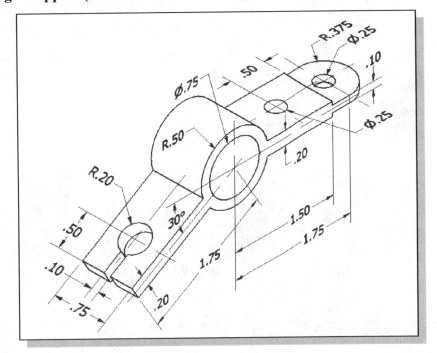

3. **Shaft coupler** (Dimensions are in inches.)

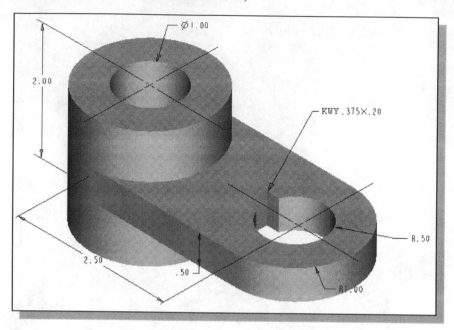

4. **Anchor Bracket** (Dimensions are in inches. Hint: Use the Profile Rib tool.)

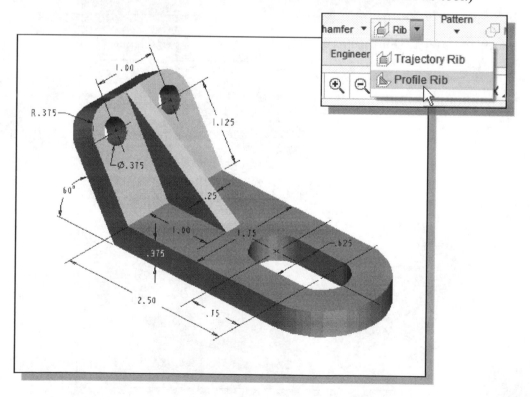

5.   Pivot Latch (Dimensions are in inches.)

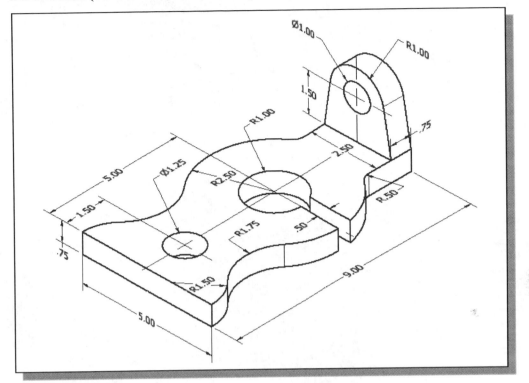

Notes:

Chapter 12
Assembly Modeling – Putting It All Together

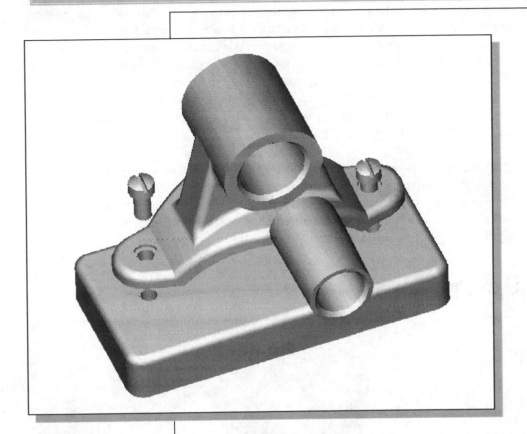

Learning Objectives

- ♦ **Understand the Assembly Modeling Methodology**
- ♦ **Place Parts in the Assembly Mode**
- ♦ **Understand and Utilize Assembly Constraints**
- ♦ **Create Subassemblies**
- ♦ **Understand the Bi-directional Full Associative Functionality**
- ♦ **Create Exploded Assemblies**

Introduction

In the previous lessons, we have gone over the fundamentals of creating basic parts and drawings. In this lesson, we will demonstrate how to create and modify assembly models. *Creo Parametric* provides full associative functionality in all *Creo Parametric* modules, including assemblies. When we change a part model, *Creo Parametric* will automatically reflect the changes in all assemblies that use the part. We can also modify a part in an assembly. The bi-directional full associative functionality allows us to make very flexible model modifications in *Creo Parametric*. Many parallels exist between assembly design and part design. The main task in creating an assembly is establishing the assembly relationships between parts. In this lesson, we will construct an assembly model using the **Bracket** part, which was created in the previous lesson.

To assemble parts into an assembly, we will need to consider the assembly relationships between parts. It is a good practice to assemble parts based on the way they would be assembled in the actual design. We should also consider breaking down the assembly into smaller subassemblies, which helps the management of parts. In *Creo Parametric*, a subassembly is treated the same way as a single part during assembling.

The BRACKET Assembly

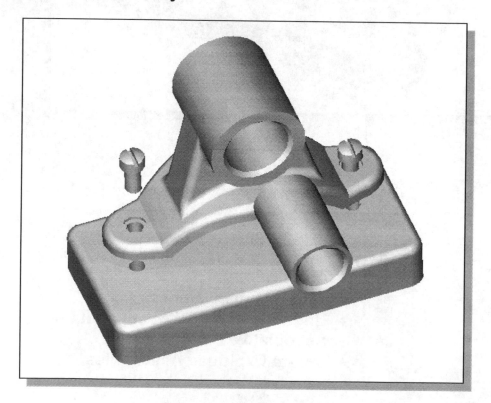

Assembly Modeling Methodology

The *Creo Parametric Assembly Modeler* provides tools and functions that allow us to create 3D parametric assembly models. An assembly model is a 3D model with any combination of multiple part models. *Parametric assembly constraints* can be used to control relationships between parts in an assembly model.

Creo Parametric can work with any of the assembly modeling methodologies:

The Bottom Up approach

The first step in the *bottom up* assembly modeling approach is to create the individual parts. The parts are then pulled together into an assembly. This approach is typically used for smaller projects with very few team members.

The Top Down approach

The first step in the *top down* assembly modeling approach is to create the assembly model of the project. Initially, individual parts are represented by names or symbolically. The details of the individual parts are added as the project gets further along. This approach is typically used for larger projects or during the conceptual design stage. Members of the project team can then concentrate on the particular section of the project to which he/she is assigned.

The Middle Out approach

The *middle out* assembly modeling approach is a mixture of the bottom-up and top-down methods. This type of assembly model is usually constructed with most of the parts already created, and additional parts are designed and created using the assembly for construction information. Some requirements are known, and some standard components are used, but new designs must also be produced to meet specific objectives. This combined strategy is a very flexible approach to creating assembly models.

The different assembly modeling approaches described above can be used as guidelines to manage design projects. Keep in mind that we can start modeling our assembly using one approach and then switch to a different approach without any problems.

In this chapter, the *bottom up* assembly modeling approach is illustrated. All of the parts (components) required to form the assembly are created first. *Creo Parametric's Assembly Modeling* tools allow us to create complex assemblies by using components that are created in separate part files or in the same part file. A component can be a subassembly or a single part, where features and parts can be modified at any time. The sketches and sections used to build part features can be fully or partially constrained. Partially constrained features may be adaptive, which means the size or shape of the associated parts are adjusted in an assembly when the parts are constrained to other parts. The basic concept and procedure of using the adaptive assembly approach is demonstrated in the tutorial.

Additional Parts

Besides the *Bracket* design from the previous chapter, we will need three additional parts: (1) *Base Plate*, (2) *Bushing* and (3) *Cap Screw*. Create the three parts as shown. Notice the positions of the parts in relation to the *datum planes*. Place all of the part files, including the Bracket part from the previous chapter, in the same folder.

(1) *Base Plate*

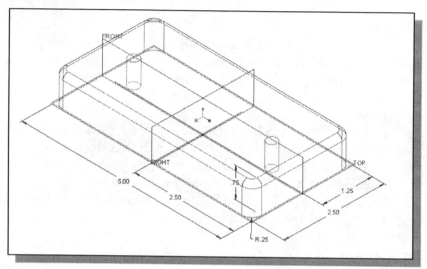

Material: Steel, Hole sizes: Ø 0.25 and 0.6 deep.
Center to center distance of the two holes: 3.5.
Rounds: R 0.125

(2) *Bushing*

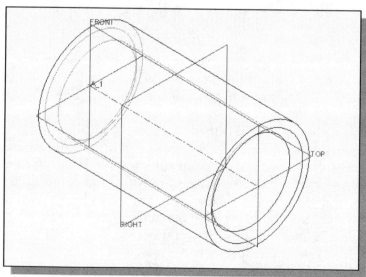

Material: Bronze, ID: Ø 0.75, OD: Ø 1.00, Length: 1.625
Chamfers: 45° × 0.0625

(3) *Cap Screw*

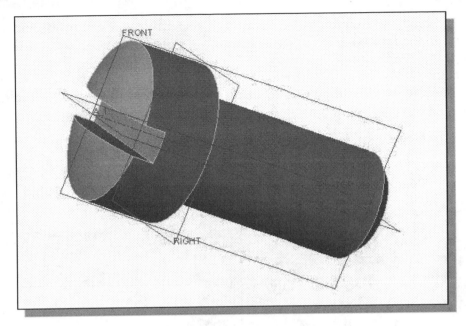

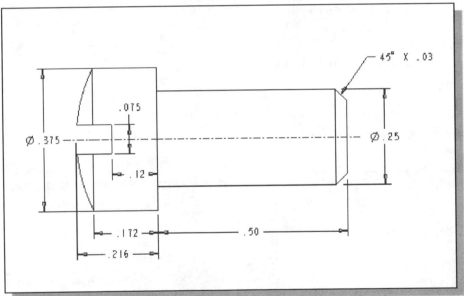

❖ We will omit the threads in this model. Threads are complex three-dimensional curves. You are encouraged to experiment with the thread option: **Insert → Helical Sweep**. Create the three parts and save the parts as separate part-files (***Base_Plate***, ***Bushing***, and ***Cap_Screw***).

● Note that many of the 3D models of standard parts are also available for download at the companies' and/or their distributors' websites.

Create a Subassembly

We are now ready to assemble the components together. We will start by assembling the **_Bracket_** and the **_Bushing_** into a subassembly.

1. Select the **Creo Parametric** option on the *Start* menu or select the **Creo Parametric** icon on the desktop to start *Creo Parametric*. The *Creo Parametric* main window will appear on the screen.

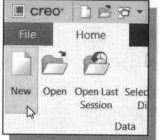

2. Click on the **New** icon, located in the *Quick Access toolbar*, as shown.

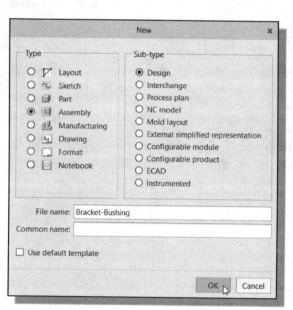

3. In the *New* form, select **Assembly** in the **Type** list.

4. Enter **_Bracket-Bushing_** as the assembly model file **Name**.

5. Turn *off* the **Use default template** option.

6. Click the **OK** button to continue. The *New File Options* window appears on the screen.

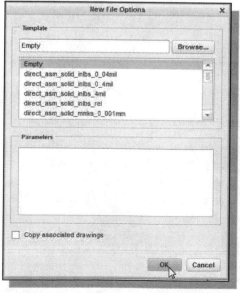

7. In the *New File Options* dialog box, select **Empty** in the option list to not use any template file.

8. Click the **OK** button to accept the settings and enter the *Assembly Modeling* mode.

Retrieve the Bracket Component

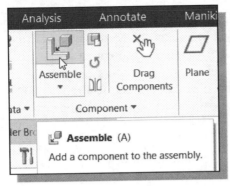

1. Pick **Assemble** in the *Component* toolbar to begin placing different parts into the assembly model.

❖ The *Open* window showing a list of part model files appears on the screen.

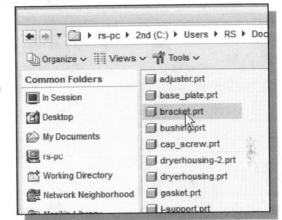

2. Select the **Bracket** (part file: *bracket.prt*) from the previous chapter in the list window and pick **Open** to retrieve the model.

❖ The **Bracket** model is placed into our assembly.

Retrieve the *Bushing* Component

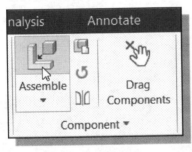

1. Pick **Assemble** in the *Component* toolbar to begin placing different parts into the assembly model.

2. Select the **Bushing** (part file: *bushing.prt*) in the list window and pick **Open** to retrieve the model.

❖ In the display area, the **Bushing** is placed next to the **Bracket** and the *Component Placement* window appears. Do not **Exit** the command.

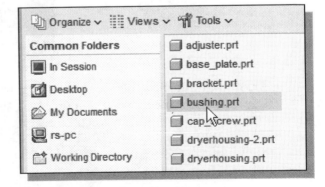

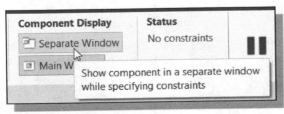

❖ The default setting of *Creo Parametric* will display the new component in the same *Assembly* window. Note that the new component can also be displayed in a separate window.

Placement Constraints

To assemble components into an assembly, we need to establish the assembly relationships between components. It is a good practice to assemble components the way they would be assembled in the actual design. *Placement constraints* create a parent/child relationship that allows us to capture the design intent of the assembly. Because the component that we are placing actually becomes a child to the already assembled components, we must use caution when choosing constraint types and references to make sure they reflect the intent.

The following is a list of the most commonly used placement constraints:

- **Coincident** – Selected surfaces point in opposite directions and become coplanar.

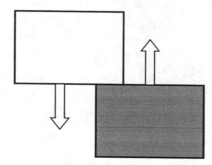

- **Coincident – Flip Direction** – Selected surfaces point in the same direction and are made coplanar.

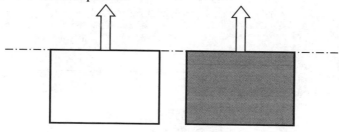

- **Offset** – Selected surfaces point in opposite directions and are offset by a specified distance. We can modify the offset dimension at any time.

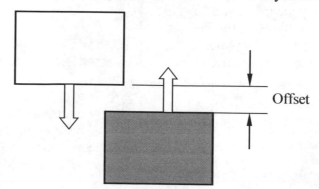

Offset

- **Angle Offset** – Selected surfaces are offset by a specified Angle. We can modify the offset dimension at any time.

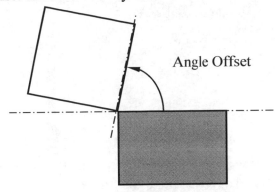

Angle Offset

- **Tangent** – Selected surfaces become Tangent. The point of contact is a tangent.

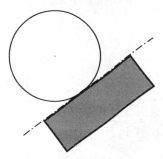

- **Fix** – This constraint will fix the current location of a component.

- **Coplanar** – This constraint will position selected surfaces to be aligned on the same plane.

❖ Note that besides surfaces, *Creo Parametric* also allows the use of points and edges when applying the placement constraints.

Placing the *Bushing*

1. In the *View* toolbar, switch *off* the display of the datum planes, datum axis and coordinate systems.

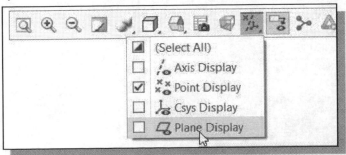

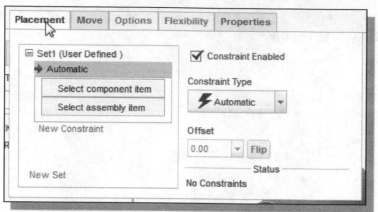

2. In the *Constraint Options Dashboard,* click the **Placement** option as shown.

3. Pick the **outer cylindrical surface** of the *Bushing.*

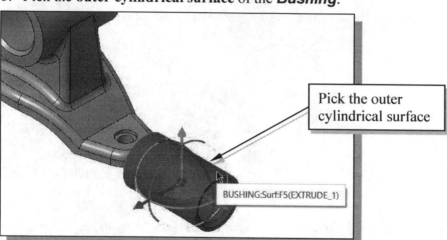

Pick the outer cylindrical surface

BUSHING:Surf:F5(EXTRUDE_1)

❖ Notice in the *Component Placement* window, the selected surface is identified.

Pick the inner cylindrical surface

BRACKET:Surf:F10(EXTRUDE_5)

4. Pick the **inner cylindrical surface** of the *Bracket.*

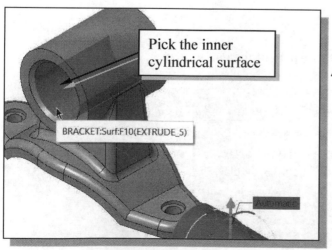

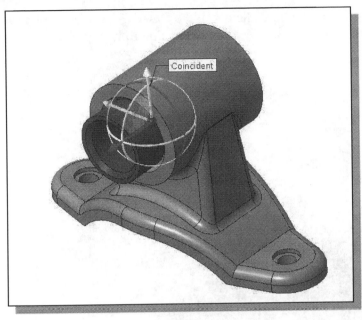

❖ In the display area, the **Bushing** is moved to a new location. Use the *Dynamic Viewing* functions to confirm that the centers are aligned.

➤ The three axes and three rings on the bushing part indicate the available degrees of freedom of the associated part.

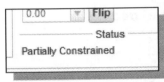

❖ At the bottom of the *Component Placement* window, *Creo Parametric* indicates the status of **Bushing** placement as **Partially Constrained**. This is because the **Bushing** can still slide along the aligned axis.

❖ In general, we need to specify two or three constraints to fully assemble a component. We will add another placement constraint to further constrain the assembly.

Create the Second Placement Constraint

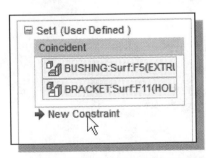

1. Click **New Constraint** to add an additional constraint.

• A new constraint, within the same constraint set of the first constraint, is activated.

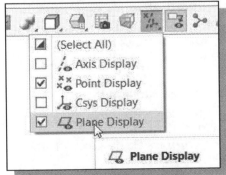

2. In the *Display Control* toolbar, switch **on** the display of the datum planes.

3. Pick the **reference plane** that is perpendicular to the *Bushing axis*, as shown in the figure below. (If necessary, set the selection filter option to help pick the proper surface.)

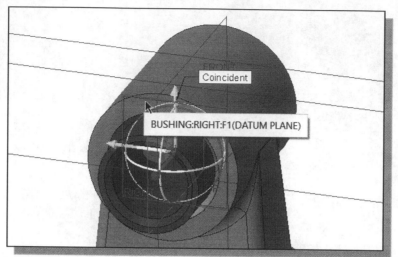

4. Pick the corresponding *reference plane* associated with the *Bracket* part.

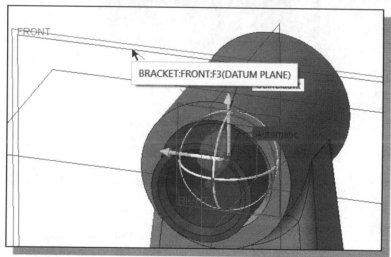

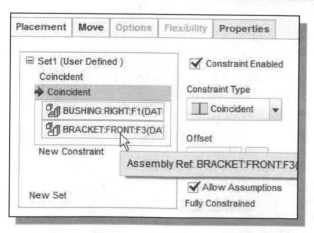

5. Set the *Constraint Type* to **Coincident** as shown. Note that the components of the placement constraints are displayed. We can redefine the selection if necessary.

• Near the bottom of the display, *Creo Parametric* indicates the status of placement of the *Bushing* as **Fully Constrained**.

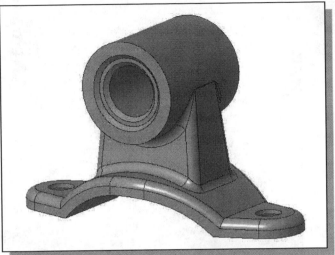

6. Pick **OK** to complete the placement of the *Bushing* component.

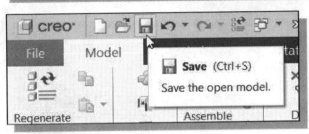

7. Pick **Save** in the toolbar.

8. Click **OK** to accept the default filename and save the subassembly on disk.

Create the Main Assembly Model

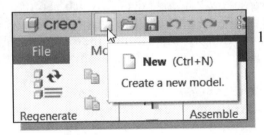

1. Pick the **New** icon in the *Quick Access* toolbar. (We can also use the key combination **Ctrl-N** to start a new object.)

2. In the *New* form, select **Assembly** in the **Type** list.

3. Enter ***Bracket-Assembly*** as the assembly model file Name.

4. Turn *off* the **Use default template** option.

5. Click the **OK** button to continue.

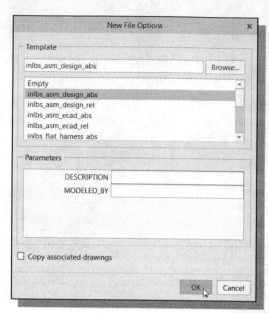

6. In the *New File Options* dialog box, select **inlbs_asm_design_abs** in the option list to use the *English Units* template file.

7. Pick **OK** to continue.

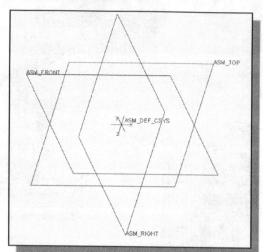

- Note that the displayed coordinate system and datum planes represent the fixed references in 3-D space for the assembly.

Base Component

In creating an assembly model in *Creo Parametric*, we will need to decide which part to use as the first component. In most cases, this *base component* should be one that is **not likely to be removed** from the assembly. For our project, we will use the **Base Plate** as the base component.

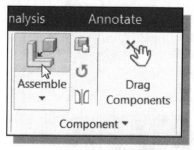

1. Pick **Assemble** in the *Component* toolbar to begin placing different parts into the assembly model.

❖ The *Open* window showing a list of part model files appears on the screen.

2. Select the **Base Plate** (part file: *base_plate.prt*) in the list window and pick **Open** to retrieve the model.

❖ The **Base Plate** model is displayed in the graphics area and the *Assembly Constraints Dashboard* is activated.

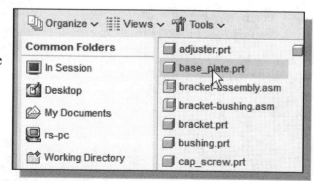

Fully Constraining the Base-Plate to the Assembly

We will remove all degrees of freedom by aligning the datum planes of the *Base Plate* part to the datum planes of the assembly.

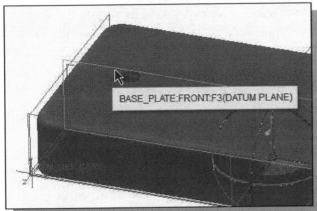

1. Pick one of the datum planes of the *Base Plate* part.

2. Select the corresponding datum plane of the assembly to align the two planes.

3. On your own, add additional constraint sets and align the other two datum plane sets. Three **coincident** constraints are used.

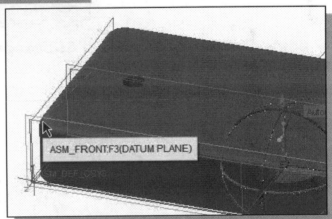

❖ Note the color of the *Base Plate* part turned orange, indicating it is fully constrained.

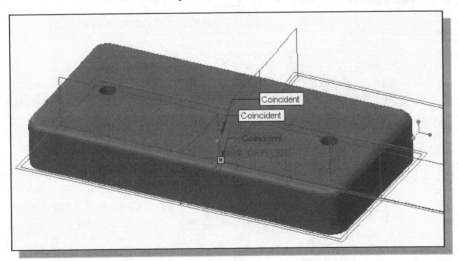

Retrieve the Bracket-Bushing Subassembly

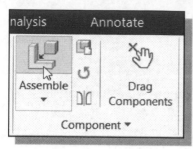

1. Pick **Assemble** in the *Component* toolbar to begin placing different parts into the assembly model.

❖ The *Open* window showing a list of part model files appears on the screen.

2. Select the **Bracket-Bushing** (assembly file: *bracket-bushing.asm*) in the list window and pick **OPEN** to retrieve the model.

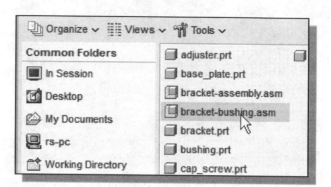

Placing the Bracket-Bushing Subassembly

1. In the toolbar, switch *off* the display of the datum planes and switch *on* the display of the datum axis.

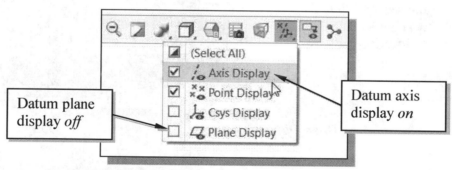

Datum plane display *off*

Datum axis display *on*

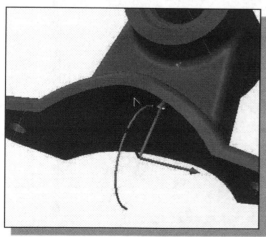

2. Drag the **red ring**, to rotate the subassembly about the x-axis, so that the front surface of the subassembly is tilted upward. (Refer to the images on the next page if necessary.)

• In *Creo Assembly*, the location and orientation of a part can be adjusted by using the **drag and drop** method.

3. Pick the **bottom surface** of the *Bracket* as shown.

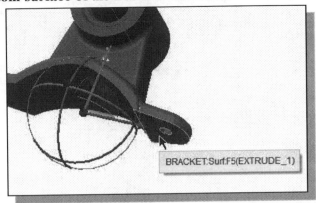

❖ Notice the selected surface is identified by a leader and the text "Automatic"; this indicates the default option is used.

4. Pick the **top plane** of the *Base Plate* as shown.

5. Click on the **Placement** tab and notice the Angle Offset constraint is applied. Note that the constraint applied with the *Automatic* option can be adjusted if necessary.

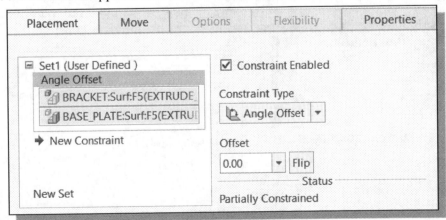

6. In the **Constraint Type** option list, choose **Coincident** as shown.

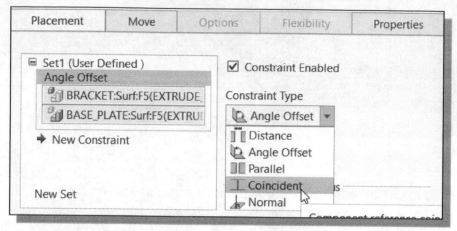

❖ Note that the applied constraint is adjusted; the two parts now align horizontally.

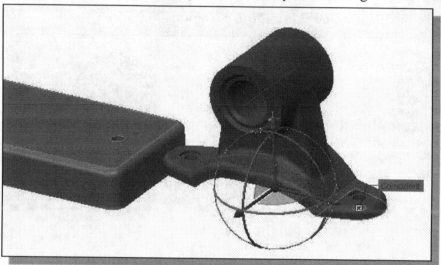

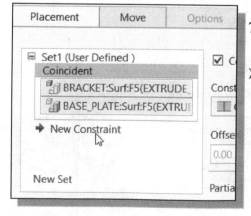

7. Click **New Constraint** to start another constraint in between the two parts.

➢ Note the new constraint is still part of the same set, **Set1**, as the previous constraint.

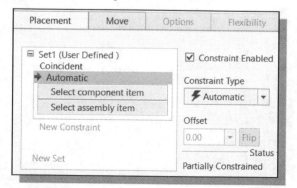

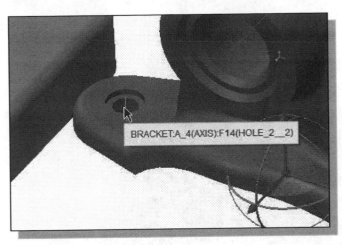

8. Pick the **left center axis**, A_4, of the *Bracket* part as shown.

9. Pick the **left axis**, A_1, of the *Base Plate* to apply the constraint between the two *bolt holes*.

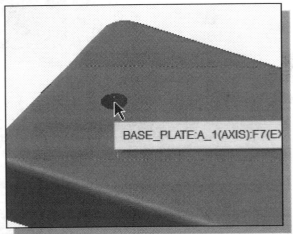

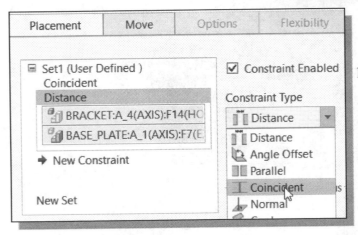

10. In the Constraint Type option list, choose **Coincident** as shown to align the two bolt holes.

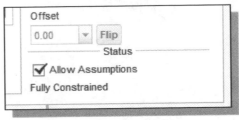

❖   At the bottom of the *Component Placement* window, *Creo Parametric* indicates the status of the placement of the ***Bracket-Bushing*** as **Fully Constrained**.

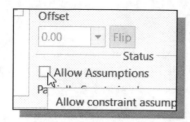

11. **Uncheck** the **Allow Assumptions** box, and notice the degrees of freedom indicator reappeared on the subassembly.

➢ In most cases, the Allow Assumptions option would apply adequate constraints to aid the alignment of parts.

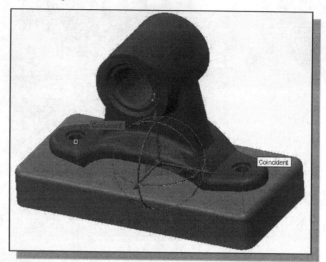

12. Drag the **green ring**, the available DOF, to rotate the subassembly in the Y-axis direction. Note that the subassembly will rotate about the left axis of the *Base Plate* while maintaining all of the applied constraints.

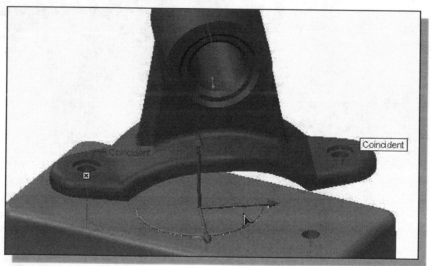

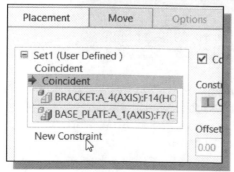

13. Click **New Constraint** to start another constraint in between the two parts.

14. On your own, apply another **Coincident** constraint to align the other set of bolt holes.

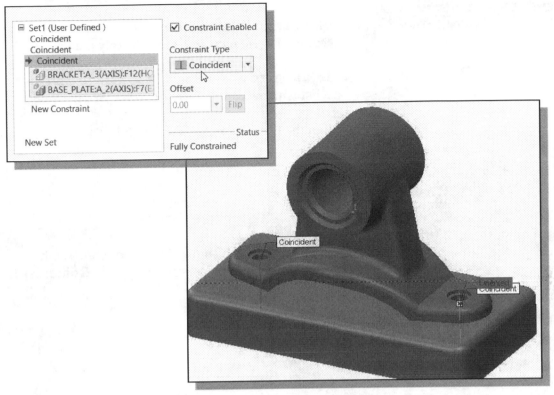

15. Click **OK** to complete the placement of the subassembly.

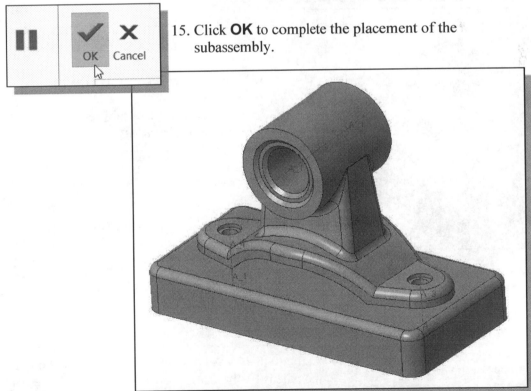

Placing the Cap Screws

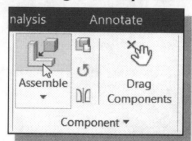

1. Pick **Assemble** in the *Component* toolbar to begin placing different parts into the assembly model.

2. Select the **Cap Screw** (part file: *cap-screw.prt*) in the list window and pick **Open** to retrieve the model.

3. Align the two *planar surfaces* as shown.

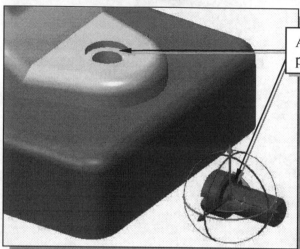

Align the two small planar surfaces

4. In the **Placement** tab, set the Constraint Type to **Coincident**.

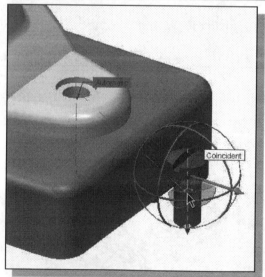

5. Click **New Constraint** to start another constraint in between the two parts.

6. Align the two **center axes** as shown.

7. In the **Placement** tab, set the Constraint Type to **Coincident**.

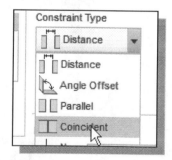

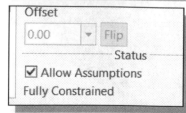

8. Use the **Flip** option if the cap-screw part is oriented up-side-down.

9. Activate the **Allow Assumptions** option and notice the part is **Fully Constrained**.

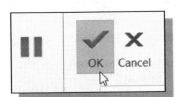

10. Pick **OK** to complete the placement of the subassembly.

11. On your own, repeat the steps and place another *Cap Screw* into the assembly.

❖ Note that in an assembly model, multiple copies of the same part can be placed individually using different constraints. As the assembled part is modified, all copies of the same part are updated automatically. This approach of assembly modeling is extremely flexible and effective.

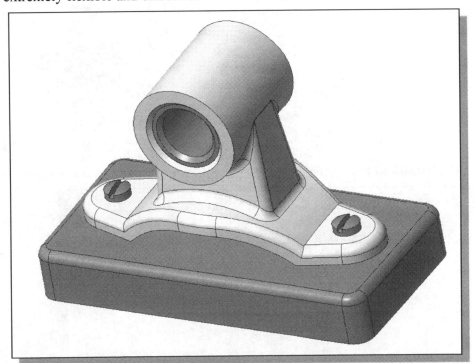

Exploding the Assembly

Exploded assemblies are often used in design presentations, catalogs, sales literature, and in the shop to show all of the parts of an assembly and how they fit together.

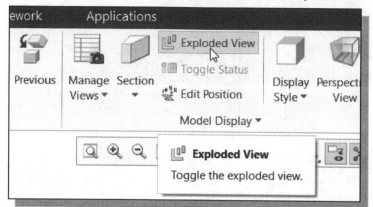

1. In the **View** tab, select **Exploded View** as shown.

❖ Note the assembly is exploded using the default settings.

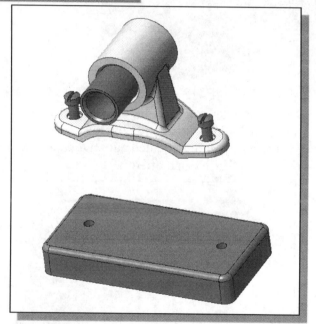

2. Click **Edit Position** to edit the positions of the different parts within the assembly.

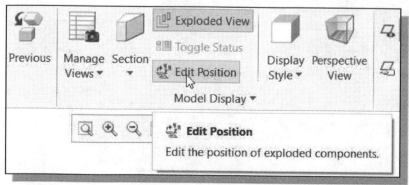

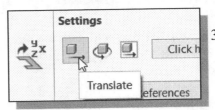

3. Confirm the *motion type* is set to **Translate** as shown.

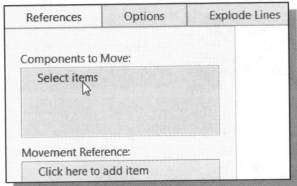

4. Click on the **References** tab to display the movement options as shown.

5. Confirm the Component to Move option is activated as shown.

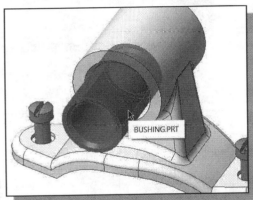

6. Pick the **Bushing** part by selecting any edge of the part. A small 3-axis system is placed at the selection point.

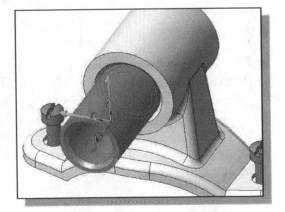

7. To move the **Bushing** part in the length direction, first touch the corresponding axis with the cursor then drag the **Bushing** part with the left-mouse-button.

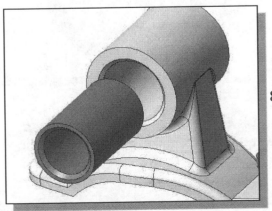

8. Reposition the **Bushing** part so that it is in front of the **Bracket** part.

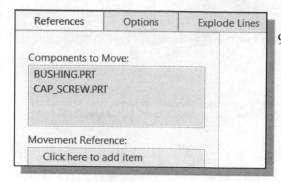

9. Hold down the [**Ctrl**] key and pick one of the **Cap Screws** and notice both pieces can be moved together.

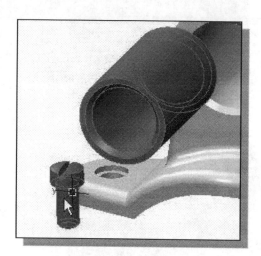

10. Hold down the [**Ctrl**] key and pick the **Bushing** part; notice the part has been de-selected.

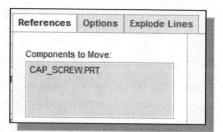

11. On your own, adjust the parts so that they appear as shown in the figure.

12. Pick **OK** to end the **Explode – Edit Position** command.

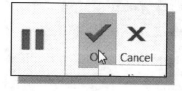

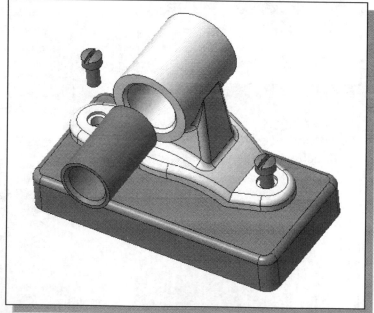

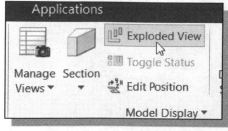

13. In the *Model Display* group, click **Exploded View** again to restore the components to their assembled status.

Bi-directional Associative Functionality

The *bi-directional associative functionality* of *Creo Parametric* allows us to change the design at any level, and the system reflects the change at all levels automatically.

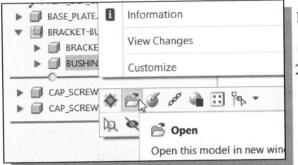

1. In the *Model History* tree, expand the Bracket bushing subassembly.

2. Choose **Open** on the Bushing part option list, through the *Model History* tree as shown.

3. On your own, modify the inside diameter of the bushing to **0.5**. Confirm the part is updated before continuing to the next step.

❖ The associative functionality of *Creo Parametric* assures the model is updated in all levels, including the assembly models.

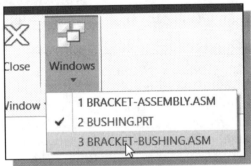

4. Select **Window** in the *View* tab and pick **BRACKET-BUSHING.ASM**. Notice the size of the **Bushing** is updated in the subassembly.

5. Select **Window** in the *View* tab and pick **BRACKET-ASSEMBLY.ASM**. Notice the size of the **Bushing** is updated in *Assembly* mode as well.

➢ We will next modify the dimension value in the *Bracket-Assembly* window.

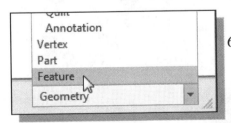

6. Pick **Features** in the *selection filter* list as shown.

7. Click on the inside cylinder of the **Bushing** part and select the *Edit dimensions* as shown.

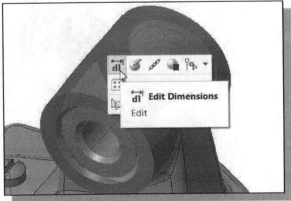

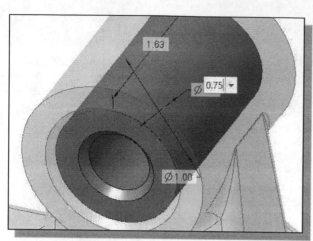

8. Modify the inside diameter by double clicking on the dimension text (**Ø 0.5**).

9. In the *value box*, enter **0.75** as the new diameter.

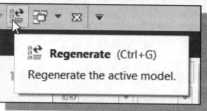

10. In the *Quick Access toolbar* area, select **Regenerate**.

11. On your own, switch to the other windows and confirm that the inside diameter of the **Bushing** part is updated in all models.

❖ With the full bi-directional associative functionality, when we **Save** an assembly in *Creo Parametric*, any changes that we made to any of the parts in the assembly are also saved. When we save a part with any changes, the assembly is automatically updated. Consequently, we can experiment with any changes in a part without being concerned about saving the changes in the assembly. The *bi-directional associative functionality* allows us greater freedom to concentrate on design while leaving the tedious tasks to *Creo Parametric*.

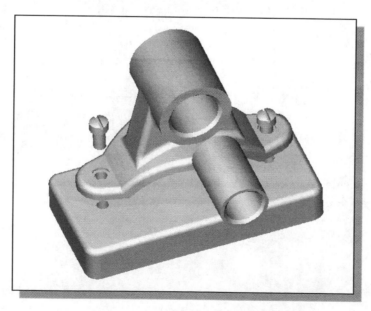

Create a Drawing of the Assembly Model

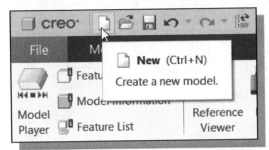

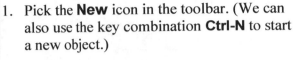

1. Pick the **New** icon in the toolbar. (We can also use the key combination **Ctrl-N** to start a new object.)

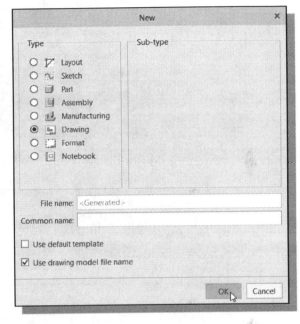

2. In the *New* form, select **Drawing** under the **Type** list.

3. Toggle *off* the **Use default template** option.

4. Activate *Use drawing model file name* so the drawing file name is the same as the model file name.

5. Pick **OK** to start a new drawing file.

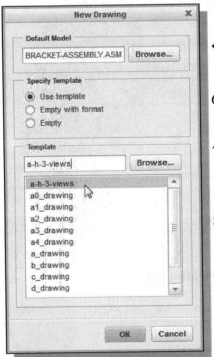

❖ In the *New Drawing* form, the currently active part is automatically selected as the drawing model.

6. In the **Specify Template** option, choose **Use Template**.

7. Pick **a-h-3views** in the **Template** list. (Use the **Browse** button to locate the *Creo Parametric template files* if necessary.)

8. Pick **OK** to proceed with the drawing template layout.

• Note that the standard three views of the assembly model are automatically generated inside the A-size Title block.

9. First select the **right view**; then select **Delete** in the option menu as shown.

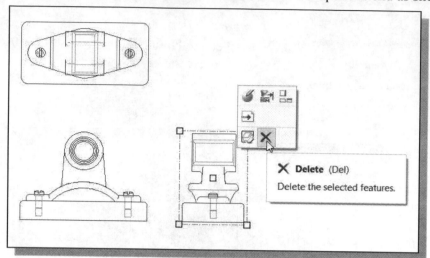

10. Repeat the above step, also delete the **top view**.

11. On your own, set the *default model orientation* to **Isometric** as shown.

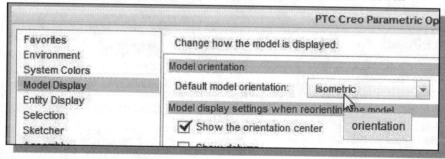

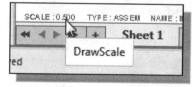

12. Near the bottom of the graphics area, double click on the **Draw Scale** text to enter the *Edit mode*.

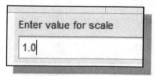

13. Enter **1.0** as the new Draw Scale.

14. Click **OK** to accept the new value.

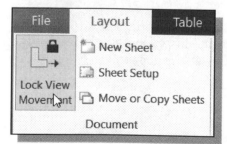

15. Turn off the **Lock View Movement** and reposition the 2D view.

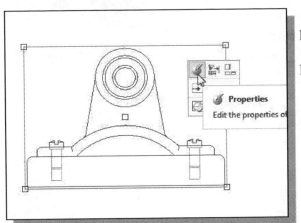

16. Select the 2D View.

17. In the option menu, select **Properties** as shown.

18. In the *View Type section*, set the view orientation to **Default Orientation** as shown.

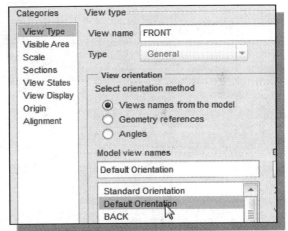

19. In the *View Display section*, set the *display type* to **No Hidden** as shown.

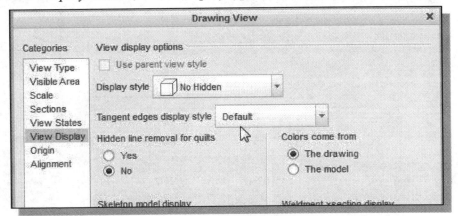

20. Also set the *Tangent edges display style* to **Default** as shown.

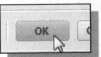

21. Click **OK** to adjust the 2D drawing.

- Note that any of the settings of the 2D drawing can be adjusted; the template file is quite helpful in providing a basic setup.

Create a Parts List

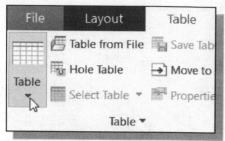

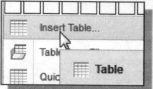

1. In the *Ribbon Toolbar*, click on the **Table** tab as shown.

2. In the *Table toolbar*, click on the *down arrow* below the **Table** button.

3. Choose **Insert Table** in the pull-down list.

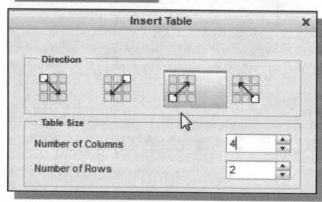

4. Set the Growth direction to start from the lower left corner as shown.

5. For the *Table Size*, set up a **4X2 Table** as shown.

6. Click **OK** to accept the settings.

7. Place the ***Parts List*** in the lower right corner, just above the *Title Block* as shown.

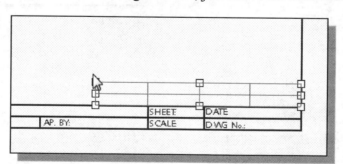

8. Zoom in and select the **first column** of the *Parts List* by moving the cursor near the top edge of the first column and select with the **left-mouse-click**.

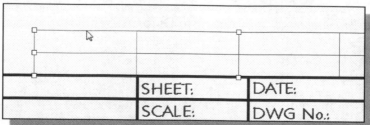

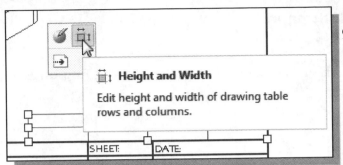

9. In the *option menu,* choose **Height and Width** as shown.

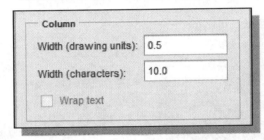

10. Set the *Column Width* to **0.5** as shown.

11. Click **OK** to accept the settings.

12. Repeat the above steps and set the width of the other columns as shown. Note that the table can be repositioned by selecting the table and drag & drop the grip points.

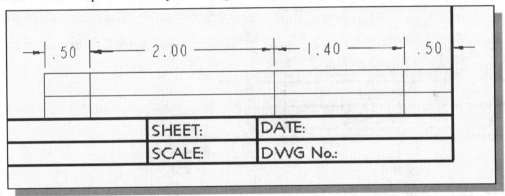

13. Double-click on the lower left cell of the table to enter the *Edit mode* and enter **ITEM** as the cell text. On your own, set the **text font** to **Font** and **height** to **0.125.**

14. Repeat the above steps and set the width of the other columns as shown.

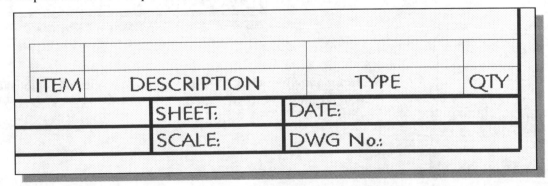

Use the Repeat Region Command

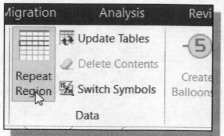

1. In the *Data toolbar*, select **Repeat Region** as shown.

2. In the *Menu Manager*, select **Add → Simple** as shown.

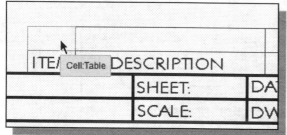

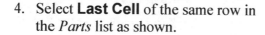

3. Select **Top Left Cell** in the *Parts* list as shown.

4. Select **Last Cell** of the same row in the *Parts* list as shown.

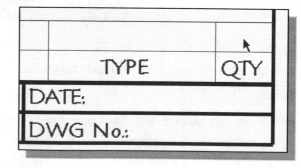

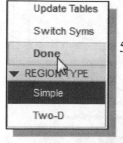

5. In the *Menu Manager*, click **Done** to accept the selected regions to be used in creating information of the assembly.

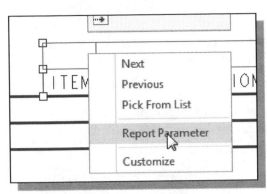

6. Select the **first cell** of the first row of the Parts List as shown.

7. Use the **right-mouse-button** to display the *option menu* and choose **Report Parameter** as shown.

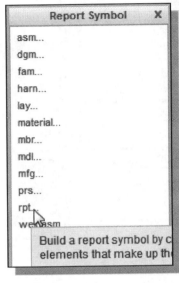

8. In the *Report Symbol* window, select **rpt...** as shown.

9. Then select **Index** as shown.

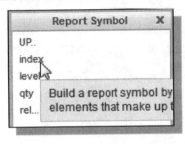

10. Double-click on the next cell to repeat the **Report Parameter** command, and set the 2nd column to **asm... → mbr... → name**

11. On your own, repeat the above steps and set the rest of the column:
    3rd column to **asm... → mbr... → type**
    4th column to **rpt... → qty**

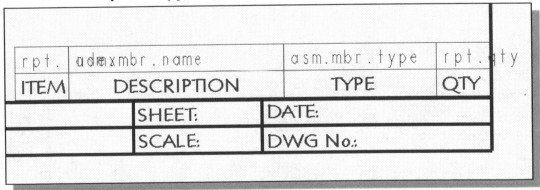

- Note that Creo will create the Parts List using the associated part/assembly properties.

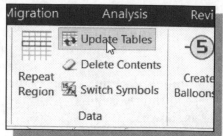

12. Click **Update Tables** to proceed to generate the Parts List for the assembly.

- Notice that the Cap_Screw part are shown twice and there is no number listed in the QTY column.

ITEM	DESCRIPTION	TYPE	QTY
4	CAP_SCREW	PART	
3	CAP_SCREW	PART	
2	BRACKET-BUSHING	ASSEMBLY	
1	BASE_PLATE	PART	

	SHEET:	DATE:
	SCALE:	DWG No.:

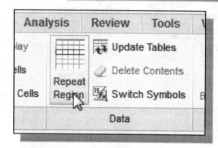

13. In the *Data toolbar*, select **Repeat Region** as shown.

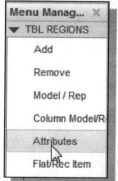

14. In the *Menu Manager*, click **Attributes** to modify the region settings.

15. Click inside the *QTY Region* to select the region.

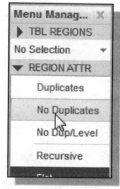

16. In the *Menu Manager*, click **No duplicates** to merge the same parts as one item.

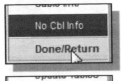

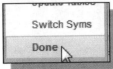

17. Click **Done/Return** to accept the settings.

18. Click **Done** to exit the *Repeat Region* command.

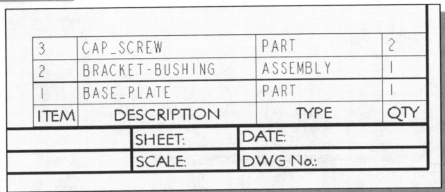

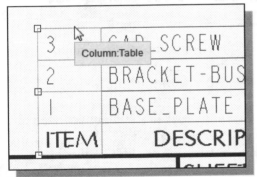

19. Inside the *Parts List*, select the first column.

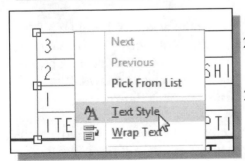

20. Right-mouse click to bring up the option menu and select **Text Style** as shown.

21. Set the *Note/Dimension alignment* to **Center** as shown.

22. Click **OK** to accept the selection.

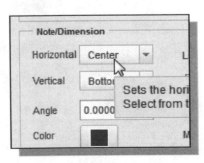

Add the Balloon Callouts

1. In the *Balloons* toolbar, click on the **Create Balloons** button and select **Create Balloons - All** as shown.

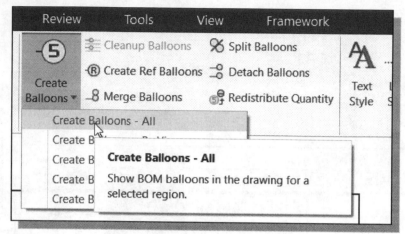

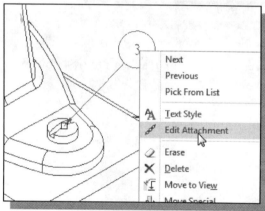

2. Select the balloon for the Cap_Screw part.

3. Use the right-mouse-button on the balloon to bring up the option menu and select **Edit Attachment** as shown.

4. On your own, reattach the balloon to the top of the cap screw part. Click **done/return** to exit the *Edit* mode.

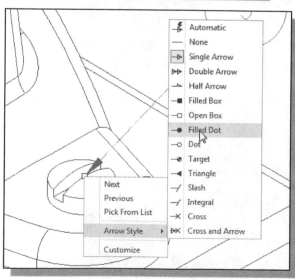

5. Select the arrow of the balloon then use the right-mouse-button on the tip of the arrow to bring up the option menu and select **Arrow Style** → **Filled Dot** as shown.

6. On your own, repeat the above steps and adjust the balloons using the same style and format.

7. Click the text of the balloon, then drag and drop to re-position if needed.

8. On your own, complete and print a hard copy of the assembly drawing of the design.

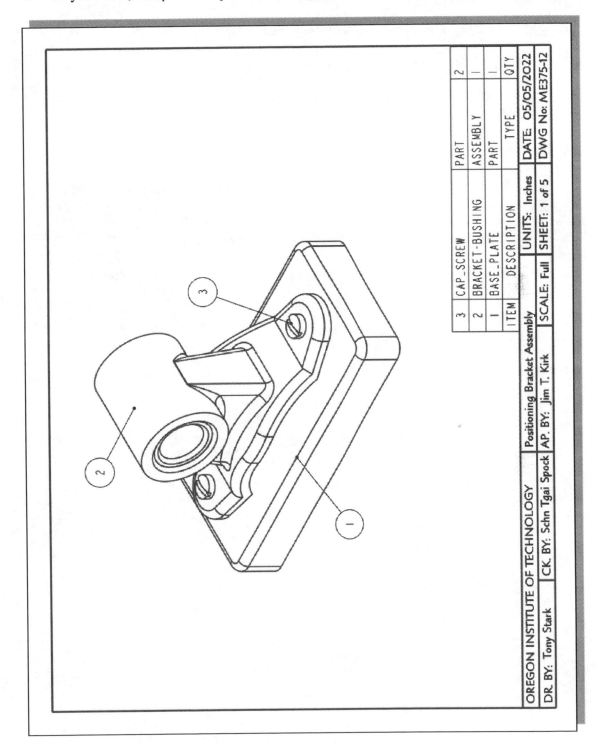

ITEM	DESCRIPTION	TYPE	QTY
3	CAP_SCREW	PART	2
2	BRACKET-BUSHING	ASSEMBLY	1
1	BASE_PLATE	PART	1

OREGON INSTITUTE OF TECHNOLOGY	Positioning Bracket Assembly	UNITS: Inches
DR. BY: Tony Stark	CK. BY: Schn Tgai Spock	AP. BY: Jim T. Kirk	SCALE: Full	SHEET: 1 of 5
		DATE: 05/05/2022
		DWG No: ME375-12

Review Questions:

1. What is the purpose of using *placement constraints*?

2. List four of the commonly used placement constraints.

3. Describe the difference between the **MATE** placement constraint and the **MATE OFFSET** placement constraint.

4. In an assembly, can we place more than one copy of a part?

5. How should we determine the assembling order of different parts in an assembly model?

6. How do we create an exploded assembly in *Creo Parametric*?

7. In *Creo Parametric*, how do we adjust the locations of parts in an exploded assembly?

Exercises:

1. **Wheel Assembly** (Create a set of detail and assembly drawings. All Dimensions are in mm.)

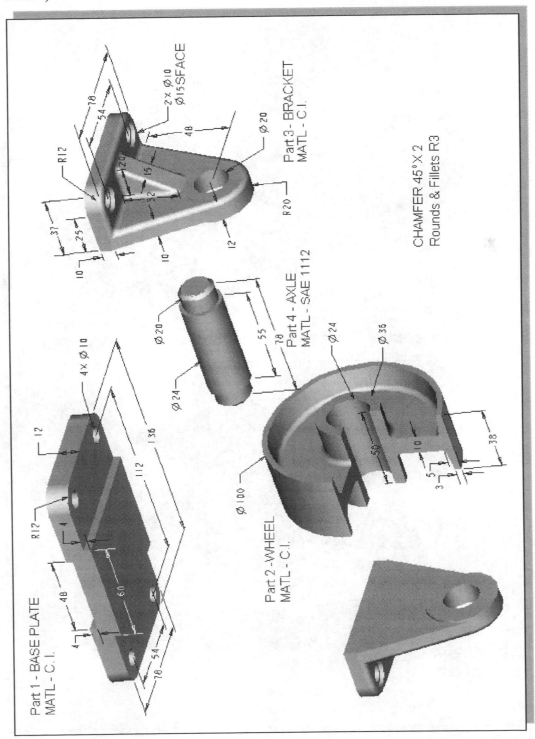

2. **Leveling Assembly** (Create a set of detail and assembly drawings. All Dimensions are in mm.)

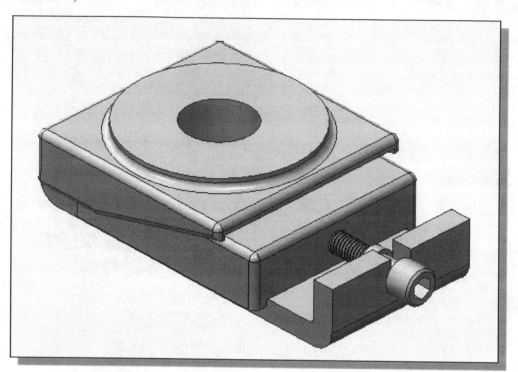

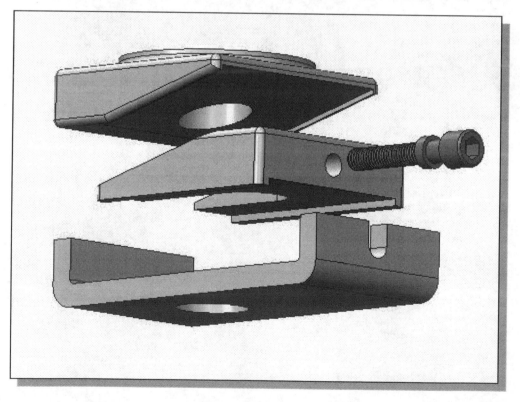

(a) Base Plate

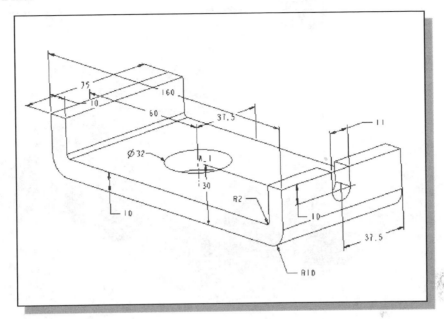

(b) Sliding Block (Rounds & Fillets: R3)

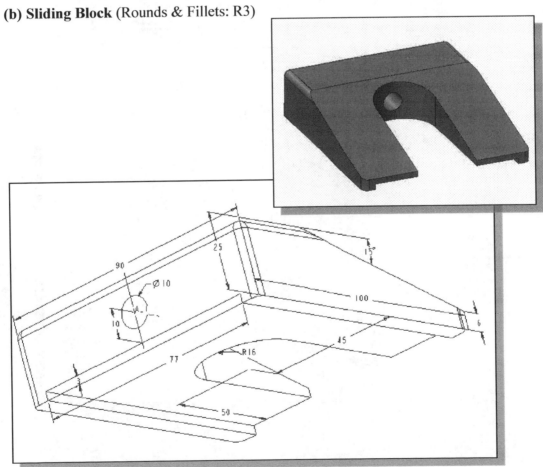

(c) Lifting Block (Rounds & Fillets: R3)

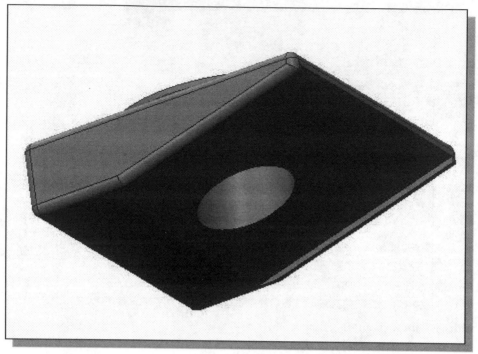

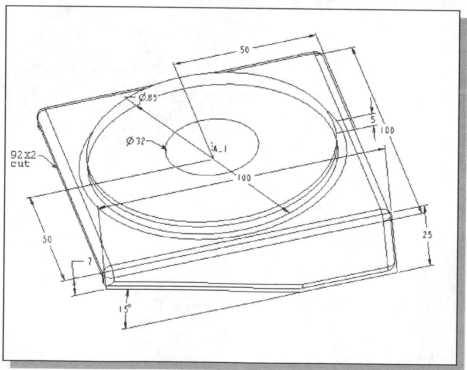

(d) Adjusting Screw (M10 × 1.5)

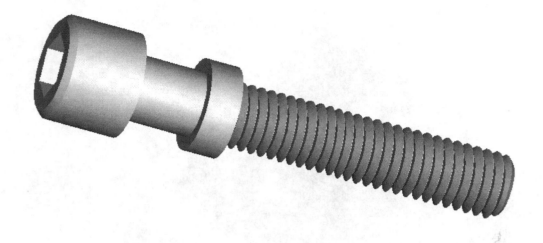

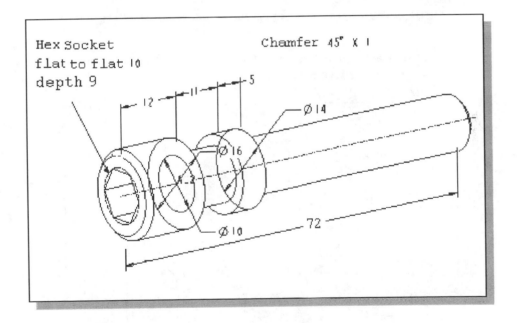

3. **Vise Assembly** (Create a set of detail and assembly drawings. All dimensions are in inches.)

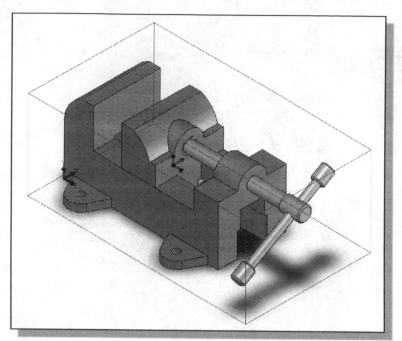

(a) **Base:** The 1.5 inch wide and 1.25 inch wide slots are cut through the entire base.
Material: **Gray Cast Iron**.

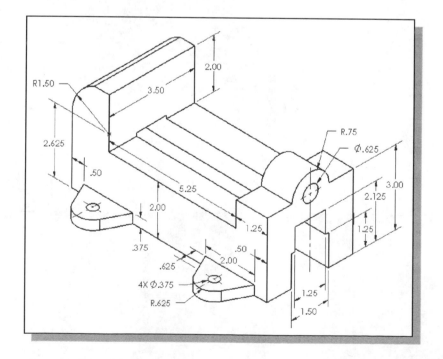

(b) Jaw: The shoulder of the jaw rests on the flat surface of the base and the jaw opening is set to 1.5 inches. Material: **Gray Cast Iron**.

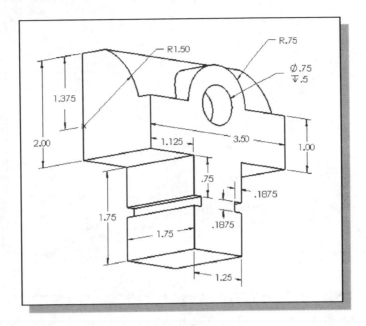

(c) Key: 0.1875 inch H x 0.375 inch W x 1.75 inch L. The keys fit into the slots on the jaw with the edge faces flush as shown in the sub-assembly to the right. Material: **Alloy Steel**.

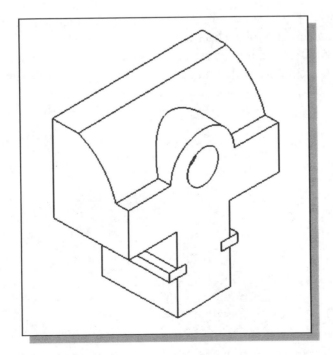

(d) Screw: There is one chamfered edge (1/16 inch x 45°). The flat ⌀ 0.75″ edge of the screw is flush with the corresponding recessed ⌀ 0.75 face on the jaw. Material: **Alloy Steel**.

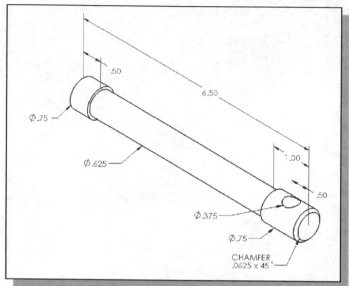

(e) Handle Rod: ⌀ 0.375″ x 5.0″ L. The handle rod passes through the hole in the screw and is rotated to an angle of 30° with the horizontal as shown in the assembly view. The flat ⌀ 0.375″ edges of the handle rod are flush with the corresponding recessed ⌀ 0.735 faces on the handle knobs. Material: **Alloy Steel**.

(f) Handle Knob: There are two chamfered edges (1/16 inch x 45°). The handle knobs are attached to each end of the handle rod. The resulting overall length of the handle with knobs is 5.50″. The handle is aligned with the screw so that the outer edge of the upper knob is 2.0″ from the central axis of the screw. Material: **Alloy Steel**.

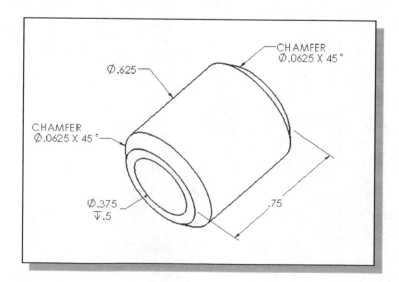

Chapter 13
Advanced Assembly Modeling and Animation

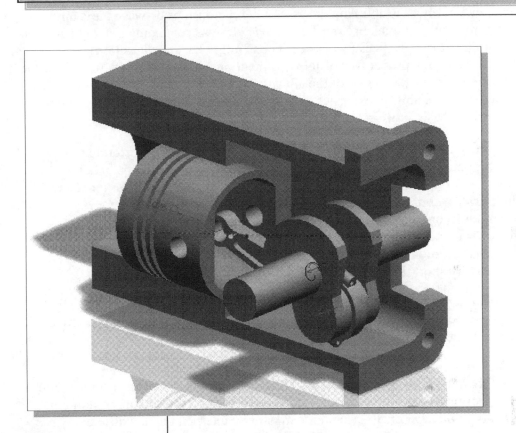

Learning Objectives

- ◆ **Understand the Basic Concepts of performing Motion Analysis**
- ◆ **Place Parts using the Joint Connections**
- ◆ **Create Animation using the Key Frame approach**
- ◆ **Use the Servo Motor to Drive the components in an Assembly**
- ◆ **Output the associated Simulation Video file**

Introduction

In the previous chapter, we went over the fundamentals of creating assembly models. In this chapter, we will examine assembly models that contain moving parts. The main task in creating an assembly is establishing the assembly relationships between parts through the use of assembly constraints. In *Creo Parametric*, designs containing moving parts can also be constrained through specially packaged constraint sets, known as **joint connections**. Placing the proper type of **joint connections** in between parts will allow the proper movements of the parts. One of the advantages of using **joint connections** in assembly models is the ability to transfer the assembly model into the *Creo Animation* module to perform basic **motion analysis**.

In designing machines with moving parts, *motion analysis* is usually performed to confirm the proper assembly of the designs, also to check for any other potential problems. In *Creo Parametric*, several options are available to perform motion analysis, such as the *Creo Mechanism* module, the *Creo Simulate* module, and the *Creo Animation* module, and we can even perform 2D analyses in *Creo Sketcher* mode. The *Creo Mechanism* module can be used to perform a very in-depth motion analysis, while the *Creo Animation* module provides a relatively simple motion analysis that can be done in a relatively short time.

One main advantage of using the *Creo Animation* module is the ability to validate the initial design concepts, check for potential problems and to explore design alternatives without actually creating a physical prototype.

The *Creo Mechanism* module does provide more *kinematics motion* analysis functionality than the *Creo Animation* tool. The *Creo Animation* module is more about showing the motion of a design concept. Motions of components are done by dragging the parts. *Creo Animation* creates animation mostly by interpolating the different positions of the components in an assembly. The *Creo Animation* module is actually a subset of the *Creo Mechanism* module, and many of the tools/procedures are the same in both modules.

In *Creo Animation*, two main approaches are commonly used to create animation: (1) construct the animation by creating *key frames* of the different positions of parts, and/or (2) the use of *servo motors* to provide more specific controls of parts.

The general procedure for creating animations using **key frames** in *Creo Animation* is similar to the creation of cartoon movies, where key frames are first created and the intermediate frames are then added. In *Creo Animation*, **servo motors** are special controls that can be used to drive parts. This allows the simulation of the real-life situation of the machines. Animations generated using the **key frames** approach typically require more work to achieve similar smoothness of motion compared with using the **servo motors** approach.

In this chapter, a crank and slider mechanism, commonly seen in engines and compressors, is used to illustrate the basic concepts and procedures of creating assembly models with connections and animations in *Creo Animation*.

Joint Connections

In *Creo Parametric*, **joint connections** are special types of packaged constraints that can be used to connect moving components. The applied joint connections will constrain the relative motion between the selected components. Each independent movement permitted by a constraint is called a *degree of freedom* (DOF). The degrees of freedom that a constraint allows can be translation and rotation about three perpendicular axes, as shown in the figure below.

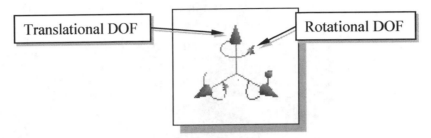

Each component in an assembly has six **degrees of freedom (DOF)**, or ways in which rigid 3D bodies can move: movement *along* the X, Y, and Z axes (translational freedom), plus rotation *around* the X, Y, and Z axes (rotational freedom).

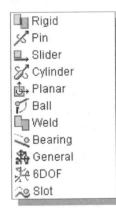

In *Creo Parametric*, the available joint connections include Rigid, Pin, Slider, Cylinder, Planar, Ball, Weld, Bearing, General, 6DOF, and Slot. Joint connections are generally created by applying several of the assembly placement constraints, such as Coincident, Offset, and Insert. For some connection types, the references must be in the same two components of the assembly. Also, note that in an assembly with moving components, instead of completely locking all movements (removing all of the *DOF*s), certain *DOF*s are kept to allow designated movement. For example, a cylinder joint is created by aligning two datum axes in their respective bodies, allowing one translational DOF and one rotational DOF.

In *Creo Parametric* (while applying the joint connection constraints) and also in *Creo Animation*, a small symbol identifying the type of connection is displayed next to the created joint connection. The symbol of a given joint also shows the translational and/or rotational DOF the joint is allowed. For an assembly with moving components, there are four basic symbols: **Pin**, **Slider**, **Planar** and **Cylinder**.

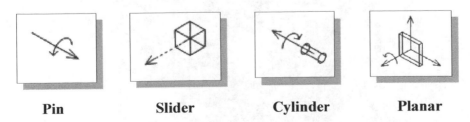

| Pin | Slider | Cylinder | Planar |

The basic four joint connections are applied as follows:

Pin

To apply this type of connection, first **Align** an axis (or a revolved surface) of the moving component to the assembly and then **Align/Mate** a flat surface (or a datum plane) of the moving component to the assembly. Note that both sets of references must be to the same two components in the assembly. This connection type allows only one rotational motion, along the aligned axis.

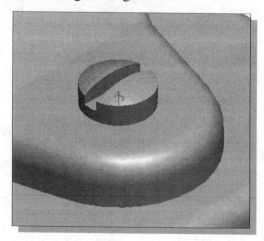

Slider

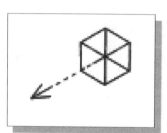

To apply this type of connection, first **Align** an axis (or an edge) of the moving component to the assembly and then **Align/Mate** a flat surface (or a datum plane) of the moving component to the assembly. Note that both sets of references must be to the same two components in the assembly. This connection type allows one rotational motion (along the aligned axis), and two translational motions on the aligned surfaces. Note that for more complex slider motions, use the *Slot connection* to move a point along a 2D or 3D path.

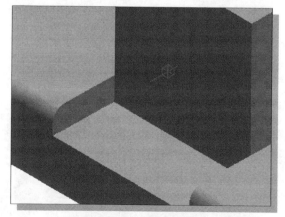

Cylinder

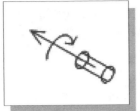

To apply this type of connection, it is necessary to **Align** an axis (or a revolved surface) of the moving component to the assembly. Note that additional constraints can also be applied to further restrict the motion of the component. This joint connection allows one rotational motion and one translational motion along the aligned axis. Note that the cylinder connection can also be used to create sliding motions in between two parts.

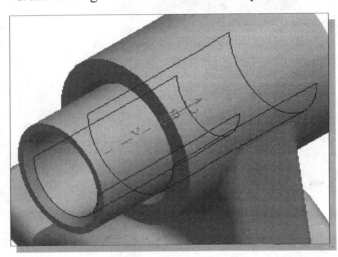

Planar

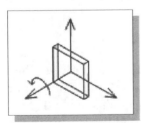

To apply this type of connection, it is necessary to **Align/Mate** a flat surface (or a datum plane) of the moving component to the assembly. Note that additional constraints can also be applied to further restrict the motion of the component. This connection type allows one rotational motion (perpendicular to the aligned surfaces) and two translational motions on the aligned surfaces.

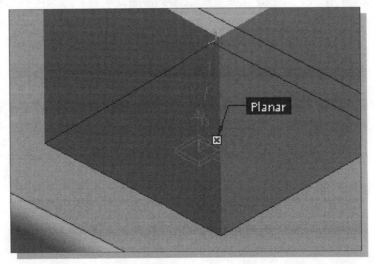

Servo Motor

In *Creo Animation*, **servo motors** are commonly used to generate a particular motion on a mechanism. *Servo motors* can be used to control a component by specifying a desired position, velocity, or acceleration as a function of time. The controlled motion can be translational or rotational motion.

Note that *Creo Animation* is using the term *servo motors* very loosely. A servo motor is typically an electronically controlled electric motor; the servo controller allows precise control of the motor shaft's position. The motor shaft can be positioned and held to specific angular positions by the servo controller. In practice, servo motors are commonly used in computer-controlled equipment, such as plotters and robots. In *Creo Animation*, the symbol of a curvy line is used to represent a servo motor as shown in the figure.

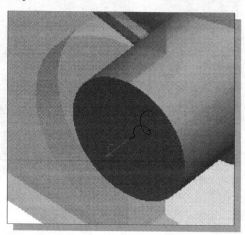

Bodies

In *Creo Animation*, a body represents a single rigid component that moves relative to the other bodies within the assembly. A body may consist of a single *Creo Parametric* part or several *Creo Parametric* parts fully constrained using placement constraints. Note that a body placed in three-dimensional space will have three translational and three rotational degrees of freedom (DOF). That is, a rigid body can translate and rotate along the X, Y, and Z-axes of a coordinate system.

Ground Body

A **ground body** (or **frame**) represents a fixed location in the three-dimensional space where the assembly is referencing its motions. The first object placed in an assembly should typically be the fixed reference of the assembly. It is generally a good habit to assemble the ground piece first when building the assembly, although it is possible to change the ground body in both *Creo E* and *Creo Animation*.

The Crank-Slider Assembly

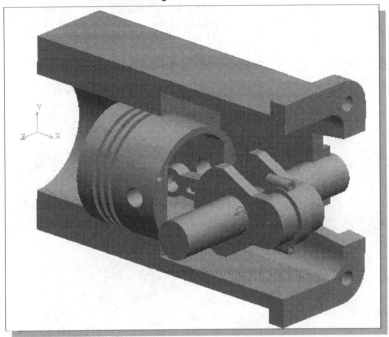

Creating the Required Parts

Five parts are required for this assembly: (1) *End Cap*, (2) *Connecting Rod*, (3) *Base Block*, (4) *Crank Shaft*, and (5) *Piston*. On your own, create the five parts shown below. Save the models as separate part files. (Note the location of the parts relative to the datum planes and close all part files after you have created the parts.)

(1) *End Cap*

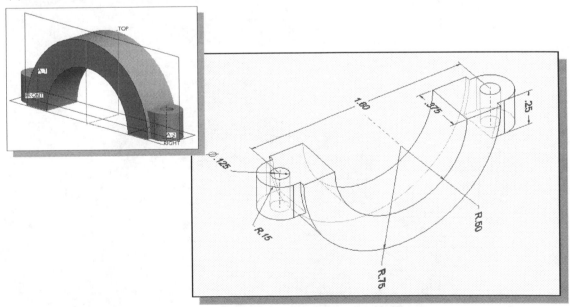

(2) *Connecting Rod*

(Construct the part with the datum planes passing through the center axes of the cylindrical surfaces. Create additional datum axes for the cylindrical surfaces if necessary.)

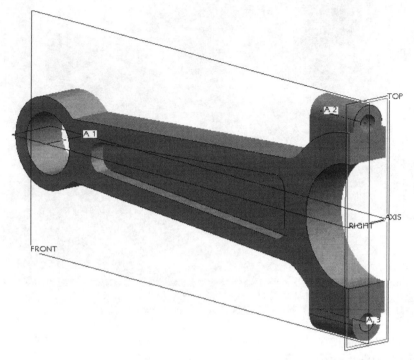

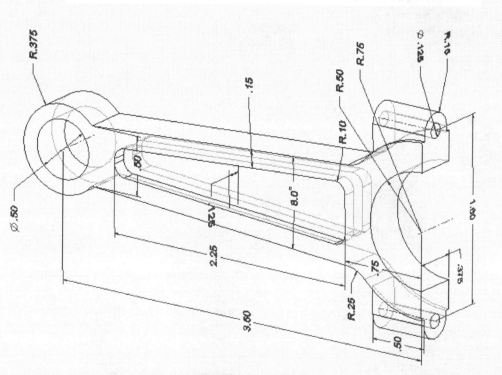

(3) *Base Block*

(Create a datum axis passing through the radius 0.5 cylindrical surface if it is not present. To make the assembling of the parts more clear, a cut feature is shown to remove the front half of the block.)

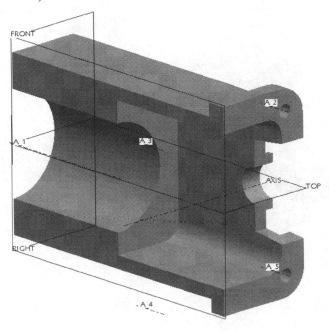

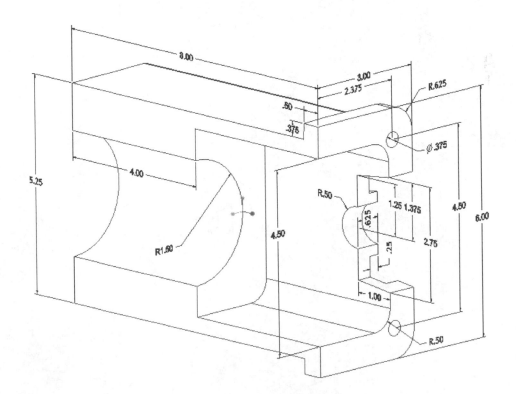

(4) *Crank Shaft*

(Construct the part with the datum planes passing through the center of the main shaft. Create additional datum axes for the cylindrical surfaces if they are not present.)

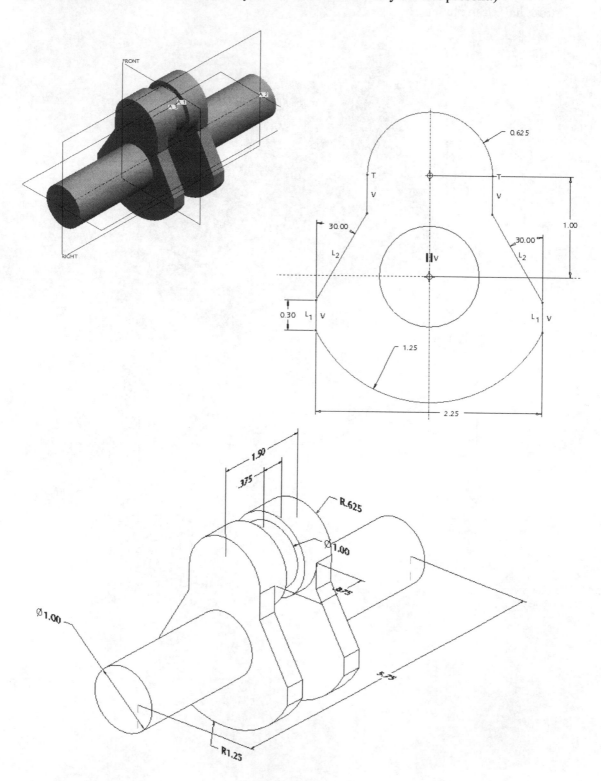

(5) *Piston*

(Construct the part with two of the datum planes passing through the center axis of the main cylinder body. Create additional datum axes for the cylindrical surfaces if not present.)

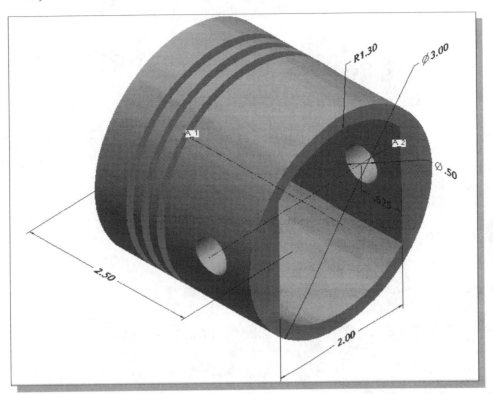

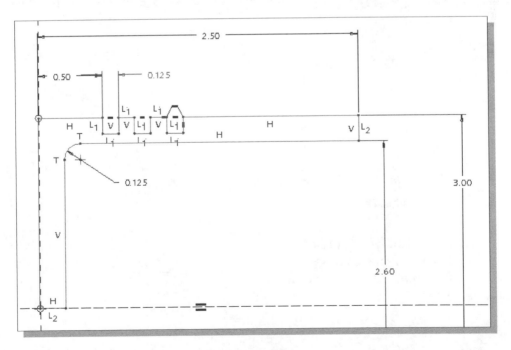

Create the Crank Slider Assembly Model

1. Pick the **New** icon in the *Quick Access* toolbar. (We can also use the key combination **Ctrl-N** to start a new object.)

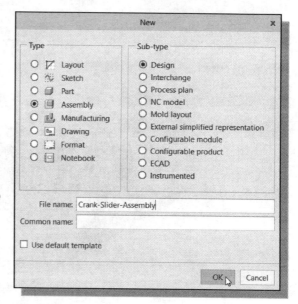

2. In the *New* form, select **Assembly** in the **Type** list.

3. Enter **Crank-Slider-Assembly** as the assembly model file **Name**.

4. Turn *off* the **Use default template** option.

5. Click the **OK** button to continue.

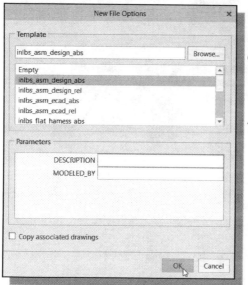

6. In the *New File Options* dialog box, select **inlbs_asm_design_abs** in the option list to use the *English Units* template file.

7. Pick **OK** to continue.

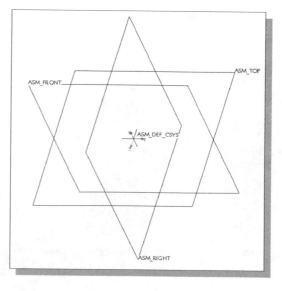

- Note that the displayed coordinate system and datum planes represent the fixed references in 3-D space for the assembly.

Base Component

In creating an assembly model in *Creo Parametric*, we will need to decide which part to use as the first component. In most cases, this *base component* should be one that is **not likely to be removed** from the assembly. For our project, we will use the **Base Block** as the base component.

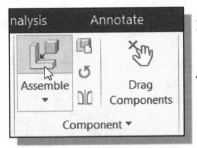

1. Pick **Assemble** in the *Component* toolbar to begin placing different parts into the assembly model.

❖ The *Open* window, showing a list of part model files, appears on the screen.

2. Select the **Base Block** (part file: *Base_Block.prt*) in the list window and pick **Open** to retrieve the model.

❖ The **Base Block** model is displayed in the graphics area and the *Assembly Constraints Dashboard* is activated.

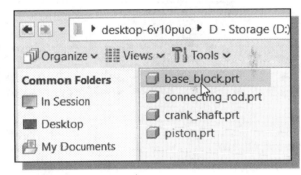

❖ We will remove all degrees of freedom by applying the **Fix** constraint; this constraint will fully constrain the part to its current location.

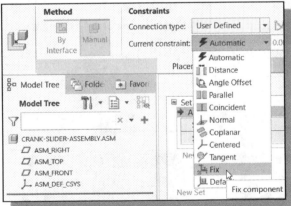

3. Select the **Fix** constraint from the Constraint Type list as shown.

4. Click on the **Placement** tab to display the constraint selection option. Note the Base-Block is fixed to the origin of the current assembly.

5. Click **OK** to accept the applied constraint.

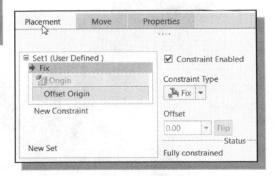

Assembling the Crank Shaft

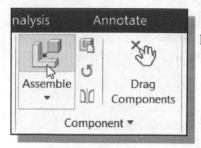

1. Pick **Assemble** in the *Component* toolbar to place more parts into the assembly model.

2. Select the ***Crank Shaft*** part (part file: *crank_shaft.prt*) in the list window and pick **Open** to retrieve the model.

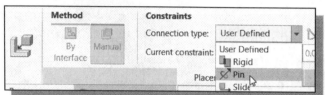

3. Select the **Pin** joint connection from the connection type list as shown.

❖ For the **Pin** joint connection, two placement constraints are required: first **Align** an axis (or a revolved surface) of the moving component to the assembly and then **Align/Mate** a flat surface (or a datum plane) of the moving component to the assembly. Note that both sets of references must be to the same two components in the assembly.

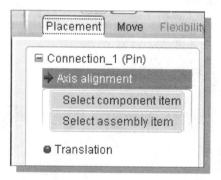

4. Pick the two *axes* of the *Crank Shaft* and the *Base Block* as shown.

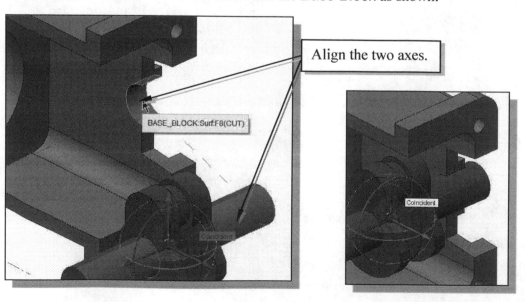

Align the two axes.

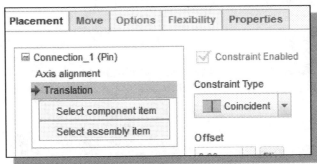

❖ Note that now the axis alignment has been completed, the **Translation** constraint is activated as shown.

5. Pick the two corresponding **vertical datum planes** of the *Base Block* and *Crank Shaft* parts as shown.

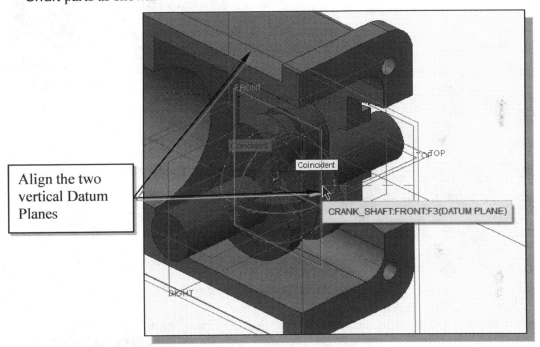

Align the two vertical Datum Planes

CRANK_SHAFT:FRONT:F3(DATUM PLANE)

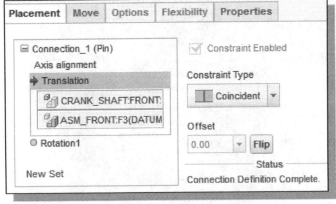

❖ At the bottom of the *Constraint Placement* control panel, *Creo Parametric* indicates the status of placement of the *Crank Shaft* as **Connection Definition Complete**.

6. Pick **OK** to complete the placement of the part.

Assembling the Connecting Rod

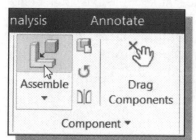

1. Pick **Assemble** in the *Component* toolbar to place another part into the assembly model.

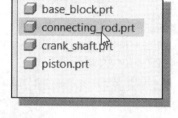

2. Select the ***Connecting Rod*** part in the list window and pick **Open** to retrieve the model.

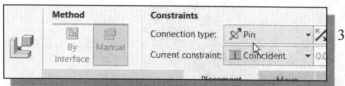

3. Select the **Pin** joint connection from the connection type list as shown.

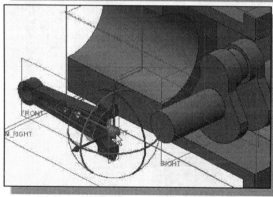

4. On your own, drag the DOF indicator attached to the connecting rod part to move the part to the left side and also adjust its orientation as shown.

5. Pick the two *surfaces* of the *Crank Shaft* and the *Connecting Rod* as shown.

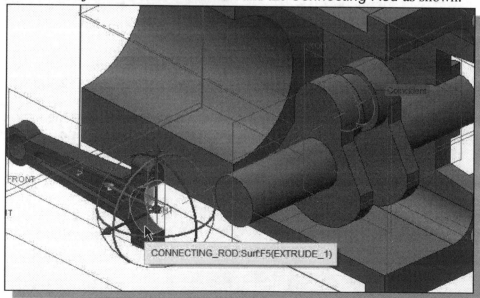

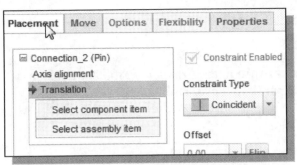

❖ Note that now the axis alignment has been completed, the **Translation** constraint is activated as shown.

6. Pick the two corresponding **vertical datum planes** of the *Connecting Rod* and *Crank Shaft* parts as shown.

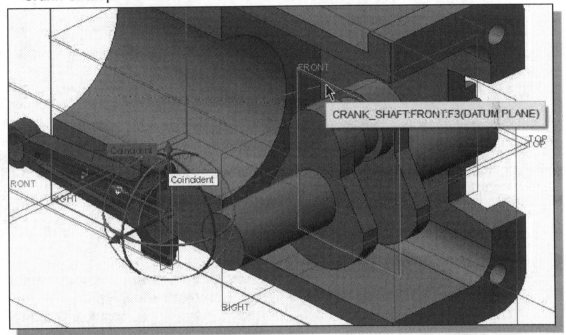

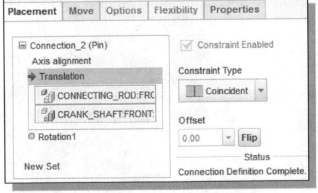

❖ At the bottom of the *Constraint Placement* control panel, *Creo Parametric* indicates the status of placement of the *Connecting Rod* as **Connection Definition Complete**.

7. Pick **OK** to complete the placement of the part.

Assembling the Piston using the Slider connection

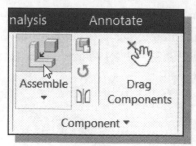

1. Pick **Assemble** in the *Component* toolbar to place a different part into the assembly model.

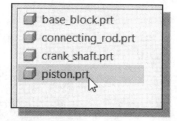

2. Select the ***Piston*** part in the list window and pick **Open** to retrieve the model.

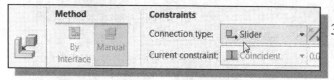

3. Select the **Slider** joint connection from the connection type list as shown.

❖ For the **Slider** joint connection, two placement constraints are required: first Align an axis (or an edge) of the moving component to the assembly, then align the planar surfaces where the sliding happens.

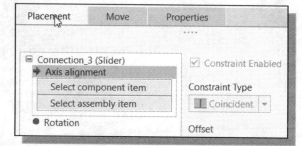

4. Pick the two center ***axes*** of the *Piston* and the *Base Block* as shown.

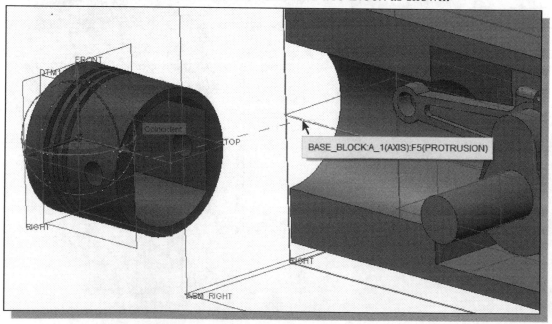

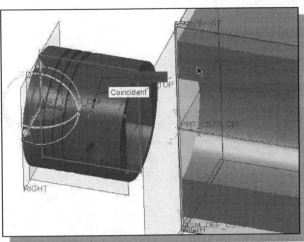

5. Pick the vertical *plane* of the *Piston* and the corresponding vertical plane of the *Base Block* as shown.

❖ At the bottom of the *Constraint Placement* control panel, *Creo Parametric* indicates the status of the Slider connection as **Fully Constrained**.

6. Click **New Set** to add another connection to also constrain the *Piston* to the *Connecting Rod*.

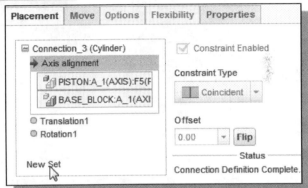

7. Select the **Cylinder** joint connection from the connection type list as shown.

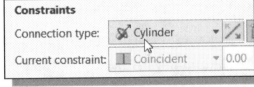

8. On your own, Align the two axes of the *Connecting Rod* and the *Piston* as shown.

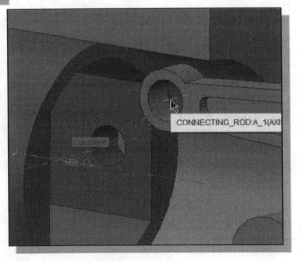

9. Pick **OK** to complete the placement of the part.

Assembling the End Cap

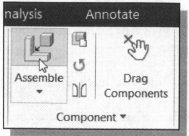

1. Pick **Assemble** in the *Component* toolbar to place another part into the assembly model.

2. Select the ***End Cap*** part in the list window and pick **Open** to retrieve the model.

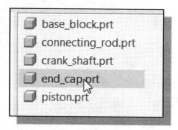

base_block.prt
connecting_rod.prt
crank_shaft.prt
end_cap.prt
piston.prt

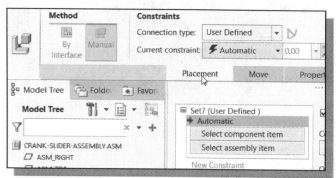

3. Click on the **Placement** tab to display the **Constraint Selection** option.

4. On your own, assemble the ***End Cap*** part by placing three coincident placement constraints.

❖ Note that several choices are feasible to fully constrain the ***End Cap*** part.

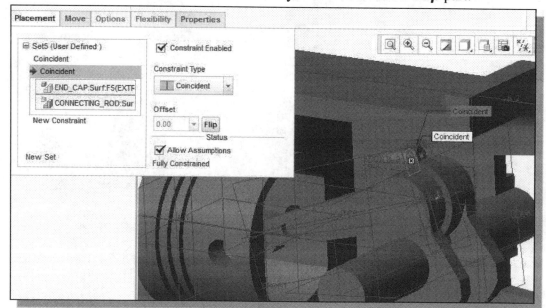

❖ In creating an assembly model, note that we can mix using the basic placement constraints and the joint connections. The key concept is to create an assembly that is ***fully constrained*** at all levels.

Starting the Creo Animation Module

The *Creo Animation* module can be used to perform a motion analysis that can be done in a relatively short time. In *Creo Animation*, two main approaches are commonly used to create an animation: (1) construct the animation by creating **key frames** of the different positions of parts, and (2) the use of **servo motors** to provide more specific controls of parts.

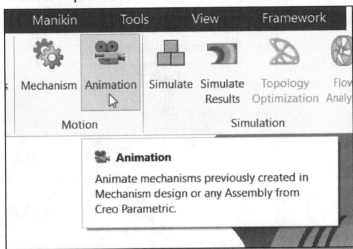

1. Select the **Applications** tab in the *Ribbon* toolbar and select **Animation**.

❖ Note the *Creo Animation* package is an integrated *Creo* module; the *Creo Animation* toolbar appears in the *Ribbon* toolbar.

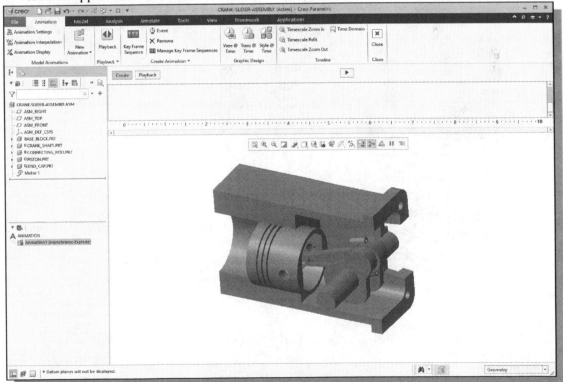

Defining an Animation in Creo Animation

With *Creo Animation*, depending on the specific objective, different steps can be used to create the desired animation. In general, *Creo Animation* is intended to provide quick and easy means for creating the following types of animation: (1) **explode animation** of assemblies (2) **snap shot animations** (3) replay analysis results from **Mechanism Dynamic Option** (MDO).

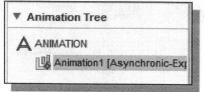

❖ In the *Animation Tree*, a default animation is set up for an *explode animation*. Note that although it is feasible to modify many elements of an animation, the animation type, once defined, cannot be changed.

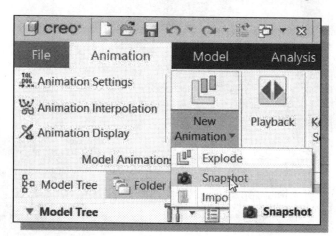

1. In the *Ribbon* toolbar, click on the **New Animation** icon to bring up the command list.

2. Select **Snapshot** to define a new *Snapshot animation*.

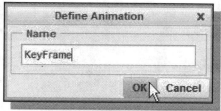

3. In the *Define Animation* dialog box, enter **KeyFrame** as the animation **Name** and click **OK** to create the new animation.

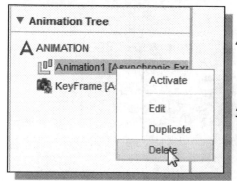

4. Inside the *Animation Tree*, press and hold down the **right-mouse-button** on top of the pre-defined animation to bring up the option menu.

5. Select **Delete** in the option menu to delete the default animation.

➢ Note that two approaches are available for *Snapshot Animation*: (1) **Key Frame** approach and (2) **Servo Motor** approach. Both methods are suitable for the *Crank-Slider-Assembly* and will be illustrated in the following sections.

The Basic Animation Body Definitions and Display Controls

For the *Crank-Slider-Assembly*, *Creo Animation* automatically organizes parts into **animation bodies**. The **Body Definition** command can be used to examine and edit the body definitions of the assembly.

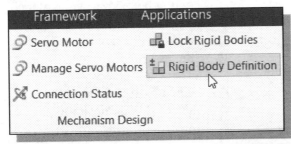

1. Select **Rigid Body Definition** in the *Mechanism Design* toolbar, which is the third icon in the toolbar.

2. Click **Ground** in the *Bodies* dialog box and notice the selected object is highlighted in the graphics area.

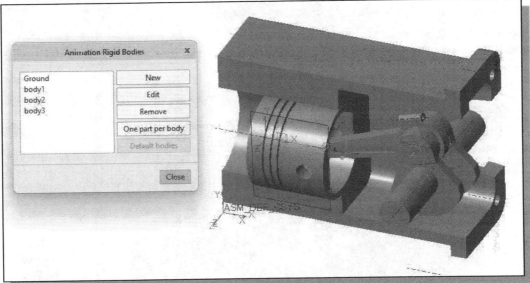

Bodies

In *Creo Animation*, a **body** represents a single rigid component that moves relative to the other bodies within the assembly. A body may consist of a single *Creo Parametric* part or several *Creo Parametric* parts fully constrained using placement constraints.

Ground Body

A **ground body** (or **frame**) represents a fixed location in the three-dimensional space where the assembly is referencing its motions. The first object placed in an assembly should typically be the fixed reference of the assembly.

3. On your own, examine the other bodies in the assembly by clicking in the *Bodies* dialog box.

❖ Note that the **Connecting Rod** and the **End Cap** parts are expected to move together and therefore they should be treated as **one body**.

Edit the Rigid Body Definitions

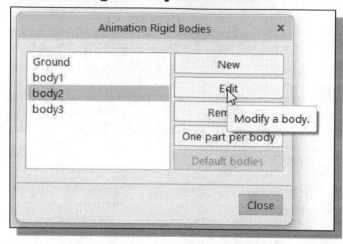

1. Select **Body2** in the assembly by clicking in the *Bodies* dialog box as shown. Note the connecting rod part and the end cap part are highlighted in the graphics window.

2. Click **Edit** to modify the selected body definition.

3. To re-define a body, hold down the **[Ctrl]** key and select the **Connecting_rod** and the **End-Cap parts**. Note that the two highlighted parts are now defined as one body.

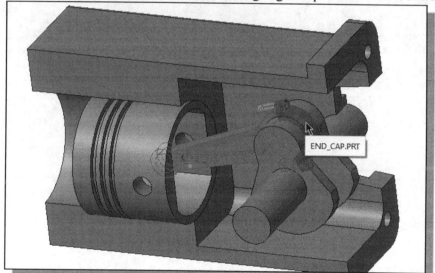

4. Click **OK** to accept the modified Body2 definition.

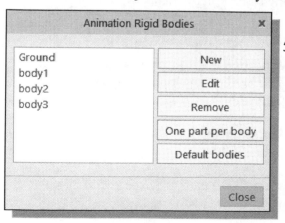

5. Note that the *Body2* definition is now containing the End-Cap part to Body2. Click **Close** to exit the dialog box.

Creating an Animation Using the Key Frame Approach

The general procedure of creating animations using *key frames* in *Creo Animation* is similar to the creation of cartoon movies, where key frames are first created, and the intermediate frames are then added by *Creo Animation*. This approach is a very flexible way to create animations.

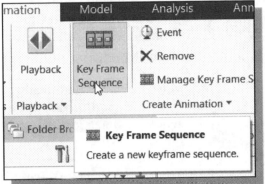

1. Select **Key Frame Sequence** in the *Create Animation* toolbar, which is the first icon in the toolbar.

❖ The **Key Frame Sequence** command uses the drag and drop method of creating key frames. In the dialog box, we have full control of taking snapshots of key frames for creating animations.

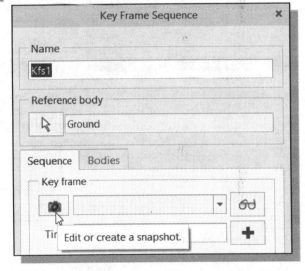

2. Click on the **Edit or Create a snapshot** icon as shown.

❖ Note the default *Key Frame Sequence* name is **Kfs1**. Note also that it is feasible to create multiple sequences to animate complex motions in an assembly.

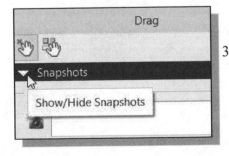

3. Click on the **Snapshots** icon to expand the control panel, which allows us to create and/or manage snapshots.

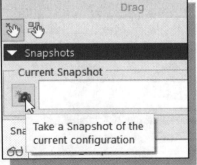

• Each time the **Camera** icon in the *Current Snapshot* section is clicked a snapshot of the current configuration will be taken.

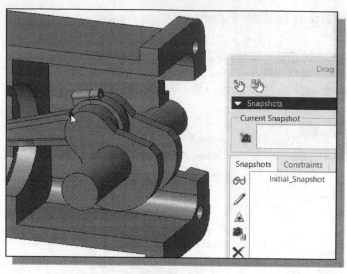

4. Left-mouse-click near the **top edge of the arc portion** of the *Crank Shaft* part as shown.

❖ Note the default option of drag and drop is set to **Drag Point** as shown in the *Drag* dialog box.

5. Move the cursor and rotate the *Crank Shaft* roughly 45 degrees counterclockwise as shown.

6. Click the **right-mouse-button** once to accept the new positions for the moving parts.

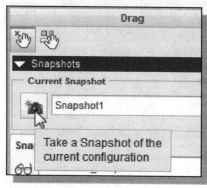

7. Click on the **Camera** icon, which will take a snapshot of the current configuration.

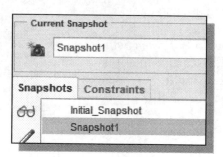

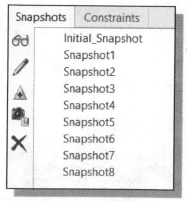

8. On your own, repeat the above steps and take additional snapshots, each at roughly 45 degrees increment of the **Crank Shaft's** position. (Hint: Use the dynamic viewing options to aid the operation.)

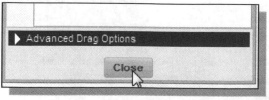

9. Pick **Close** to end the Create Snapshot command.

❖ Note that the default time increment between the snapshots is set to 1 second; this is also shown in the animation dashboard near the top of the graphics area.

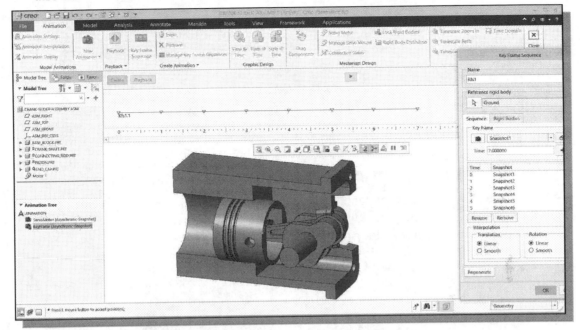

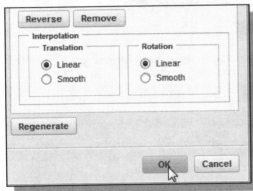

❖ Near the bottom of the *Key Frame Sequence* dialog box, note the settings for interpolating the intermediate steps of the moving parts. *Creo Animation* will automatically generate additional frames for our animation.

10. Click **OK** to accept the current setup.

11. Click **Generate**, in the *Animation* dashboard, to start the calculations and create the animation.

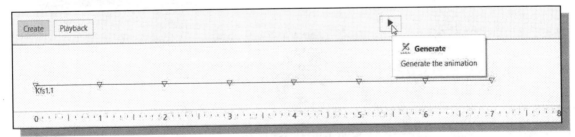

❖ Note that *Creo Animation* will use the key frames we have created and also generate additional frames for the intermediate positions for a smoother transition of the animation.

12. Once the calculations are completed, click on the **Playback** icon to open the *Playback* control panel.

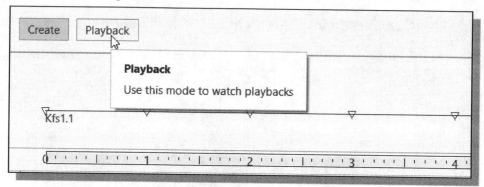

❖ Note the *Playback* control panel acts almost as if it is a tape recorder. We can play forward, as well as play backward. We can adjust the playback speed and have the playback in a continuous loop or reverse the play direction.

13. Click on the **Play** button to watch the animation.

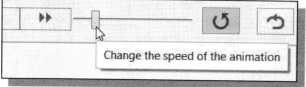

14. On your own, experiment with the different controls available in the control panel.

15. Click on the **Stop** button to stop the animation.

16. Click on the **Create** button to exit the *Playback* mode.

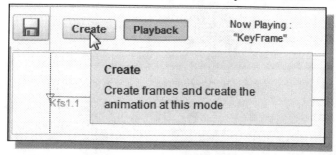

Adjusting the Time Domain of the Animation

The default time domain is set to 10 seconds, but our snapshots only covered the first seven seconds. We will adjust the timeline to match the generated animation.

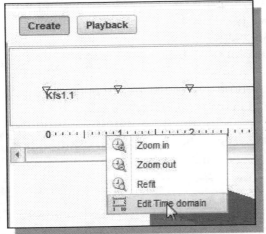

1. Click once with the **right-mouse-button** in the *Timeline* area near the bottom of the digit display area to bring up the option list.

2. Select **Edit Time domain** as shown.

3. In the *Animation Time Domain* form, set the **End Time** to **7** as shown.

4. Click **OK** to accept the new settings.

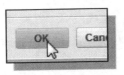

5. Click **Generate**, in the *Animation* dashboard, to start the calculations and create the animation.

6. Click **Yes** to overwrite the previous result set.

❖ Note that *Creo Animation* will regenerate the animation, still using the four key frames we have created, but with the new time domain we have specified.

7. Click **Playback** to open the *Playback* control panel.

8. Click on the **Play** button to watch the animation.

9. On your own, experiment with the different settings.

❖ Also note that several other options are also available to adjust the snapshots in the timeline; we could use the mouse control and adjust the key frames in the timeline or adjust them through the *Key Frame Sequence* dialog box.

Creating another Animation by Using a Servo Motor

Besides using the **key frames** approach, we can also create an animation of the *Crank-Slider* mechanism through the use of a **servo motor**. In *Creo Animation*, a *servo motor* can be used to control a component by specifying a desired position, velocity, or acceleration as a function of time. This approach is very useful if the position, or speed, or acceleration of the driving component is known or can be estimated. In this section, we will create another animation using this approach for our *Crank-Slider* assembly.

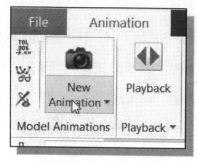

1. Select **New Animation** in the *Model Animations* toolbar located in the *Ribbon* toolbar.

2. Enter **ServoMotor** as the animation **Name** and click **OK** to accept the default settings as shown.

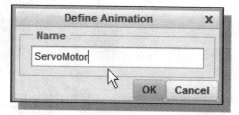

❖ Note the timeline is reset back to the default 10 seconds, near the bottom of the graphics area.

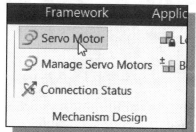

3. Select **Servo Motor** in the *Animation* toolbar located to the right side of the graphics area.

4. Select the **connection axis** that passes through the shaft of the *Crank Shaft* part. This will attach the servo motor to the shaft of the part.

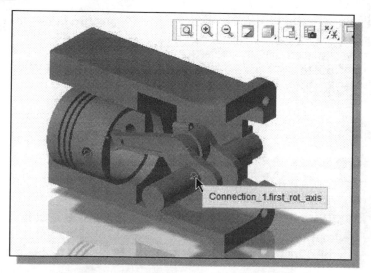

5. Click the **Profile Detail** tab to view/adjust the settings of the position functions for the new servo motor.

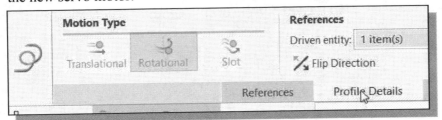

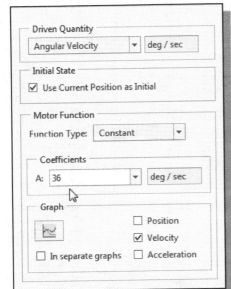

6. Set the *Driven Quantity* to **Angular Velocity** as shown.

7. Set the **Motor Function** option to **Constant** and enter **36** as the *coefficients value*.

❖ Note that the current timeline is set to 10 seconds. It is therefore necessary to set the velocity to 36 degrees per second to achieve a full revolution in 10 seconds.

8. Click **OK** to accept the settings.

❖ Note that the servo motor is attached to the **Crank Shaft** and the animation definition also appears in the timeline.

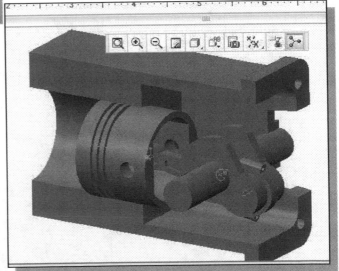

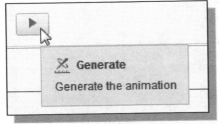

9. Click **Generate**, in the *Animation* dashboard, to start the calculations and create the animation.

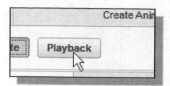

10. Once the calculations are completed, click on the **Playback** icon to open the *Playback* control panel.

11. Click on the **Play** button to watch the animation.

12. Click **Stop** to end the playback.

❖ Note that *Creo Animation* automatically generated all the necessary key frames. The animation of the parts appears to be better than the animation done through the key frame approach.

Output the Animation as a Video file

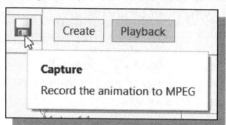

1. Click on the **Capture** button in the *dashboard* as shown.

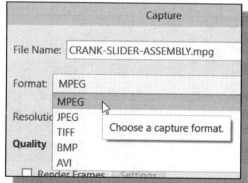

2. Click on the **Type** list to view the different output formats that are available.

3. Accept the default settings and click **OK** to save the animation as an **MPEG** video file.

4. On your own, view the captured movie file.

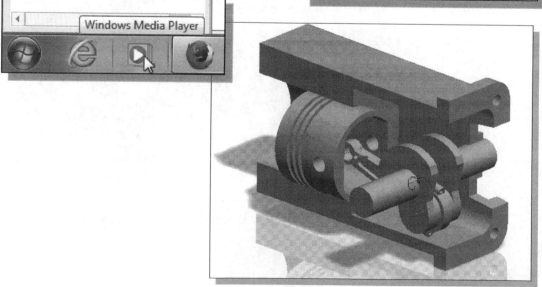

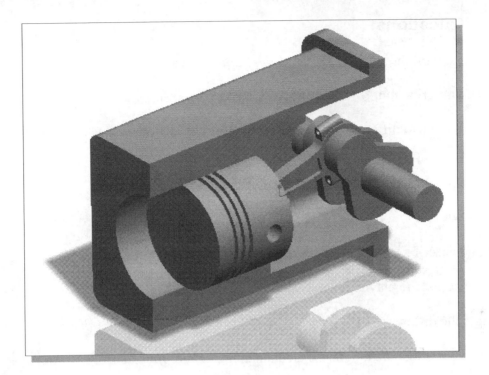

Conclusion

Design includes all activities involved from the original concept to the finished product. Design is the process by which products are created and modified. For many years designers sought ways to describe and analyze three-dimensional designs without building physical models. With the advancements in computer technology, the creation of parametric models on computers offers a wide range of benefits. Parametric models are easier to interpret and can be easily altered. Parametric models can be analyzed using finite element analysis software, and simulation of real-life loads can be applied to the models and the results graphically displayed. The finalized solid models can also be used directly by manufacturing equipment to manufacture the product.

Throughout this text, various modeling techniques have been presented. Mastering these techniques will enable you to create intelligent and flexible solid models. The goal is to make use of the tools provided by *Creo Parametric* and to successfully capture the ***design intent*** of the product. In many instances, only a single approach to the modeling tasks was presented; you are encouraged to repeat all of the tutorials and develop different ways of thinking in accomplishing the same tasks. We have only scratched the surface of *Creo Parametric's* functionality. The more time you spend using the system, the easier it will be to perform parametric modeling with *Creo Parametric*.

Review Questions:

1. When and why should we use *joint connections* in an assembly?

2. List and describe four of the commonly used *joint connections*.

3. Describe the two different approaches available to create animations in *Creo Animation*.

4. Describe the procedure to apply a Pin connection in an assembly.

5. In what video file formats can we save an animation generated in *Creo Animation*?

6. List and describe the procedure to attach a servo motor in *Creo Animation*.

7. Can we adjust the playback speed of the animation in *Creo Animation*?

8. What is the difference between a *body* and a *ground body* in *Creo Animation*?

9. How do we adjust the timeline in an animation in *Creo Animation*?

10. How did we constrain the *Base Block* part in the *Crank-Slider* assembly? Which joint connection did we use to accomplish that?

Exercises: (Create Assembly models and animations in *Creo Animation.*)

1. **Geneva-Cam** (Design and create the necessary parts to complete the assembly. All Dimensions are in inches. Create an animation using the *Key Frames* approach.)

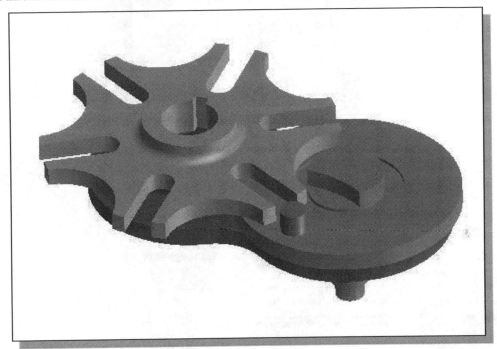

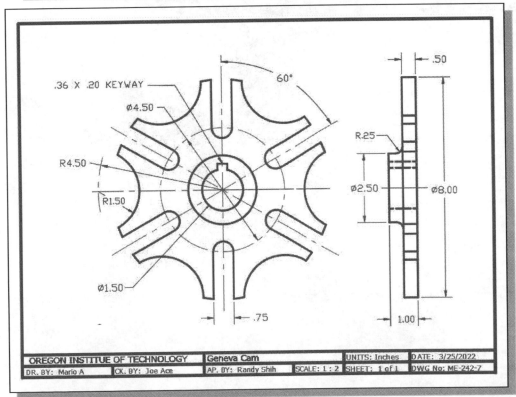

| OREGON INSTITUE OF TECHNOLOGY | Geneva Cam | | UNITS: Inches | DATE: 3/25/2022 |
| DR. BY: Mario A | CK. BY: Joe Ace | AP. BY: Randy Shih | SCALE: 1 : 2 | SHEET: 1 of 1 | DWG No: ME-242-7 |

2. Quick Return Mechanism (Design and create the necessary parts. Animations can be created using either the *Key Frames* approach or the Servo Motor approach.)

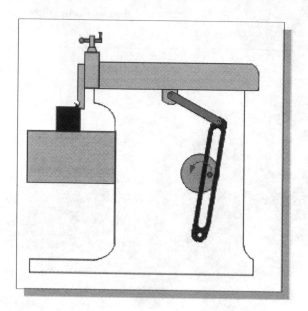

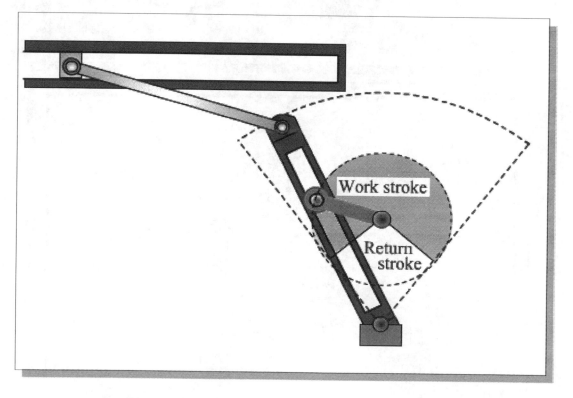

Appendix A

Sec. 206. Standard gauge for sheet and plate iron and steel

For the purpose of securing uniformity, the following is established as the only standard gauge for sheet and plate iron and steel in the United States of America, namely:

| Gauge | Thickness | | | Weight | | | | |
	Frac. Inch	Dec. Inch	mm	oz/ft^2	lb/ft^2	kg/ft^2	kg/m^2	lb/m^2
0000000	1/2	.5	12.7	320	20.00	9.072	97.65	215.28
000000	15/32	.46875	11.90625	300	18.75	8.505	91.55	201.82
00000	7/16	.4375	11.1125	280	17.50	7.983	85.44	188.37
0000	13/32	.40625	10.31875	260	16.25	7.371	79.33	174.91
000	3/8	.375	9.525	240	15	6.804	73.24	161.46
00	11/32	.34375	8.73125	220	13.75	6.237	67.13	148.00
0	5/16	.3125	7.9375	200	12.50	5.67	61.03	134.55
1	9/32	.28125	7.14375	180	11.25	5.103	54.93	121.09
2	17/64	.265625	6.746875	170	10.625	4.819	51.88	114.37
3	1/4	.25	6.35	160	10	4.536	48.82	107.64
4	15/64	.234375	5.953125	150	9.375	4.252	45.77	100.91
5	7/32	.21875	5.55625	140	8.75	3.969	42.72	94.18
6	13/64	.203125	5.159375	130	8.125	3.685	39.67	87.45
7	3/16	.1875	4.7625	120	7.5	3.402	36.62	80.72
8	11/64	.171875	4.365625	110	6.875	3.118	33.57	74.00
9	5/32	.15625	3.96875	100	6.25	2.835	30.52	67.27
10	9/64	.140625	3.571875	90	5.625	2.552	27.46	60.55
11	1/8	.125	3.175	80	5	2.268	24.41	53.82
12	7/64	.109375	2.778125	70	4.375	1.984	21.36	47.09
13	3/32	.09375	2.38125	60	3.75	1.701	18.31	40.36
14	5/64	.078125	1.984375	50	3.125	1.417	15.26	33.64
15	9/128	.0703125	1.7859375	45	2.8125	1.276	13.73	30.27
16	1/16	.0625	1.5875	40	2.5	1.134	12.21	26.91
17	9/160	.05625	1.42875	36	2.25	1.021	10.99	24.22
18	1/20	.05	1.27	32	2	.9072	9.765	21.53
19	7/160	.04375	1.11125	28	1.75	.7938	8.544	18.84
20	3/80	.0375	.9525	24	1.50	.6804	7.324	16.15
21	11/320	.034375	.873125	22	1.375	.6237	6.713	14.80
22	1/32	.03125	.793750	20	1.25	.567	6.103	13.46
23	9/320	.028125	.714375	18	1.125	.5103	5.493	12.11
24	1/40	.025	.635	16	1	.4536	4.882	10.76
25	7/320	.021875	.555625	14	.875	.3969	4.272	9.42

26	3/160	.01875	.47625	12	.75	.3402	3.662	8.07
27	11/640	.0171875	.4365625	11	.6875	.3119	3.357	7.40
28	1/64	.015625	.396875	10	.625	.2835	3.052	6.73
29	9/640	.0140625	.3571875	9	.5625	.2551	2.746	6.05
30	1/80	.0125	.3175	8	.5	.2268	2.441	5.38
31	7/640	.0109375	.2778125	7	.4375	.1984	2.136	4.71
32	13/1280	.01015625	.25796875	6 1/2	.40625	.1843	1.983	4.37
33	3/320	.009375	.238125	6	.375	.1701	1.831	4.04
34	11/1280	.00859375	.21828125	5 1/2	.34375	.1559	1.678	3.70
35	5/640	.0078125	.1984375	5	.3125	.1417	1.526	3.36
36	9/1280	.00703125	.17859375	4 1/2	.28125	.1276	1.373	3.03
37	17/2560	.006640625	.168671875	4 1/4	.265625	.1205	1.297	2.87
38	1/160	.00625	.15875	4	.25	.1134	1.221	2.69

Reformatted from **http://www4.law.cornell.edu/uscode/15/206.html**

INDEX

Notes: